Probability and Its Applications

Published in association with the Applied Probability Trust

Editors: J. Gani, C.C. Heyde, P. Jagers, T.G. Kurtz

Probability and Its Applications

Anderson: Continuous-Time Markov Chains.
Azencott/Dacunha-Castelle: Series of Irregular Observations.
Bass: Diffusions and Elliptic Operators.
Bass: Probabilistic Techniques in Analysis.
Chen: Eigenvalues, Inequalities and Ergodic Theory
Choi: ARMA Model Identification.
Costa, Fragoso and Marques: Discrete-Time Markov Jump Linear Systems
Daley/Vere-Jones: An Introduction to the Theory of Point Processes. Volume I: Elementary Theory and Methods. Second Edition.
de la Peña/Giné: Decoupling: From Dependence to Independence.
Durrett: Probability Models for DNA Sequence Evolution.
Galambos/Simonelli: Bonferroni-type Inequalities with Applications.
Gani (Editor): The Craft of Probabilistic Modelling.
Grandell: Aspects of Risk Theory.
Gut: Stopped Random Walks.
Guyon: Random Fields on a Network.
Kallenberg: Foundations of Modern Probability, Second Edition.
Last/Brandt: Marked Point Processes on the Real Line.
Leadbetter/Lindgren/Rootzén: Extremes and Related Properties of Random Sequences and Processes.
Molchanov: Theory of Random Sets
Nualari: The Malliavin Calculus and Related Topics.
Rachev/Rüschendorf: Mass Transportation Problems. Volume I: Theory.
Rachev/Rüschendorf: Mass Transportation Problems. Volume II: Applications.
Resnick: Extreme Values, Regular Variation and Point Processes.
Shedler: Regeneration and Networks of Queues.
Silvestrov: Limit Theorems for Randomly Stopped Stochastic Processes
Thorisson: Coupling, Stationarity, and Regeneration.
Todorovic: An Introduction to Stochastic Processes and Their Applications.

Nils Berglund and Barbara Gentz

Noise-Induced Phenomena in Slow-Fast Dynamical Systems

A Sample-Paths Approach

With 57 Figures

 Springer

Nils Berglund, PhD
Centre de Physique Theorique
Campus de Luminy
F13288 Marseille Cedex 9
France

Barbara Gentz, PhD
Weierstrasse Institute for Applied
Analysis and Stochastics
Mohrenstrasse 39
D-10117 Berlin
Germany

Series Editors
J Gani
Stochastic Analysis Group, CMA
Australian National University
Canberra ACT 0200
Australia

C.C. Heyde
Stochastic Analysis Group, CMA
Australian National University
Canberra ACT 0200
Australia

Peter Jagers
Mathematical Statistics
Chalmers University of Technology
S-412 96 Göteborg
Sweden

T.G. Kurtz
Department of Mathematics
University of Wisconsin
480 Lincoln Drive
Madison, WI 53706
USA

Mathematics Subject Classification (2000): 34E15, 37H20, 60F10, 60H10, 60J60

British Library Cataloguing in Publication Data
Berglund, Nils
  Noise-induced phenomena in slow-fast dynamical systems : a
  sample-paths approach. - (Probability and its applications)
    1. Stochastic differential equations 2. Noise - Mathematical models
    I. Title II. Gentz, Barbara
  519.2
ISBN-10: 1846280389

Library of Congress Control Number: 2005931125

ISBN-10: 1-84628-038-9   e-ISBN 1-84628-186-5   Printed on acid-free paper
ISBN-13: 978-1-84628-038-2

© Springer-Verlag London Limited 2006

Apart from any fair dealing for the purposes of research or private study, or criticism or review, as permitted under the Copyright, Designs and Patents Act 1988, this publication may only be reproduced, stored or transmitted, in any form or by any means, with the prior permission in writing of the publishers, or in the case of reprographic reproduction in accordance with the terms of licences issued by the Copyright Licensing Agency. Enquiries concerning reproduction outside those terms should be sent to the publishers.

The use of registered names, trademarks, etc. in this publication does not imply, even in the absence of a specific statement, that such names are exempt from the relevant laws and regulations and therefore free for general use.

The publisher makes no representation, express or implied, with regard to the accuracy of the information contained in this book and cannot accept any legal responsibility or liability for any errors or omissions that may be made.

Printed in the United States of America   (EB)

9 8 7 6 5 4 3 2 1

Springer Science+Business Media
springeronline.com

To Erwin Bolthausen and Hervé Kunz

Preface

When constructing a mathematical model for a problem in natural science, one often needs to combine methods from different fields of mathematics. Stochastic differential equations, for instance, are used to model the effect of noise on a system evolving in time. Being at the same time differential equations and random, their description naturally involves methods from the theory of dynamical systems, and from stochastic analysis.

Much progress has been made in recent years in the quantitative description of the solutions to stochastic differential equations. Still, it seems to us that communication between the various communities interested in these models (mathematicians in probability theory and analysis, physicists, biologists, climatologists, and so on) is not always optimal. This relates to the fact that researchers in each field have developed their own approaches, and their own jargon. A climatologist, trying to use a book on stochastic analysis, in the hope of understanding the possible effects of noise on the box model she is studying, often experiences similar difficulties as a probabilist, trying to find out what his latest theorem on martingales might imply for action-potential generation in neurons.

The purpose of this book is twofold. On the one hand, it presents a particular approach to the study of stochastic differential equations with two timescales, based on a characterisation of typical sample paths. This approach, which combines ideas from singular perturbation theory with probabilistic methods, is developed in Chapters 3 and 5 in a mathematically rigorous way. On the other hand, Chapters 4 and 6 attempt to bridge the gap between abstract mathematics and concrete applications, by illustrating how the method works, in a few specific examples taken from physics, biology, and climatology. The choice of applications is somewhat arbitrary, and was mainly influenced by the field of speciality of the modellers we happened to meet in the course of the last years.

The present book grew out of a series of articles on the effect of noise on dynamical bifurcations. In the process of writing, however, we realised that many results could be improved, generalised, and presented in a more concise way. Also, the bibliographical research for the chapters on applications revealed many new interesting problems, that we were tempted to discuss as well. Anyone who has written a book knows that there is no upper bound on the amount of time and work one can put into the writing. For the sake of timely publishing, however, one has to stop somewhere, even though the feeling remains to leave something unfinished. The reader will thus find many open problems, and our hope is that they will stimulate future research.

We dedicate this book to Erwin Bolthausen and Hervé Kunz on the occasion of their sixtieth birthdays. As our thesis advisors in Zürich and Lausanne, respectively, they left us with a lasting impression of their way to use mathematics to solve problems from physics in an elegant way. Our special thanks go to Anton Bovier, who supported and encouraged our collaboration from the very beginning. Without his sharing our enthusiasm and his insightful remarks, our research would not have led us that far.

Many more people have provided help by giving advice, by clarifying subtleties of models used in applications, by serving on our habilitation committees, or simply by showing their interest in our work. We are grateful to Ludwig Arnold, Gérard Ben Arous, Jean-Marie Barbaroux, Dirk Blömker, Fritz Colonius, Predrag Cvitanović, Jean-Dominique Deuschel, Werner Ebeling, Bastien Fernandez, Jean-François Le Gall, Jean-Michel Ghez, Maria Teresa Giraudo, Peter Hänggi, Peter Imkeller, Alain Joye, Till Kuhlbrodt, Robert Maier, Adam Monahan, Khashayar Pakdaman, Etienne Pardoux, Cecile Penland, Pierre Picco, Arkady Pikovsky, Claude-Alain Pillet, Michael Rosenblum, Laura Sacerdote, Michael Scheutzow, Lutz Schimansky-Geier, Igor Sokolov, Dan Stein, Alain-Sol Sznitman, Peter Talkner, Larry Thomas, and Michael Zaks.

Our long-lasting collaboration would not have been possible without the gracious support of a number institutions. We thank the Weierstraß-Institut für Angewandte Analysis und Stochastik in Berlin, the Forschungsinstitut für Mathematik at the ETH in Zürich, the Université du Sud Toulon–Var, and the Centre de Physique Théorique in Marseille-Luminy for kind hospitality.

Financial support of the research preceding the endeavour of writing this book was provided by the Weierstraß-Institut für Angewandte Analysis und Stochastik, the Forschungsinstitut für Mathematik at the ETH Zürich, the Université du Sud Toulon–Var, the Centre de Physique Théorique, the French Ministry of Research by way of the *Action Concertée Incitative (ACI) Jeunes Chercheurs, Modélisation stochastique de systèmes hors équilibre*, and the ESF Programme *Phase Transitions and Fluctuation Phenomena for Random Dynamics in Spatially Extended Systems (RDSES)* and is gratefully acknowledged.

During the preparation of the manuscript, we enjoyed the knowledgeable and patient collaboration of Stephanie Harding and Helen Desmond at Springer, London.

Finally, we want to express our thanks to Ishin and Tian Fu in Berlin for providing good reasons to spend the occasional free evening with friends, practising our skill at eating with chopsticks.

Marseille and Berlin, *Nils Berglund*
June 2005 *Barbara Gentz*

Contents

1 Introduction 1
   1.1 Stochastic Models and Metastability 1
   1.2 Timescales and Slow–Fast Systems 6
   1.3 Examples 8
   1.4 Reader's Guide 13
   Bibliographic Comments 15

2 Deterministic Slow–Fast Systems 17
   2.1 Slow Manifolds 18
      2.1.1 Definitions and Examples 18
      2.1.2 Convergence towards a Stable Slow Manifold 22
      2.1.3 Geometric Singular Perturbation Theory 24
   2.2 Dynamic Bifurcations 27
      2.2.1 Centre-Manifold Reduction 27
      2.2.2 Saddle–Node Bifurcation 28
      2.2.3 Symmetric Pitchfork Bifurcation and Bifurcation Delay 34
      2.2.4 How to Obtain Scaling Laws 37
      2.2.5 Hopf Bifurcation and Bifurcation Delay 43
   2.3 Periodic Orbits and Averaging 45
      2.3.1 Convergence towards a Stable Periodic Orbit 45
      2.3.2 Invariant Manifolds 47
   Bibliographic Comments 48

3 One-Dimensional Slowly Time-Dependent Systems 51
   3.1 Stable Equilibrium Branches 53
      3.1.1 Linear Case 56
      3.1.2 Nonlinear Case 62
      3.1.3 Moment Estimates 66
   3.2 Unstable Equilibrium Branches 68
      3.2.1 Diffusion-Dominated Escape 71
      3.2.2 Drift-Dominated Escape 78

XII   Contents

      3.3   Saddle–Node Bifurcation 84
            3.3.1   Before the Jump 87
            3.3.2   Strong-Noise Regime........................... 90
            3.3.3   Weak-Noise Regime............................ 96
      3.4   Symmetric Pitchfork Bifurcation 97
            3.4.1   Before the Bifurcation 99
            3.4.2   Leaving the Unstable Branch 101
            3.4.3   Reaching a Stable Branch 103
      3.5   Other One-Dimensional Bifurcations 105
            3.5.1   Transcritical Bifurcation 105
            3.5.2   Asymmetric Pitchfork Bifurcation 108
   Bibliographic Comments 110

4   **Stochastic Resonance** 111
      4.1   The Phenomenon of Stochastic Resonance 112
            4.1.1   Origin and Qualitative Description 112
            4.1.2   Spectral-Theoretic Results......................116
            4.1.3   Large-Deviation Results........................124
            4.1.4   Residence-Time Distributions 126
      4.2   Stochastic Synchronisation: Sample-Paths Approach 132
            4.2.1   Avoided Transcritical Bifurcation...................132
            4.2.2   Weak-Noise Regime............................135
            4.2.3   Synchronisation Regime 138
            4.2.4   Symmetric Case 139
   Bibliographic Comments 141

5   **Multi-Dimensional Slow–Fast Systems** 143
      5.1   Slow Manifolds...144
            5.1.1   Concentration of Sample Paths...................145
            5.1.2   Proof of Theorem 5.1.6.........................151
            5.1.3   Reduction to Slow Variables 164
            5.1.4   Refined Concentration Results 166
      5.2   Periodic Orbits...172
            5.2.1   Dynamics near a Fixed Periodic Orbit 172
            5.2.2   Dynamics near a Slowly Varying Periodic Orbit 175
      5.3   Bifurcations 178
            5.3.1   Concentration Results and Reduction 178
            5.3.2   Hopf Bifurcation 185
   Bibliographic Comments 190

6   **Applications** 193
      6.1   Nonlinear Oscillators....................................194
            6.1.1   The Overdamped Langevin Equation 194
            6.1.2   The van der Pol Oscillator......................196
      6.2   Simple Climate Models..................................199

		6.2.1	The North-Atlantic Thermohaline Circulation..........200

        6.2.1 The North-Atlantic Thermohaline Circulation..........200
        6.2.2 Ice Ages and Dansgaard–Oeschger Events............204
   6.3 Neural Dynamics...207
        6.3.1 Excitability..209
        6.3.2 Bursting 212
   6.4 Models from Solid-State Physics.............................214
        6.4.1 Ferromagnets and Hysteresis........................214
        6.4.2 Josephson Junctions 219

A A Brief Introduction to Stochastic Differential Equations ..223
   A.1 Brownian Motion..223
   A.2 Stochastic Integrals..225
   A.3 Strong Solutions..229
   A.4 Semigroups and Generators 230
   A.5 Large Deviations..232
   A.6 The Exit Problem 234
   Bibliographic Comments 236

B Some Useful Inequalities...239
   B.1 Doob's Submartingale Inequality and a Bernstein Inequality ..239
   B.2 Using Tail Estimates...240
   B.3 Comparison Lemma 241
   B.4 Reflection Principle..242

C First-Passage Times for Gaussian Processes..................243
   C.1 First Passage through a Curved Boundary 243
   C.2 Small-Ball Probabilities for Brownian Motion 247
   Bibliographic Comments 248

References..249

List of Symbols and Acronyms..................................263

Index...271

1
Introduction

1.1 Stochastic Models and Metastability

Stochastic models play an important rôle in the mathematical description of systems whose degree of complexity does not allow for deterministic modelling. For instance, in ergodic theory and in equilibrium statistical mechanics, the invariant probability measure has proved an extremely successful concept for the characterisation of a system's long-time behaviour.

Dynamical stochastic models, or *stochastic processes*, play a similarly important rôle in the description of *non-equilibrium* properties. An important class of stochastic processes are *Markov processes*, that is, processes whose future evolution only depends on the present state, and not on the past evolution. Important examples of Markov processes are

- *Markov chains* for discrete time;
- *Markovian jump processes* for continuous time;
- *Diffusion processes* given by solutions of *stochastic differential equations* (SDEs) for continuous time.

This book concerns the sample-path behaviour of a particular class of Markov processes, which are solutions to either non-autonomous or coupled systems of stochastic differential equations. This class includes in particular time-inhomogeneous Markov processes.

There are different approaches to the quantitative study of Markov processes. A first possibility is to study the properties of the system's invariant probability measure, provided such a measure exists. However, it is often equally important to understand transient phenomena. For instance, the speed of convergence to the invariant measure may be extremely slow, because the system spends long time spans in so-called *metastable* states, which can be very different from the asymptotic equilibrium state.

A classical example of metastable behaviour is a glass of water, suddenly exposed to an environment of below-freezing temperature: Its content may stay liquid for a very long time, unless the glass is shaken, in which case the water freezes instantly. Here the asymptotic equilibrium state corresponds to the ice phase, and the metastable state is called *supercooled water*. Such a non-equilibrium behaviour is quite common near first-order phase transition points. It has been established mathematically in various lattice models with Markovian dynamics (e.g., Glauber or Kawasaki dynamics). See, for example, [dH04] for a recent review.

In continuous-time models, metastability occurs, for instance, in the case of a Brownian particle moving in a multiwell potential landscape. For weak noise, the typical time required to escape from a potential well (called *activation time* or *Kramers' time*) is exponentially long in the inverse of the noise intensity. In such situations, it is natural to try to describe the dynamics hierarchically, on two (or perhaps more) different levels: On a coarse-grained level, the rare transitions between potential wells are described by a jump process, while a more precise local description is used for the metastable intrawell dynamics.

Different techniques have been used to achieve such a separation of time and length scales. Let us mention two of them. The first approach is more analytical in nature, while the second one is more probabilistic, as it brings into consideration the notion of *sample paths*.

- **Spectral theory:** The eigenvalues and eigenfunctions of the diffusion's generator yield information on the relevant timescales. Metastability is characterised by the existence of a *spectral gap* between exponentially small eigenvalues, corresponding to the long timescales of interwell transitions, and the rest of the spectrum, associated with the intrawell fluctuations.
- **Large-deviation theory:** The exponential asymptotics of the probability of rare events (e.g., interwell transitions) can be obtained from a variational principle, by minimising a so-called *rate function* (or *action functional*) over the set of all possible escape paths. This yields information on the distribution of random transition times between attractors of the deterministic system.

To be more explicit, let us consider an n-dimensional diffusion process, given by the SDE
$$\mathrm{d}x_t = f(x_t)\,\mathrm{d}t + \sigma F(x_t)\,\mathrm{d}W_t\;, \qquad (1.1.1)$$
where f takes values in $\mathbb{R}^n$, W_t is a k-dimensional Brownian motion, and F takes values in the set $\mathbb{R}^{n\times k}$ of $(n\times k)$-matrices. The state $x_t(\omega)$ of the system depends both on time t, and on the realisation ω of the noise via the sample path $W_t(\omega)$ of the Brownian motion. Depending on the point of view, one may rather be inclined to consider

- either properties of the random variable $\omega \mapsto x_t(\omega)$ for fixed time t, in particular its distribution;
- or properties of the sample path $t \mapsto x_t(\omega)$ for each realisation ω of the noise.

The density of the distribution is connected to the *generator* of the diffusion process, which is given by the differential operator

$$L = \sum_{i=1}^{n} f_i(x)\frac{\partial}{\partial x_i} + \frac{\sigma^2}{2}\sum_{i,j=1}^{n} d_{ij}(x)\frac{\partial^2}{\partial x_i \partial x_j}, \qquad (1.1.2)$$

where $d_{ij}(x)$ are the elements of the square matrix $D(x) = F(x)F(x)^{\mathrm{T}}$. Denote by $p(x,t|y,s)$ the transition probabilities of the diffusion, considered as a Markov process. Then $(y,s) \mapsto p(x,t|y,s)$ satisfies *Kolmogorov's backward equation*

$$-\frac{\partial}{\partial s}u(y,s) = Lu(y,s), \qquad (1.1.3)$$

while $(x,t) \mapsto p(x,t|y,s)$ satisfies *Kolmogorov's forward equation* or *Fokker–Planck equation*

$$\frac{\partial}{\partial t}v(x,t) = L^* v(x,t), \qquad (1.1.4)$$

where L^* is the adjoint operator of L.

Under rather mild assumptions on f and F, the spectrum of L is known to consist of isolated non-positive reals $0 = -\lambda_0 > -\lambda_1 > -\lambda_2 > \ldots$. If $q_k(y)$ and $p_k(x)$ denote the (properly normalised) eigenfunctions of L and $L^\star$ respectively, the transition probabilities take the form

$$p(x,t|y,s) = \sum_{k \geqslant 0} \mathrm{e}^{-\lambda_k(t-s)} q_k(y) p_k(x). \qquad (1.1.5)$$

In particular, the transition probabilities approach the invariant density $p_0(x)$ as $t - s \to \infty$ (because $q_0(x) = 1$). However if, say, the first p eigenvalues of L are very small, then all terms up to $k = p - 1$ contribute significantly to the sum (1.1.5) as long as $t - s \leqslant \lambda_k^{-1}$. Thus the inverses λ_k^{-1} are interpreted as metastable lifetimes, and the corresponding eigenfunctions $p_k(x)$ as "quasi-invariant densities". The eigenfunctions $q_k(y)$ are typically close to linear combinations of indicator functions $1_{\mathcal{A}_k}(y)$, where the sets $\mathcal{A}_k$ are interpreted as supports of metastable states.

In general, determining the spectrum of the generator is a difficult task. Results exist in particular cases, for instance perturbative results for small noise intensity σ.[1] If the drift term f derives from a potential and $F = \mathbb{1}$ is the identity matrix, the small eigenvalues λ_k behave like $\mathrm{e}^{-2h_k/\sigma^2}$, where the h_k are the depths of the potential wells.

One of the alternative approaches to spectral theory is based on the concept of *first-exit times*. Let $\mathcal{D}$ be a bounded, open subset of $\mathbb{R}^n$, and fix an initial condition $x_0 \in \mathcal{D}$. Typically, one is interested in situations where $\mathcal{D}$ is positively invariant under the deterministic dynamics, as a candidate for the support of a metastable state. The *exit problem* consists in characterising the laws of first-exit time

$$\tau_{\mathcal{D}} = \tau_{\mathcal{D}}(\omega) = \inf\{t > 0 \colon x_t(\omega) \notin \mathcal{D}\} \qquad (1.1.6)$$

---

[1] Perturbation theory for generators of the form (1.1.2) for small σ is closely related to semiclassical analysis.

and first-exit location $x_{\tau_\mathcal{D}} \in \partial\mathcal{D}$. For each realisation ω of the noise, $\tau_\mathcal{D}(\omega)$ is the first time at which the sample path $x_t(\omega)$ leaves $\mathcal{D}$. Metastability is associated with the fact that for certain domains $\mathcal{D}$, the first-exit time $\tau_\mathcal{D}$ is likely to be extremely large. Like the transition probabilities, the laws of $\tau_\mathcal{D}$ and of the first-exit location $x_{\tau_\mathcal{D}}$ can be linked to partial differential equations (PDEs) involving the generator L (cf. Appendix A.6).

Another approach to the exit problem belongs to the theory of large deviations. The basic idea is that, although the sample paths of an SDE are in general nowhere differentiable, they tend to concentrate near certain differentiable curves as σ tends to zero. More precisely, let $\mathcal{C} = \mathcal{C}([0,T], \mathbb{R}^n)$ denote the space of continuous functions $\varphi : [0,T] \to \mathbb{R}^n$. The *rate function* or *action functional* of the SDE (1.1.1) is the function $J : \mathcal{C} \to \mathbb{R}^+ \cup \{\infty\}$ defined by

$$J(\varphi) = J_{[0,T]}(\varphi) = \frac{1}{2}\int_0^T (\dot\varphi_t - f(\varphi_t))^\mathrm{T} D(\varphi_t)^{-1}(\dot\varphi_t - f(\varphi_t))\,dt \qquad (1.1.7)$$

for all φ in the Sobolev space $H_1 = H_1([0,T], \mathbb{R}^n)$ of absolutely continuous functions $\varphi : [0,T] \to \mathbb{R}^n$ with square-integrable derivative $\dot\varphi$.[2] By definition, J is infinite for all other paths φ in $\mathcal{C}$. The rate function measures the "cost" of forcing a sample path to track the curve φ — observe that it vanishes if and only if φ_t satisfies the deterministic limiting equation $\dot x = f(x)$. For other φ, the probability of sample paths remaining near φ decreases roughly like $e^{-J(\varphi)/\sigma^2}$. More generally, the probability of x_t belonging to a set of paths Γ, behaves like e^{-J^*/σ^2}, where J^* is the infimum of J over Γ. This can be viewed as an infinite-dimensional variant of the Laplace method.

One of the results of the Wentzell–Freidlin theory is that, for fixed time $t \in [0,T]$,

$$\lim_{\sigma \to 0} \sigma^2 \log \mathbb{P}^{x_0}\{\tau_\mathcal{D} < t\} \qquad (1.1.8)$$
$$= -\inf\{J_{[0,t]}(\varphi) : \varphi \in \mathcal{C}([0,t], \mathbb{R}^n),\ \varphi_0 = x_0,\ \exists s \in [0,t)\ \text{s.t.}\ \varphi_s \notin \mathcal{D}\}.$$

If $\mathcal{D}$ is positively invariant under the deterministic flow, the right-hand side is strictly negative. We can rewrite (1.1.8) as

$$\mathbb{P}^{x_0}\{\tau_\mathcal{D} < t\} \sim e^{-J_{[0,t]}(\varphi^*)/\sigma^2}, \qquad (1.1.9)$$

where $\sim$ stands for logarithmic equivalence, and φ^* denotes a minimiser of $J_{[0,t]}$. The minimisers of $J_{[0,t]}$ are interpreted as "most probable exit paths" of the diffusion from $\mathcal{D}$ in time t; their cost is decreasing in t, and typically tends to a positive limit as $t \to \infty$. Since the limit in (1.1.8) is taken for fixed t, (1.1.9) is not sufficient to estimate the typical exit time, though it suggests that this time grows exponentially fast in $1/\sigma^2$.

---

[2] Note that in the general case, when $D(x)$ is not positive-definite, the rate function J cannot be represented by (1.1.7), but is given by a variational principle.

1.1 Stochastic Models and Metastability

Under additional assumptions on the deterministic dynamics, more precise results have been established. Assume for instance that the closure of $\mathcal{D}$ is contained in the basin of attraction of an asymptotically stable equilibrium point $x^\star$. Then

$$\lim_{\sigma \to 0} \sigma^2 \log \mathbb{E}^{x_0}\{\tau_\mathcal{D}\}$$
$$= \inf_{t > 0,\, y \in \partial \mathcal{D}} \{J_{[0,t]}(\varphi) : \varphi \in \mathcal{C}([0,t], \mathbb{R}^n),\ \varphi_0 = x^\star,\ \varphi_t = y\}, \quad (1.1.10)$$

and thus the expected first-exit time satisfies $\mathbb{E}^{x_0}\{\tau_\mathcal{D}\} \sim e^{J^\star/\sigma^2}$, where $J^\star$ is the right-hand side of (1.1.10).

In the case of gradient systems, that is, for $f(x) = -\nabla U(x)$ and $F(x) = \mathbb{1}$, the cost for leaving a potential well is determined by the minimum of the rate function, taken over all paths starting at the bottom $x^\star$ of the well and occasionally leaving the well. This minimum is attained for a path satisfying $\lim_{t \to -\infty} \varphi_t = x^\star$ and moving against the deterministic flow, that is, $\dot{\varphi}_t = -f(\varphi_t)$. The resulting cost is equal to twice the potential difference H to be overcome, which implies the exponential asymptotics of Kramers' law on the expected escape time behaving like e^{2H/σ^2}.

Applying the theory of large deviations is only a first step in the study of the exit problem, and many more precise results have been obtained, using other methods. Day has proved that the law of the first-exit time from a set $\mathcal{D}$, contained in the basin of attraction of an equilibrium point $x^\star$, converges, if properly rescaled, to an exponential distribution as $\sigma \to 0$:

$$\lim_{\sigma \to 0} \mathbb{P}^{x_0}\{\tau_\mathcal{D} > s \mathbb{E}^{x_0}\{\tau_\mathcal{D}\}\} = e^{-s}. \quad (1.1.11)$$

There are also results on more complex situations arising when the boundary of $\mathcal{D}$ contains a saddle point, or, in dimension two, when $\partial \mathcal{D}$ is an unstable periodic orbit. In the case of gradient systems, the subexponential asymptotics of the expected first-exit time has also been studied in detail. In particular, the leading term of the prefactor $e^{-J^\star/\sigma^2} \mathbb{E}^{x_0}\{\tau_\mathcal{D}\}$ of the expected first-exit time $\mathbb{E}^{x_0}\{\tau_\mathcal{D}\}$ can be expressed in terms of eigenvalues of the Hessian of the potential at $x^\star$ and at the saddle.

All these results allow for quite a precise understanding of the dynamics of transitions between metastable equilibrium states. They also show that between these transitions, sample paths spend long time spans in neighbourhoods of the attractors of the deterministic system. Apart from the timescales associated with these transitions between metastable states, dynamical systems typically have additional inherent timescales or might be subject to slow external forcing. The interplay between these various timescales leads to many interesting phenomena, which are the main focus of the present book. The description of these phenomena will often require to develop methods going beyond spectral analysis and large deviations.

1.2 Timescales and Slow–Fast Systems

Most nonlinear ordinary differential equations (ODEs) are characterised by several different timescales, given, for instance, by relaxation times to equilibrium, periods of periodic orbits, or Lyapunov times, which measure the exponential rate of divergence of trajectories in chaotic regimes.

There are many examples of systems in which such different timescales are well-separated, for instance

- slowly forced systems, such as the Earth's atmosphere, subject to periodic changes in incoming Solar radiation;
- interacting systems with different natural timescales, e.g., predator–prey systems in which the prey reproduces much faster than the predators, or interacting particles of very different mass;
- systems near instability thresholds, in which the modes losing stability evolve on a longer timescale than stable modes; a typical example is Rayleigh–Bénard convection near the appearance threshold of convection rolls.

Such a separation of timescales allows to write the system as a *slow–fast ODE*, of the form

$$\begin{aligned} \varepsilon \frac{\mathrm{d}x_t}{\mathrm{d}t} &= f(x_t, y_t)\,, \\ \frac{\mathrm{d}y_t}{\mathrm{d}t} &= g(x_t, y_t)\,, \end{aligned} \qquad (1.2.1)$$

where ε is a small parameter. Here x_t contains the fast degrees of freedom, and y_t the slow ones. Equivalently, the system can be written in *fast time* $s = t/\varepsilon$ as

$$\begin{aligned} \frac{\mathrm{d}x_s}{\mathrm{d}s} &= f(x_s, y_s)\,, \\ \frac{\mathrm{d}y_s}{\mathrm{d}s} &= \varepsilon g(x_s, y_s)\,. \end{aligned} \qquad (1.2.2)$$

It is helpful to think of (1.2.2) as a perturbation of the parameter-dependent ODE

$$\frac{\mathrm{d}x_s}{\mathrm{d}s} = f(x_s, \lambda)\,, \qquad (1.2.3)$$

in which the parameter λ slowly varies in time. Depending on the dynamics of the associated system (1.2.3), different situations can occur:

- If the associated system admits an asymptotically stable equilibrium point $x^\star(\lambda)$ for each value of λ, the fast degrees of freedom can be eliminated, at least locally, by projection onto the set of equilibria, called *slow manifold*. This yields the effective system

$$\frac{\mathrm{d}y_t}{\mathrm{d}t} = g(x^\star(y_t), y_t) \qquad (1.2.4)$$

for the slow degrees of freedom, which is simpler to analyse than the full system. More generally, if the associated system admits an attractor depending smoothly on λ, such a reduction is obtained by averaging over the invariant measure of the attractor.

- In other situations, most notably when the associated system admits bifurcation points, new phenomena may occur, which cannot be captured by an effective slow system, for instance
  - *jumps* between different (parts of) slow manifolds;
  - *relaxation oscillations*, that is, periodic motions in which fast and slow phases alternate;
  - *hysteresis loops*, e.g., for periodically forced systems, where the state for a given value of the forcing can depend on whether the forcing increases or decreases;
  - *bifurcation delay*, when solutions keep tracking a branch of unstable equilibria for some time after a bifurcation has occurred;
  - solutions *scaling* in a nontrivial way with the adiabatic parameter ε, for instance like ε^ν, where ν is a fractional exponent.

Deterministic slow–fast systems may thus display a rich behaviour, which can only be analysed by a combination of different methods, depending on whether the system operates near a slow manifold, near a bifurcation point, near a periodic orbit, etc.

Adding noise to a slow–fast ODE naturally yields a slow–fast SDE. This adds one or several new timescales to the dynamics, namely the metastable lifetimes (or Kramers' times) mentioned above. The dynamics will depend in an essential way on the relative values of the deterministic system's intrinsic timescales, and the Kramers times introduced by the noise:

- In cases where the Kramers time is much longer than all relevant deterministic timescales, the system is likely to follow the deterministic dynamics for very long time spans, with rare transitions between attractors.
- In cases where the Kramers time is much shorter than all relevant deterministic timescales, noise-induced transitions are frequent, and thus the system's behaviour is well captured by its invariant density.
- The most interesting situations occurs for Kramers' times lying somewhere between the typical slow and fast deterministic timescales. In these cases, noise-induced transitions are frequent for the slow system, but rare for the fast one. As we shall see, this can yield remarkable, and sometimes unexpected phenomena, a typical example being the effect known as *stochastic resonance*.

1.3 Examples

Let us now illustrate the different approaches, and the interplay of timescales, on a few simple examples.

We start with the two-parameter family of ODEs

$$\frac{dx_s}{ds} = \mu x_s - x_s^3 + \lambda, \qquad (1.3.1)$$

which is a classical example in bifurcation theory.[3] In dimension one, all systems can be considered as gradient systems. In the present case, the right-hand side derives from the potential

$$U(x) = -\frac{1}{2}\mu x^2 + \frac{1}{4}x^4 - \lambda x. \qquad (1.3.2)$$

The potential has two wells and a saddle if $\mu^3 > (27/4)\lambda^2$. For further reference, let us denote the bottoms of the wells by $x_\pm^\star$ and the position of the saddle by $x_0^\star$. If $\mu^3 < (27/4)\lambda^2$, the potential only has one well. When $\mu^3 = (27/4)\lambda^2$ and $\lambda \neq 0$, there is a *saddle–node bifurcation* between the saddle and one of the wells. The point $(x, \lambda, \mu) = (0, 0, 0)$ is a *pitchfork bifurcation* point, where all equilibria meet.

To model the effect of noise on this system, we replace now the deterministic family of ODEs (1.3.1) by the family of SDEs

$$dx_s = \left[\mu x_s - x_s^3 + \lambda\right] ds + \sigma\, dW_s. \qquad (1.3.3)$$

The invariant density of the stochastic process x_s is given by

$$p_0(x) = \frac{1}{N} e^{-2U(x)/\sigma^2}, \qquad (1.3.4)$$

where N is the normalisation. The generator of the diffusion is essentially self-adjoint on $L^2(\mathbb{R}, p_0(x)\,dx)$, which implies in particular that the eigenfunctions q_k and p_k of L and L^* are related by $p_k(x) = q_k(x) p_0(x)$. For small noise intensities σ, $p_0(x)$ is strongly concentrated near the local minima of the potential. If the potential has two wells of different depths, the invariant density favours the deeper well.

In situations where the potential has a single minimum of positive curvature, the first non-vanishing eigenvalue $-\lambda_1$ of the generator is bounded away from zero, by a constant independent of σ. This implies that the distribution of x_s relaxes to the invariant distribution $p_0(x)$ in a time of order one.

In the double-well situation, on the other hand, the leading eigenvalue has the expression

$$-\lambda_1 = -\frac{\omega_0}{2\pi}\left[\omega_+ e^{-2h_+/\sigma^2} + \omega_- e^{-2h_-/\sigma^2}\right]\left[1 + \mathcal{O}(\sigma^2)\right], \qquad (1.3.5)$$

---

[3] The right-hand side of (1.3.1) is the universal unfolding of the simplest possible singular vector field of codimension 2, namely $\dot{x} = -x^3$.

where $h_\pm = U(x_0^\star) - U(x_\pm^\star)$ are the depths of the potential wells, and $\omega_i = |U''(x_i^\star)|^{1/2}$, $i \in \{-, 0, +\}$, are the square roots of the curvatures of the potential at its stationary points. The other eigenvalues are again bounded away from zero by some constant $c > 0$. As a consequence, the transition probabilities satisfy

$$p(x, s|y, 0) = \left[1 + e^{-\lambda_1 s} q_1(y) q_1(x) + \mathcal{O}(e^{-cs})\right] p_0(x) . \tag{1.3.6}$$

It turns out that $q_1(y)q_1(x)$ is close to a positive constant (compatible with the normalisation) if x and y belong to the same well, and to -1 if they belong to different wells. As a consequence, the distribution of x_s is concentrated in the starting well for $s \ll \lambda_1^{-1}$, and approaches $p_0(x)$ for $s \gg \lambda_1^{-1}$.

The theory of large deviations confirms this picture, but is more precise, as it shows that between interwell transitions, x_t spends on average times of order e^{2h_-/σ^2} in the left-hand well, and times of order e^{2h_+/σ^2} in the right-hand well. This naturally leads to the following hierarchical description:

- On a coarse-grained level, the dynamics is described by a two-state Markovian jump process, with transition rates $e^{-2h_\pm/\sigma^2}$.
- The dynamics between transitions inside each well can be approximated by ignoring the other well; as the process starting, say, in the left-hand well spends most of the time near the bottom in $x_-^\star$, one may for instance approximate the dynamics of the deviation $x_t - x_-^\star$ by the linearisation

$$dy_s = -\omega_-^2 y_s \, ds + \sigma \, dW_s , \tag{1.3.7}$$

whose solution is an *Ornstein–Uhlenbeck process* of asymptotic variance $\sigma^2/2\omega_-^2$.

Let us now turn to situations in which the potential $U(x) = U(x, \varepsilon s)$ depends slowly on time. That is, we consider SDEs of the form

$$dx_s = -\frac{\partial U}{\partial x}(x_s, \varepsilon s) \, ds + \sigma \, dW_s , \tag{1.3.8}$$

which can also be written, on the scale of the slow time $t = \varepsilon s$ as

$$dx_t = -\frac{1}{\varepsilon} \frac{\partial U}{\partial x}(x_t, t) \, dt + \frac{\sigma}{\sqrt{\varepsilon}} \, dW_t . \tag{1.3.9}$$

The potential-well depths $h_\pm = h_\pm(t)$ naturally also depend on time, and may even vanish if one of the bifurcation curves is crossed. As a result, the "instantaneous" Kramers timescales $e^{2h_\pm(t)/\sigma^2}$ are no longer fixed quantities. If the timescale ε^{-1}, at which the potential changes shape, is longer than the maximal Kramers time of the system, one can expect the dynamics to be a slow modulation of the dynamics for frozen potential. Otherwise, the interplay between the timescales of modulation and of noise-induced transitions becomes nontrivial. Let us discuss this in a few examples.

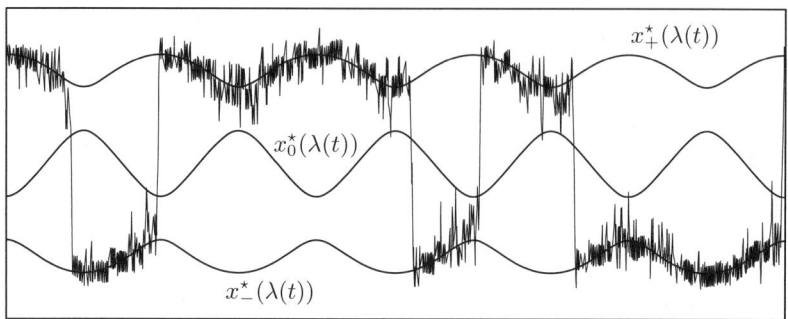

Fig. 1.1. A sample path of Equation (1.3.10) for parameter values $\varepsilon = 0.02$, $\sigma = 0.2$, $A = 0.3$. *Heavy curves* denote locations $x_{\pm}^{\star}(\lambda(t))$ of the potential minima and $x_0^{\star}(\lambda(t))$ of the saddle.

Example 1.3.1. Assume $\mu > 0$ is fixed, say $\mu = 1$, and $\lambda = \lambda(\varepsilon s)$ changes periodically, say $\lambda(\varepsilon s) = A \cos \varepsilon s$. The SDE in slow time reads

$$\mathrm{d}x_t = \frac{1}{\varepsilon}\left[x_t - x_t^3 + A \cos t\right] \mathrm{d}t + \frac{\sigma}{\sqrt{\varepsilon}} \mathrm{d}W_t \ . \qquad (1.3.10)$$

If the amplitude A of the modulation is smaller than the critical value $A_{\mathrm{c}} = \sqrt{4/27}$, the drift term always has two stable and one unstable equilibria, i.e., it derives from a double-well potential. Since the depths $h_{\pm}(t)$ of the two potential wells vary periodically between two values $h_{\min} < h_{\max}$ (and half a period out of phase), so do the instantaneous transition rates. Transitions from left to right are more likely when the left-hand well is shallowest, and vice versa for transitions back from right to left (Fig. 1.1). This yields transition times which are not uniformly distributed within each period, an effect known as *stochastic resonance* (SR). If the modulation period ε^{-1} is larger than twice the maximal Kramers time $\mathrm{e}^{2h_{\max}/\sigma^2}$, transitions are likely to occur at least twice per period, and one is in the so-called *synchronisation regime*.

If the amplitude A of the modulation is larger than the critical value $A_{\mathrm{c}} = \sqrt{4/27}$, the potential periodically changes from double-well to single-well, each time the saddle–node bifurcation curve is crossed. As a result, transitions between potential wells occur even in the deterministic case $\sigma = 0$. Solutions represented in the (λ, x)-plane have the shape of *hysteresis cycles*: When $|\lambda(t)| < A$, the well the particle x_t is "inhabiting" depends on whether $\lambda(t)$ is decreasing or increasing (Fig. 1.2a). The area of the cycles is known to scale like $\mathcal{A}_0 + \varepsilon^{2/3}(A - A_{\mathrm{c}})^{1/3}$, where $\mathcal{A}_0$ is the "static hysteresis area" which is of order 1. The main effect of noise is to enable earlier transitions, already when there is still a double-well configuration. This modifies the hysteresis cycles. We shall see in Section 6.4.1 that for sufficiently large noise intensities, the typical area of cycles decreases by an amount proportional to $\sigma^{4/3}$.

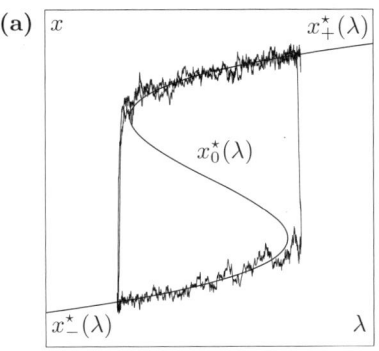 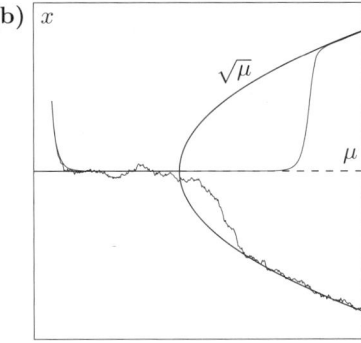

Fig. 1.2. (a) Sample paths of Equation (1.3.10) displaying hysteresis loops in the (λ, x)-plane. Parameter values are $\varepsilon = 0.05$, $\sigma = 0.1$, $A = 0.45$. (b) Sample paths of Equation (1.3.11), describing a dynamic pitchfork bifurcation. The deterministic solution ($\sigma = 0$) displays a bifurcation delay of order 1; noise decreases this delay to order $\sqrt{\varepsilon|\log \sigma|}$. Parameter values are $\varepsilon = 0.02$, $\sigma = 0.02$.

Example 1.3.2. Assume that $\lambda = 0$, and $\mu = \mu(\varepsilon s) = \mu_0 + \varepsilon s$ is slowly growing, starting from some initial value $\mu_0 < 0$. The SDE in slow time thus reads

$$\mathrm{d}x_t = \frac{1}{\varepsilon}\left[\mu(t)x_t - x_t^3\right]\mathrm{d}t + \frac{\sigma}{\sqrt{\varepsilon}}\mathrm{d}W_t . \qquad (1.3.11)$$

When $\mu(t)$ becomes positive, the potential changes from single-well to double-well. In the deterministic case $\sigma = 0$, the system displays what is called *bifurcation delay*: Even if the initial condition x_0 is nonzero, solutions approach the bottom of the well at $x = 0$ exponentially closely. Thus, when $\mu(t)$ becomes positive, the solution continues to stay close to $x = 0$, which is now the position of a saddle, for a time of order 1 before falling into one of the wells.

For positive noise intensity σ, the fluctuations around the saddle help to push the paths away from it, which decreases the bifurcation delay. We shall show in Section 3.4.2 that the delay in the presence of noise is of order $\sqrt{\varepsilon|\log \sigma|}$ (Fig. 1.2b).

Another interesting situation arises when $\mu(t)$ remains positive but approaches zero periodically. Near the moments of low barrier height, sample paths may reach the saddle with a probability that can approach 1. Owing to the symmetry of the potential, upon reaching the saddle the process has probability 1/2 to settle for the other well. The coarse-grained interwell dynamics thus resembles a Bernoulli process.

These examples already reveal some questions about the sample-path behaviour that we would like to answer:

- How long do sample paths remain concentrated near stable equilibrium branches, that is, near the bottom of slowly moving potential wells?

(a) 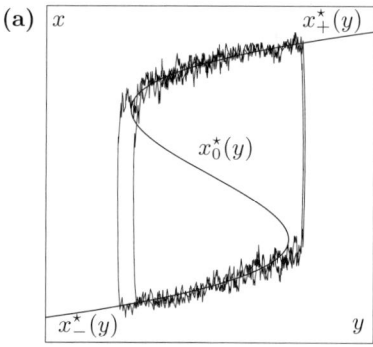 (b)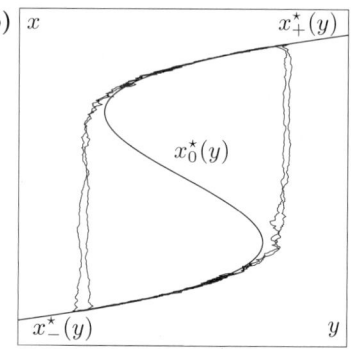

Fig. 1.3. Sample paths of the van der Pol Equation (1.3.12) **(a)** with noise added to the fast variable x, and **(b)** with noise added to the slow variable y. The *heavy curves* have the equation $y = x^3 - x$. Parameter values are $\varepsilon = 0.01$ and **(a)** $\sigma = 0.1$, **(b)** $\sigma = 0.2$.

- How fast do sample paths depart from unstable equilibrium branches, that is, from slowly moving saddles?
- What happens near bifurcation points, when the number of equilibrium branches changes?
- What can be said about the dynamics far from equilibrium branches?

In the two examples just given, the separation of slow and fast timescales manifests itself in a slowly varying parameter, or a slow forcing. We now mention two examples of slow–fast systems, in which the slow variables are coupled dynamically to the fast ones.

Example 1.3.3 (Van der Pol oscillator). The deterministic slow–fast system

$$\varepsilon \dot{x} = x - x^3 + y ,$$
$$\dot{y} = -x , \qquad\qquad (1.3.12)$$

describes the dynamics of an electrical oscillating circuit with a nonlinear resistance.[4] Small values of ε correspond to large damping. The situation resembles the one of Example 1.3.1, except that the periodic driving is replaced by a new dynamic variable y. The term $x - x^3 + y$ can be considered as deriving from the potential $-\frac{1}{2}x^2 + \frac{1}{4}x^4 - yx$, which has one or two wells depending on the value of y. For positive x, y slowly decreases until the right-hand well disappears, and x falls into the left-hand well. From there on, y starts increasing again until the left-hand well disappears. The system thus displays self-sustained periodic motions, so-called *relaxation oscillations*.

---

[4]See Example 2.1.6 for the more customary formulation of the van der Pol oscillator as a second-order equation, and the time change yielding the above slow–fast system.

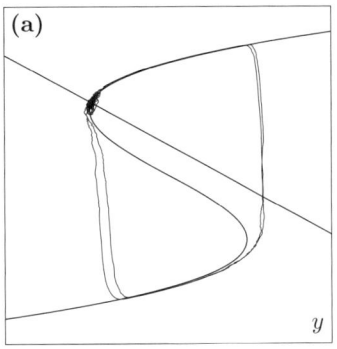 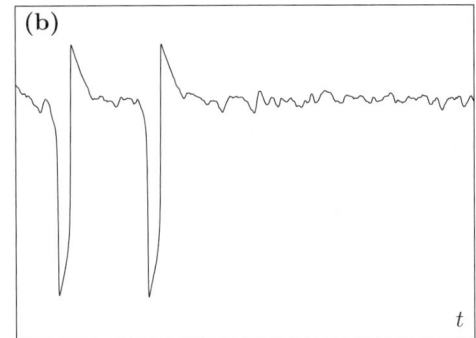

Fig. 1.4. Sample paths of the Fitzhugh–Nagumo Equation (1.3.13) with noise added to the y-variable, **(a)** in the (y,x)-plane, and **(b)** in the (t,x)-plane. The curve $y = x^3 - x$ and the line $y = \alpha - \beta x$ are indicated. Parameter values are $\varepsilon = 0.03$, $\alpha = 0.27$, $\beta = 1$ and $\sigma = 0.03$.

Adding noise to the system changes the shape and period of the cycles. In particular, if noise is added to the fast variable x, paths may cross the curve $y = x^3 - x$, which results, as in Example 1.3.1, in smaller cycles (Fig. 1.3a).

Example 1.3.4 (Excitability). The *Fitzhugh–Nagumo equations* are a simplification of the *Hodgkin–Huxley equations* modelling electric discharges across cell membranes, and generalise the van der Pol equations. They can be written in the form

$$\varepsilon \dot{x} = x - x^3 + y \,,$$
$$\dot{y} = \alpha - \beta x - y \,. \tag{1.3.13}$$

The difference to the previous situation is that $\dot{y}$ changes sign on the line $\beta x = \alpha - y$. Depending on how this line intersects the curve $y = x^3 - x$, the system admits a stable equilibrium point instead of a limit cycle. Noise, however, may drive the system near the saddle–node bifurcation point, and induce a pulse, before the system returns to equilibrium (Fig. 1.4). This phenomenon is called *excitability*.

1.4 Reader's Guide

The examples we just discussed show that the solutions of slow–fast SDEs typically display an alternation of small fluctuations near and quick transitions between attractors. The phases between transitions can last very long, a characteristic feature of metastable systems. The invariant distribution of the stochastic process, if existing, is not able to capture this non-equilibrium behaviour.

In this book, we develop an approach to slow–fast SDEs based on a characterisation of typical sample paths. We aim at constructing sets of optimal shape, in which the sample paths are concentrated for long time spans. Then we proceed to characterise the distribution of first-exit times from these sets by providing rather precise concentration estimates for the sample paths.

Our point of view will be to assume that the deterministic dynamics is sufficiently well known, meaning for instance that we know the invariant sets of the associated system, and have some information on their basins of attraction. We are then interested in what happens when noise is added, particularly when the noise intensity is such that the Kramers transition times and the typical timescales of the deterministic system are of comparable length. We are mainly interested in the following questions:

- What is the shape of the domains in which sample paths are concentrated between transitions?
- What can we say about the distribution of waiting times between transitions?
- How much time is required for a typical transition?
- When a transition occurs, where does it lead the system?
- Do certain quantities exhibit a scaling-law behaviour as a function of parameters such as ε, measuring the difference in timescales, and the noise intensity σ? Is there an easy way to determine these scaling laws?

The approach we develop provides a very precise description in the case of one-dimensional systems. For instance, we know that the probability to leave the neighbourhood of a stable equilibrium branch grows linearly with time during long time spans. In the general, n-dimensional case, the results are not yet as precise: For instance, we only obtain upper and lower bounds on the probability of leaving the neighbourhood of a stable slow manifold, with different time-dependences. Moreover, we treat only some of the most generic bifurcations.

We have endeavoured to present the material gradually, starting with simpler particular cases before discussing the general case, and intermingling abstract parts with more applied ones. Here is a brief overview of the contents of the chapters.

- **Chapter 2** contains an overview of the results on *deterministic* slow–fast systems that will be needed in the sequel. We start by giving standard results on the motion near asymptotically stable *slow manifolds*, that is, near collections of equilibrium points of the associated system. Then we discuss the most generic *dynamic bifurcations*, which arise when a slow manifold loses stability. Finally, we briefly describe the case of an associated system admitting stable periodic orbits.
- **Chapter 3** considers the effect of noise on a particular class of slow–fast systems, namely *one-dimensional* equations with slowly varying parameters. In this case, we can give precise estimates on the time spent by sample paths near stable and unstable equilibrium branches. We then discuss in detail the dynamic saddle–node, pitchfork and transcritical bifurcations with noise.

- **Chapter 4** is concerned with a special case of such slowly-driven, one-dimensional systems, namely systems which display *stochastic resonance*. The first part of the chapter gives an overview of mathematical results based on spectral theory, and on the theory of large deviations. In the second part, we apply the methods of Chapter 3 to the *synchronisation regime*, in which typical sample paths make regular transitions between potential wells.
- In **Chapter 5**, we turn to the general case of multidimensional, fully coupled slow–fast systems with noise. A substantial part of the discussion is concerned with the dynamics near *asymptotically stable* slow manifolds. We prove results on the concentration of sample paths in an explicitly constructed neighbourhood of the manifold, and on the reduction of the dynamics to an effective equation involving only slow variables. The remainder of the chapter is concerned with the dynamics near periodic orbits, and with dynamic bifurcations, in particular the Hopf bifurcation.
- Finally, **Chapter 6** illustrates the theory by giving some applications to problems from physics, biology, and climatology. In particular, we discuss bistable models of the North-Atlantic thermohaline circulation, the phenomena of excitability and bursting in neural dynamics, and the effect of noise on hysteresis cycles in ferromagnets.

The appendices gather some information on stochastic processes as needed for the purpose of the book. Appendix A gives a general introduction to stochastic integration, Itô calculus, SDEs, and the related large-deviation estimates provided by the Wentzell–Freidlin theory. Appendix B collects some useful inequalities, and Appendix C discusses some results on first-passage-time problems for Gaussian processes.

The mathematical notations we use are rather standard, except for some unavoidable compromises in the choice of symbols for certain variables. We shall introduce the set-up in each chapter, as soon as it is first needed. The list of symbols at the end of the book gives an overview of the most commonly used notations, with the place of their first appearance.

Bibliographic Comments

The use of stochastic models as simplified descriptions of complex deterministic systems is very widespread. However, its justification is a difficult problem, which has so far only been solved in a number of particular situations:

- For systems coupled to a (classical) heat reservoir, the effective description by a stochastic differential equation goes back to Ford, Kac and Mazur for a linear model [FKM65], and was developed by Spohn and Lebowitz [SL77]. This approach has then been extended to more general nonlinear systems (chains of nonlinear oscillators) in [EPRB99]; see, for instance, [RBT00, RBT02] for recent developments.

- For quantum heat reservoirs, the situation is still far from a complete understanding. See, for instance, [Mar77, Att98, vK04] for different approaches.
- Another class of systems for which an effective stochastic description has been proposed are slow–fast systems of differential equations. Khasminskii [Kha66] seems to have been the first to suggest that the effect of fast degrees of freedom on the dynamics of slow variables can be approximated by a noise term, an idea that Hasselmann applied to climate modelling [Has76]. See, for instance, [Arn01, Kif03, JGB+03] for extensions of these ideas.

Concerning the mathematical foundations of the theory of stochastic differential equations, we refer to the bibliographical comments at the end of Appendix A.

We mentioned different mathematical approaches to the description of SDEs. A (very) partial list of references is the following:

- For large-deviation results for the solutions of SDEs, the standard reference is the book by Freidlin and Wentzell [FW98], see also the corresponding chapters in [DZ98].
- For a review of spectral-theoretic results, see, for instance, [Kol00].
- Precise relations between spectral properties of the generator, metastable lifetimes, and potential theory for reversible diffusions have recently been established in [BEGK04, BGK05], and [HKN04, HN05].
- Another approach to SDEs is based on the concept of random dynamical systems, see [Arn98, Sch89].

Concerning deterministic slow–fast differential equations, we refer to the bibliographical comments at the end of Chapter 2.

2
Deterministic Slow–Fast Systems

A *slow–fast system* involves two kinds of dynamical variables, evolving on very different timescales. The ratio between the fast and slow timescale is measured by a small parameter ε. A slow–fast ordinary differential equation (ODE) is customarily written in the form[1]

$$\begin{aligned} \varepsilon \dot{x} &= f(x,y)\,, \\ \dot{y} &= g(x,y)\,, \end{aligned} \qquad (2.0.1)$$

where the components of $x \in \mathbb{R}^n$ are called *fast variables*, while those of $y \in \mathbb{R}^m$ are called *slow variables*. Rather than considering ε as a fixed parameter, it is of interest to study how the dynamics of the system (2.0.1) depends on ε, for all values of ε in an interval $(0, \varepsilon_0]$.

The particularity of a slow–fast ODE such as (2.0.1) is that, instead of remaining a system of coupled differential equations in the limit $\varepsilon \to 0$, it becomes an algebraic–differential system. Such equations are called *singularly perturbed*. Of course, it is always possible to convert (2.0.1) into a regular perturbation problem: The derivatives x' and y' of x and y with respect to the *fast time* $s = t/\varepsilon$ satisfy the system

$$\begin{aligned} x' &= f(x,y)\,, \\ y' &= \varepsilon g(x,y)\,, \end{aligned} \qquad (2.0.2)$$

which can be considered as a small perturbation of the *associated system* (or *fast system*)

$$x' = f(x, \lambda)\,, \qquad (2.0.3)$$

in which λ plays the rôle of a parameter. However, standard methods from perturbation theory only allow one to control deviations of the solutions of (2.0.2)

---

[1] In many applications, the right-hand side of (2.0.1) shows an explicit dependence on ε. We refrain from introducing this set-up here, but refer to Remark 2.1.4 concerning an equivalent reformulation.

from those of the associated system (2.0.3) for fast times s of order 1 at most, that is, for t of order ε. The dynamics on longer timescales has to be described by other methods, belonging to the field of *singular perturbation theory*.

The behaviour of the slow–fast system (2.0.1) is nonetheless strongly linked to the dynamics of the associated system (2.0.3). We recall in this chapter results from singular perturbation theory corresponding to the following situations.

- In Section 2.1, we consider the simplest situation, occurring when the associated system admits a hyperbolic equilibrium point $x^\star(\lambda)$ for all parameters λ in some domain. The set of all points $(x^\star(y), y)$ is called a *slow manifold* of the system. We state classical results by Tihonov and Fenichel describing the dynamics near such a slow manifold.
- In Section 2.2, we consider situations in which a slow manifold ceases to be hyperbolic, giving rise to a so-called *dynamic bifurcation*. These singularities can cause new phenomena such as bifurcation delay, relaxation oscillations, and hysteresis. We summarise results on the most generic dynamic bifurcations, including saddle–node, pitchfork and Hopf bifurcations.
- In Section 2.3, we turn to situations in which the associated system admits a stable periodic orbit, depending on the parameter λ. In this case, the dynamics of the slow variables is well approximated by averaging the system over the fast motion along the periodic orbit.

We will not consider more difficult situations, arising for more complicated asymptotic dynamics of the associated system (e.g., quasiperiodic or chaotic). Such situations have been studied in the deterministic case, but the analysis of their stochastic counterpart lies beyond the scope of this book.

2.1 Slow Manifolds

2.1.1 Definitions and Examples

We consider the slow–fast system (2.0.1), where we assume that f and g are twice continuously differentiable in a connected, open set $\mathcal{D} \subset \mathbb{R}^n \times \mathbb{R}^m$. The simplest situation occurs when the associated system admits one or several hyperbolic equilibrium points, i.e., points on which f vanishes, while the Jacobian matrix $\partial_x f$ of $x \mapsto f(x, \lambda)$ has no eigenvalue on the imaginary axis. Collections of such points define *slow manifolds* of the system. More precisely, we will distinguish between the following types of slow manifolds:

Definition 2.1.1 (Slow manifold). *Let $\mathcal{D}_0 \subset \mathbb{R}^m$ be a connected set of nonempty interior, and assume that there exists a continuous function $x^\star : \mathcal{D}_0 \to \mathbb{R}^n$ such that $(x^\star(y), y) \in \mathcal{D}$ and*

$$f(x^\star(y), y) = 0 \tag{2.1.1}$$

for all $y \in \mathcal{D}_0$. Then the set $\mathcal{M} = \{(x,y)\colon x = x^\star(y), y \in \mathcal{D}_0\}$ is called a *slow manifold* of the system (2.0.1).

Let $A^\star(y) = \partial_x f(x^\star(y), y)$ denote the *stability matrix* of the associated system at $x^\star(y)$. The slow manifold $\mathcal{M}$ is called

- *hyperbolic* if all eigenvalues of $A^\star(y)$ have nonzero real parts for all $y \in \mathcal{D}_0$;
- *uniformly hyperbolic* if all eigenvalues of $A^\star(y)$ have real parts which are uniformly bounded away from zero for $y \in \mathcal{D}_0$;
- *asymptotically stable* if all eigenvalues of $A^\star(y)$ have negative real parts for all $y \in \mathcal{D}_0$;
- *uniformly asymptotically stable* if all eigenvalues of $A^\star(y)$ have negative real parts which are uniformly bounded away from zero for $y \in \mathcal{D}_0$.

Finally, a hyperbolic slow manifold $\mathcal{M}$ is called *unstable* if at least one of the eigenvalues of the stability matrix $A^\star(y)$ has positive real part for some $y \in \mathcal{D}_0$.

In the particular case of a one-dimensional slow variable, a slow manifold is also called *equilibrium branch* (the terminology comes from bifurcation theory). The graph of all equilibrium branches versus y is called *bifurcation diagram*.

In the adiabatic limit $\varepsilon \to 0$, the dynamics on the slow manifold is described by the so-called *reduced system* (or *slow system*)

$$\dot{y} = g(x^\star(y), y) \,. \tag{2.1.2}$$

However, it remains yet to be proved that the reduced system indeed gives a good approximation of the original system's dynamics. Before developing the general theory, let us consider a few examples.

Example 2.1.2 (Slowly varying parameters). Consider a dynamical system

$$x' = f_\mu(x) \,, \tag{2.1.3}$$

depending on a set of parameters $\mu \in \mathbb{R}^p$ (the prime always indicates derivation with respect to a fast time s). In an experimental set-up, it is often possible to modify one or several parameters at will, for instance an energy supply, an adjustable resistor, etc. Modifying parameters sufficiently slowly (in comparison to the relaxation time of the system) may allow measurements for different parameter values to be taken in the course of a single experiment. This procedure can be described mathematically by setting $\mu = h(\varepsilon s)$, for a given function $h \colon \mathbb{R} \to \mathbb{R}^p$. Introducing the slow time $t = \varepsilon s$, one arrives at the slow–fast system

$$\begin{aligned} \varepsilon \dot{x} &= f_{h(y)}(x) =: f(x,y) \,, \\ \dot{y} &= 1 \,. \end{aligned} \tag{2.1.4}$$

Equation (2.1.3) is of course the associated system of (2.1.4). If f_μ vanishes on some set of equilibrium points $x = x^\star_\mu$, then the slow manifold is given by the equation $x = x^\star_{h(y)} =: x^\star(y)$. Note that the reduced dynamics is trivial, since it is always given by the equation $\dot y = 1$.

Example 2.1.3 (Overdamped motion of a particle in a potential). The dynamics of a particle in $\mathbb{R}^d$, subject to a field of force deriving from a potential $U(x)$, and a viscous drag, is governed by the second-order equation

$$x'' + \gamma x' + \nabla U(x) = 0 \,. \tag{2.1.5}$$

One possible way to write this equation as a first-order system is

$$\begin{aligned} x' &= \gamma(y - x)\,, \\ y' &= -\frac{1}{\gamma}\nabla U(x)\,. \end{aligned} \tag{2.1.6}$$

Let us assume that the friction coefficient γ is large, and set $\varepsilon = 1/\gamma^2$. With respect to the slow time $t = \sqrt{\varepsilon}s$, the dynamics is governed by the slow–fast system

$$\begin{aligned} \varepsilon \dot x &= y - x\,, \\ \dot y &= -\nabla U(x)\,. \end{aligned} \tag{2.1.7}$$

The slow manifold is given by $x^\star(y) = y$. The stability matrix being simply $A^\star(y) \equiv -\mathbb{1}$, the slow manifold is uniformly asymptotically stable, and the reduced dynamics is governed by the equation

$$\dot y = -\nabla U(y)\,, \tag{2.1.8}$$

or, equivalently, $\dot x = -\nabla U(x)$. This relation is sometimes called *Aristotle's law*, since it reflects the fact that at large friction, velocity is proportional to force, as if inertia were absent.

Remark 2.1.4. For simplicity, we assumed in (2.0.1) that the right-hand side does not explicitly depend on ε. This is no real constraint as we can always introduce a dummy variable for ε. We rewrite the slow–fast system

$$\begin{aligned} \varepsilon \dot x &= f(x, y, \varepsilon)\,, \\ \dot y &= g(x, y, \varepsilon)\,, \end{aligned} \tag{2.1.9}$$

as

$$\begin{aligned} \varepsilon \dot x &= f(x, y, z)\,, \\ \dot y &= g(x, y, z)\,, \\ \dot z &= 0\,, \end{aligned} \tag{2.1.10}$$

and consider z as an additional slow variable. Thus slow manifolds for (2.1.9) are of the form $x^\star(y, \varepsilon)$.[2]

---

[2] Some authors do not allow for ε-dependent slow manifolds and consider $x^\star(y, 0)$, obtained by setting $\varepsilon = 0$, as the slow manifold.

The following example provides an application with ε-dependent right-hand side.

Example 2.1.5 (Stommel's box model). Simple climate models, whose dynamic variables are averaged values of physical quantities over some large volumes, or *boxes*, are called box models. Stommel's model gives a simple qualitative description of the North Atlantic thermohaline circulation. The fast variable is proportional to the temperature difference between a low- and a high-latitude box, and the slow variable is proportional to the salinity difference (see Section 6.2 for a more detailed description). Their dynamics is governed by the system

$$\begin{aligned} \varepsilon \dot{x} &= -(x-1) - \varepsilon x Q(x-y) \,, \\ \dot{y} &= \mu - y Q(x-y) \,, \end{aligned} \qquad (2.1.11)$$

where the small parameter ε reflects the fact that the relaxation time for the temperature difference is much shorter than the one for the salinity difference. The parameter μ is proportional to the freshwater flux, while the function Q describes the Fickian mass exchange. Typical choices are $Q(z) = 1 + \eta |z|$ [Sto61][3] or $Q(z) = 1 + \eta^2 z^2$ [Ces94], where η, which depends on the volume of the boxes, is of order one.

The slow manifold is of the form $x^*(y,\varepsilon) = 1 - \varepsilon Q(1-y) + \mathcal{O}(\varepsilon^2)$ for small ε and y from a bounded set, and is obviously uniformly asymptotically stable. The reduced dynamics is given by an equation of the form

$$\dot{y} = \mu - y Q(1-y) + \mathcal{O}(\varepsilon) \,. \qquad (2.1.12)$$

For the above-mentioned choices of Q, depending on the values of μ and η, there can be up to three equilibrium points, one of which is unstable.

Example 2.1.6 (Van der Pol oscillator). The van der Pol oscillator is an electric circuit including a current-dependent resistor (see Section 6.1.2). Its (scalar) equation is

$$x'' + \gamma(x^2 - 1)x' + x = 0 \,. \qquad (2.1.13)$$

For $\gamma = 0$, it reduces to an harmonic oscillator. For large γ, however, the dynamics becomes very far from harmonic. Proceeding as in Example 2.1.3, (2.1.13) can be transformed into the slow–fast system

$$\begin{aligned} \varepsilon \dot{x} &= y + x - \frac{x^3}{3} \,, \\ \dot{y} &= -x \,, \end{aligned} \qquad (2.1.14)$$

where again $\varepsilon = 1/\gamma^2$ and $t = \sqrt{\varepsilon} s$. The associated system has up to three equilibria, and the slow manifold is a curve, given implicitly by the equation

---
[3] Note that Stommel's choice of Q does not satisfy our differentiability assumption. Since left and right derivatives exist, the system can nevertheless be studied by patching together solutions for $x < y$ and $x > y$.

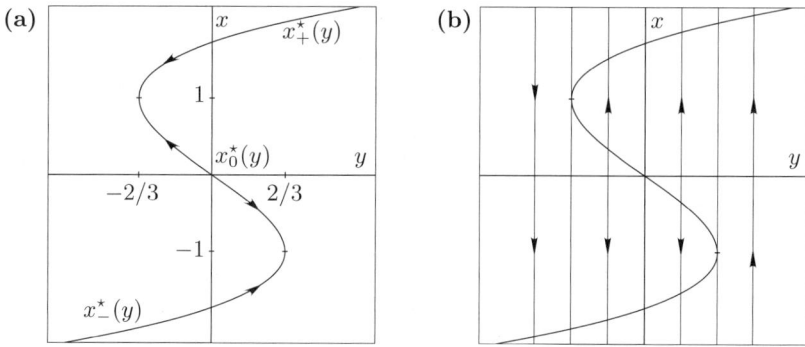

Fig. 2.1. Behaviour of the van der Pol equation in the singular limit $\varepsilon \to 0$: (a) the reduced dynamics (2.1.17) on the slow manifold, (b) the fast dynamics of the associated system $x' = y + x - \frac{1}{3}x^3$.

$x^3/3 - x = y$. The stability matrix, which is a scalar in this case, has value $1 - x^2$, showing that points with $|x| > 1$ are stable while points with $|x| < 1$ are unstable. The slow manifold is thus divided into three equilibrium branches (Fig. 2.1),

$$\begin{aligned} x_-^\star &: (-\infty, \tfrac{2}{3}) \to (-\infty, -1) \ , \\ x_0^\star &: (-\tfrac{2}{3}, \tfrac{2}{3}) \to (-1, 1) \ , \\ x_+^\star &: (-\tfrac{2}{3}, \infty) \to (1, \infty) \ , \end{aligned} \qquad (2.1.15)$$

meeting at two *bifurcation points* $\pm(1, -\tfrac{2}{3})$.

The reduced system is best expressed with respect to the variable x. Since on the slow manifold one has

$$-x = \dot{y} = (x^2 - 1)\dot{x} \ , \qquad (2.1.16)$$

the reduced dynamics is governed by the equation

$$\dot{x} = -\frac{x}{x^2 - 1} \ , \qquad (2.1.17)$$

which becomes singular in $x = \pm 1$. We will see below (Example 2.2.3 in Section 2.2.2) that the true dynamics for $\varepsilon > 0$ involves *relaxation oscillations*, in which slow and fast motions alternate (Fig. 2.5).

2.1.2 Convergence towards a Stable Slow Manifold

The first results showing that the reduced equation on a slow manifold may indeed give a good approximation to the full dynamics are due to Tihonov [Tih52] and Gradšteǐn [Gra53]. In particular, the following theorem on exponentially fast convergence of solutions to an ε-neighbourhood of a uniformly asymptotically stable slow manifold is contained in their results.

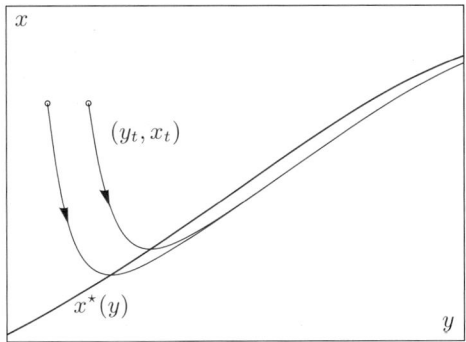

Fig. 2.2. Two orbits approaching a uniformly asymptotically stable slow manifold.

Theorem 2.1.7 (Convergence towards a slow manifold). *Let $\mathcal{M} = \{(x, y)\colon x = x^\star(y), y \in \mathcal{D}_0\}$ be a uniformly asymptotically stable slow manifold of the slow–fast system (2.0.1). Let f and g, as well as all their derivatives up to order two, be uniformly bounded in norm in a neighbourhood $\mathcal{N}$ of $\mathcal{M}$. Then there exist positive constants $\varepsilon_0, c_0, c_1, \kappa = \kappa(n), M$ such that, for $0 < \varepsilon \leqslant \varepsilon_0$ and any initial condition $(x_0, y_0) \in \mathcal{N}$ satisfying $\|x_0 - x^\star(y_0)\| \leqslant c_0$, the bound*

$$\|x_t - x^\star(y_t)\| \leqslant M\|x_0 - x^\star(y_0)\| e^{-\kappa t/\varepsilon} + c_1 \varepsilon \qquad (2.1.18)$$

holds as long as $y_t \in \mathcal{D}_0$.

This result shows that after a time of order $\varepsilon|\log \varepsilon|$, all orbits starting in a neighbourhood of order 1 of the slow manifold $\mathcal{M}$ will have reached a neighbourhood of order ε, where they stay as long as the slow dynamics permits (Fig. 2.2). This phenomenon is sometimes called "slaving principle" or "adiabatic reduction".

A simple proof of Theorem 2.1.7 uses Lyapunov functions. Since, for any $y \in \mathcal{D}_0$, $x^\star(y)$ is an asymptotically stable equilibrium point of the associated system, there exists a Lyapunov function $x \mapsto V(x, y)$, admitting a non-degenerate minimum in $x^\star(y)$, with $V(x^\star(y), y) = 0$, and satisfying

$$\langle \nabla_x V(x, y), f(x, y) \rangle \leqslant -2\kappa V(x, y) \qquad (2.1.19)$$

in a neighbourhood of $x^\star(y)$, see, e.g., [Arn92, § 22]. One can choose $V(\cdot, y)$ to be a quadratic form in $x - x^\star(y)$, with $V(x, y)/\|x - x^\star(y)\|^2$ uniformly bounded above and below by positive constants. Taking into account the y-dependence of V and $x^\star$, one arrives at the relation

$$\varepsilon \frac{\mathrm{d}}{\mathrm{d}t} V(x_t, y_t) =: \varepsilon \dot{V}_t \leqslant -2\kappa V_t + \mathrm{const}\, \varepsilon \sqrt{V_t}\,, \qquad (2.1.20)$$

which implies (2.1.18).

2.1.3 Geometric Singular Perturbation Theory

While Tihonov's theorem is useful to describe the orbits' approach to a small neighbourhood of a uniformly asymptotically stable slow manifold, it does not provide a very precise picture of the dynamics in this neighbourhood. Fenichel has initiated a geometrical approach [Fen79], which allows for a description of the dynamics in terms of invariant manifolds. The simplest result of this kind concerns the existence of an invariant manifold near a hyperbolic (not necessarily stable) slow manifold.

Theorem 2.1.8 (Existence of an adiabatic manifold). *Let the slow manifold $\mathcal{M} = \{(x,y) \colon x = x^\star(y), y \in \mathcal{D}_0\}$ of the slow–fast system (2.0.1) be uniformly hyperbolic. Then there exists, for sufficiently small ε, a locally invariant manifold*

$$\mathcal{M}_\varepsilon = \{(x,y) \colon x = \bar{x}(y, \varepsilon), y \in \mathcal{D}_0\}, \qquad (2.1.21)$$

where $\bar{x}(y, \varepsilon) = x^\star(y) + \mathcal{O}(\varepsilon)$. In other words, if the initial condition is taken on $\mathcal{M}_\varepsilon$, that is, $x_0 = \bar{x}(y_0, \varepsilon)$, then $x_t = \bar{x}(y_t, \varepsilon)$ as long as $y_t \in \mathcal{D}_0$.

We shall call $\mathcal{M}_\varepsilon$ an *adiabatic manifold*. The dynamics on $\mathcal{M}_\varepsilon$ is governed by the equation

$$\dot{y} = g(\bar{x}(y, \varepsilon), y), \qquad (2.1.22)$$

which reduces to (2.1.2) in the limit $\varepsilon \to 0$. By extension, we also call it *reduced system*. The deviations from the limiting system can now be treated by standard methods of regular perturbation theory.

Fenichel has in fact proved more general results, in particular on the existence of invariant manifolds associated with the stable and unstable manifolds of a family of hyperbolic equilibria of the fast system. These results, together with their many refinements, are known as *geometric singular perturbation theory*. See, for instance, [Jon95] for a review.

Theorem 2.1.8 can be proved by using the centre-manifold theorem (see, for instance, [Car81]). Indeed, by again viewing ε as a dummy dynamic variable, and using the fast time $s = t/\varepsilon$, the slow–fast system can be rewritten as

$$\begin{aligned} x' &= f(x,y), \\ y' &= \varepsilon g(x,y), \\ \varepsilon' &= 0. \end{aligned} \qquad (2.1.23)$$

Any point of the form $(x^\star(y), y, 0)$ is an equilibrium point of this system.[4] The linearisation of (2.1.23) around such a point has the structure

$$\begin{pmatrix} A^\star(y) & \partial_y f(x^\star(y), y) & 0 \\ 0 & 0 & g(x^\star(y), y) \\ 0 & 0 & 0 \end{pmatrix}. \qquad (2.1.24)$$

---

[4] There might also exist equilibrium points $(x^\star(y), y, \varepsilon)$ with $\varepsilon > 0$, namely if $g(x^\star(y), y) = 0$. At these points, slow and adiabatic manifold coïncide for *all* ε.

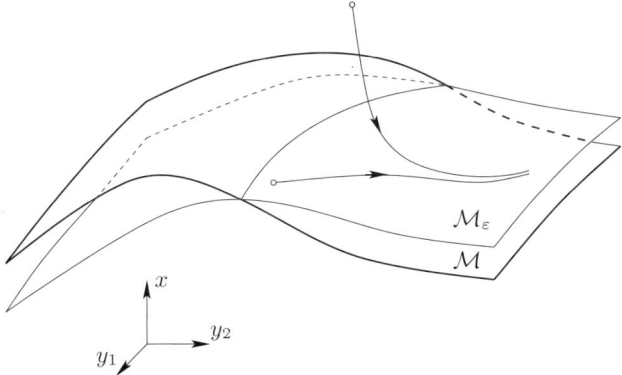

Fig. 2.3. An adiabatic manifold $\mathcal{M}_\varepsilon$ associated with a uniformly asymptotically stable slow manifold. Orbits starting in its vicinity converge exponentially fast to an orbit on the adiabatic manifold.

Hence it admits $m+1$ vanishing eigenvalues, while the eigenvalues of $A^\star(y)$ are bounded away from the imaginary axis by assumption. Thus the centre-manifold theorem (cf. [Car81, Theorem 1, p. 4]) implies the existence of an invariant manifold.

Two important properties of centre manifolds carry over to this special case. Firstly, if the slow manifold $\mathcal{M}$ is uniformly asymptotically stable, then the centre manifold $\mathcal{M}_\varepsilon$ is locally attractive, cf. [Car81, Theorem 2, p. 4]. This means that for each initial condition (x_0, y_0) sufficiently close to $\mathcal{M}_\varepsilon$, there exists a particular solution $\widehat{y}_t$ of the reduced equation (2.1.22) such that

$$\|(x_t, y_t) - (\widehat{x}_t, \widehat{y}_t)\| \leqslant M \|(x_0, y_0) - (\widehat{x}_0, \widehat{y}_0)\| e^{-\kappa t/\varepsilon} \qquad (2.1.25)$$

for some $M, \kappa > 0$, where $\widehat{x}_t = \bar{x}(\widehat{y}_t, \varepsilon)$. Thus, after a short time, the orbit (x_t, y_t) has approached an orbit on the invariant manifold.

The second property concerns approximations of $\bar{x}(y, \varepsilon)$. Using invariance, one easily sees from (2.0.1) that $\bar{x}(y, \varepsilon)$ must satisfy the partial differential equation

$$\varepsilon \partial_y \bar{x}(y, \varepsilon) g(\bar{x}(y, \varepsilon), y) = f(\bar{x}(y, \varepsilon), y) \ . \qquad (2.1.26)$$

This equation is difficult to solve in general. However, the approximation theorem for centre manifolds (cf. [Car81, Theorem 3, p. 5]) states that, if f and g are sufficiently differentiable, then any power series in ε or y, satisfying (2.1.26) to order ε^k or y^k, is indeed an approximation of $\bar{x}(y, \varepsilon)$ to that order. It is thus possible to construct an approximate solution of (2.1.26) by inserting a power series in ε and equating like powers. The first orders of the expansion in ε are

$$\begin{aligned}\bar{x}(y, \varepsilon) &= x^\star(y) + \varepsilon A^\star(y)^{-1} \partial_y x^\star(y) g(x^\star(y), y) + \mathcal{O}(\varepsilon^2) \\ &= x^\star(y) - \varepsilon A^\star(y)^{-2} \partial_y f(x^\star(y), y) g(x^\star(y), y) + \mathcal{O}(\varepsilon^2) \ , \end{aligned} \qquad (2.1.27)$$

where the last line follows from the implicit-function theorem, applied to the equation $f(x^\star(y), y) = 0$.

Example 2.1.9 (Overdamped motion of a particle in a potential – continued). Consider again the slow–fast system of Example 2.1.3. The adiabatic manifold satisfies the equation

$$\varepsilon \partial_y \bar{x}(y, \varepsilon) \nabla U(\bar{x}(y, \varepsilon)) = -y + \bar{x}(y, \varepsilon), \qquad (2.1.28)$$

and admits the expansion

$$\bar{x}(y, \varepsilon) = y + \varepsilon \nabla U(y) + 2\varepsilon^2 \partial_{yy} U(y) \nabla U(y) + \mathcal{O}(\varepsilon^3), \qquad (2.1.29)$$

where $\partial_{yy} U(y)$ denotes the Hessian matrix of U. The dynamics of $x = \bar{x}(y, \varepsilon)$ on the adiabatic manifold is thus governed by the equation

$$\begin{aligned} \dot{x} &= \frac{y - x}{\varepsilon} \\ &= -\partial_y \bar{x}(y, \varepsilon) \nabla U(x) \\ &= -\big[\mathbb{1} + \varepsilon \partial_{xx} U(x) + \mathcal{O}(\varepsilon^2)\big] \nabla U(x) . \end{aligned} \qquad (2.1.30)$$

This is indeed a small correction to the limiting equation $\dot{x} = -\nabla U(x)$.

One should note two facts concerning the power-series expansion of adiabatic manifolds, in the case where the fast and slow vector fields f and g are analytic. First, centre manifolds are in general not unique. However, all centre manifolds of a non-hyperbolic equilibrium share the same expansions, and thus differ typically only by exponentially small terms. Second, although these expansions are asymptotic expansions, they do not in general provide convergent series.

Example 2.1.10. Consider the slow–fast system

$$\begin{aligned} \varepsilon \dot{x} &= -x + h(y) , \\ \dot{y} &= 1 , \end{aligned} \qquad (2.1.31)$$

where h is analytic. The slow manifold is given by $x^\star(y) = h(y)$ and is uniformly asymptotically stable. The adiabatic manifold satisfies the equation

$$\varepsilon \partial_y \bar{x}(y, \varepsilon) = -\bar{x}(y, \varepsilon) + h(y) , \qquad (2.1.32)$$

and admits the asymptotic expansion

$$\bar{x}(y, \varepsilon) = h(y) - \varepsilon h'(y) + \varepsilon^2 h''(y) - \ldots . \qquad (2.1.33)$$

This series will not converge in general, since the norm of the kth derivative of h may grow like $k!$ by Cauchy's formula. In fact, (2.1.31) admits for each k the particular solution

$$x_t = \sum_{j=0}^{k-1}(-\varepsilon)^j h^{(j)}(y_t) - (-\varepsilon)^{k-1}\int_0^t e^{-(t-s)/\varepsilon} h^{(k)}(y_s)\,\mathrm{d}s\;. \qquad (2.1.34)$$

If $|h|$ is bounded by a constant M in a strip of width $2R$ around the real axis, Cauchy's formula shows that $|h^{(k)}| \leqslant M R^{-k} k!$, and the last term in (2.1.34) is bounded by $(\varepsilon/R)^k M k!$. Using Stirling's formula, one finds that this remainder is smallest for $k \simeq R/\varepsilon$, and that for this value of k it is of order $e^{-1/\varepsilon}$. This situation is quite common in analytic singularly perturbed systems.

2.2 Dynamic Bifurcations

An essential assumption in the previous section was the (uniform) hyperbolicity of the slow manifold, that is, the fact that no eigenvalue of the Jacobian matrix $A^\star(y) = \partial_x f(x^\star(y), y)$ approaches the imaginary axis. It can happen, however, that the slow dynamics of y on the adiabatic manifold takes the orbit to a region where this assumption is violated, that is, to a *bifurcation point* of the associated system. We have seen such a situation in the case of the van der Pol oscillator (Example 2.1.6), where the points $\pm(1, -\tfrac{2}{3})$ correspond to saddle–node bifurcations.

We will start, in Section 2.2.1, by showing how the dynamics near such a bifurcation point can be reduced to a suitable centre manifold, involving only bifurcating directions. Then we turn to a detailed description of the most generic bifurcations.

2.2.1 Centre-Manifold Reduction

We say that $(\widehat{x}, \widehat{y})$ is a *bifurcation point* of the slow–fast system

$$\begin{aligned}\varepsilon \dot{x} &= f(x,y)\;,\\ \dot{y} &= g(x,y)\;,\end{aligned} \qquad (2.2.1)$$

if $f(\widehat{x}, \widehat{y}) = 0$ and $\partial_x f(\widehat{x}, \widehat{y})$ has q eigenvalues on the imaginary axis, with $1 \leqslant q \leqslant n$. We consider here the situation where $q < n$, because otherwise no reduction is possible, and assume that the other $n-q$ eigenvalues of $\partial_x f(\widehat{x}, \widehat{y})$ have strictly negative real parts (in order to obtain a locally attracting invariant manifold).

We introduce coordinates $(x^-, z) \in \mathbb{R}^{n-q} \times \mathbb{R}^q$ in which the matrix $\partial_x f(\widehat{x}, \widehat{y})$ becomes block-diagonal, with a block $A^- \in \mathbb{R}^{(n-q)\times(n-q)}$ having eigenvalues in the left half-plane, and a block $A^0 \in \mathbb{R}^{q \times q}$ having eigenvalues on the imaginary axis. We consider again ε as a dummy dynamic variable, and write the system in slow time as

$$\begin{aligned}
(x^-)' &= f^-(x^-, z, y) \,, \\
z' &= f^0(x^-, z, y) \,, \\
y' &= \varepsilon g(x^-, z, y) \,, \\
\varepsilon' &= 0 \,.
\end{aligned} \qquad (2.2.2)$$

This system admits $(\widehat{x}^-, \widehat{z}, \widehat{y}, 0)$ as an equilibrium point, with a linearisation having $q + m + 1$ eigenvalues on the imaginary axis (counting multiplicity), which correspond to the directions z, y and ε. In other words, z has become a slow variable near the bifurcation point. We can thus apply the centre-manifold theorem to obtain the existence, for sufficiently small ε and in a neighbourhood $\mathcal{N}$ of $(\widehat{z}, \widehat{y})$, of a locally attracting invariant manifold

$$\widehat{\mathcal{M}}_\varepsilon = \{(x^-, z, y) \colon x^- = \bar{x}^-(z, y, \varepsilon), (z, y) \in \mathcal{N}\} \,. \qquad (2.2.3)$$

The function $\bar{x}^-(z, y, \varepsilon)$ satisfies $\bar{x}^-(\widehat{z}, \widehat{y}, 0) = \widehat{x}^-$, and is a solution of the partial differential equation

$$\begin{aligned}
f^-(\bar{x}^-(z, y, \varepsilon), z, y) &= \partial_z \bar{x}^-(z, y, \varepsilon) f^0(\bar{x}^-(z, y, \varepsilon), z, y) \\
&\quad + \varepsilon \partial_y \bar{x}^-(z, y, \varepsilon) g(\bar{x}^-(z, y, \varepsilon), z, y) \,. \quad (2.2.4)
\end{aligned}$$

In particular, $\bar{x}^-(z, y, 0)$ corresponds to a centre manifold of the associated system.

By local attractivity, it is sufficient to study the $(q+m)$-dimensional reduced system

$$\begin{aligned}
\varepsilon \dot{z} &= f^0(\bar{x}^-(z, y, \varepsilon), z, y) \,, \\
\dot{y} &= g(\bar{x}^-(z, y, \varepsilon), z, y) \,,
\end{aligned} \qquad (2.2.5)$$

which contains only bifurcating fast variables.

2.2.2 Saddle–Node Bifurcation

Assume that $\partial_x f(\widehat{x}, \widehat{y})$ has one vanishing eigenvalue, all its other eigenvalues having negative real parts, that is, $q = 1$. The most generic bifurcation then is the *saddle–node bifurcation*.

We will ease notations by choosing $\widehat{x} = 0$ and $\widehat{y} = 0$, writing x instead of z, and omitting the ε-dependence on the right-hand side of (2.2.5).[5] We thus arrive at the reduced system

$$\begin{aligned}
\varepsilon \dot{x} &= \tilde{f}(x, y) \,, \\
\dot{y} &= \tilde{g}(x, y) \,,
\end{aligned} \qquad (2.2.6)$$

---

[5] This ε-dependence is harmless for the saddle–node bifurcation, which is structurally stable, but may play a nontrivial rôle for less generic bifurcations. We will examine examples of such situations, involving avoided bifurcations, in Chapter 4.

where now $x \in \mathbb{R}$. With a slight abuse of notation we also drop the tilde and write (2.2.6) as

$$\varepsilon \dot{x} = f(x,y), \qquad \dot{y} = g(x,y). \qquad (2.2.7)$$

The fast vector field f satisfies the bifurcation conditions

$$f(0,0) = 0 \quad \text{and} \quad \partial_x f(0,0) = 0. \qquad (2.2.8)$$

We will discuss here the case of a one-dimensional slow variable $y \in \mathbb{R}$, that is, $m = 1$, as for the van der Pol oscillator in Example 2.1.6. A saddle–node bifurcation occurs if

$$\partial_{xx} f(0,0) \neq 0 \quad \text{and} \quad \partial_y f(0,0) \neq 0. \qquad (2.2.9)$$

For the sake of definiteness, we will choose the variables in such a way that

$$\partial_{xx} f(0,0) < 0 \quad \text{and} \quad \partial_y f(0,0) < 0 \qquad (2.2.10)$$

hold. This implies in particular that the slow manifold exists for $y < 0$. Finally, the additional assumption

$$g(0,0) > 0 \qquad (2.2.11)$$

guarantees that trajectories starting near the stable slow manifold are driven towards the bifurcation point. The simplest example of this kind is the system

$$\varepsilon \dot{x} = -x^2 - y, \qquad \dot{y} = 1, \qquad (2.2.12)$$

where the slow manifold consists of a stable branch $\mathcal{M}_- = \{(x,y): x = \sqrt{-y}, y < 0\}$ and an unstable branch $\mathcal{M}_+ = \{(x,y): x = -\sqrt{-y}, y < 0\}$. The linearisation of f at these branches is given by $\mp 2\sqrt{-y}$.

For general systems satisfying (2.2.8), (2.2.10) and (2.2.11), a qualitatively similar behaviour holds in a neighbourhood of the bifurcation point $(0,0)$. By rescaling x and y, we may arrange for

$$\partial_{xx} f(0,0) = -2 \quad \text{and} \quad \partial_y f(0,0) = -1. \qquad (2.2.13)$$

Using the implicit-function theorem, one easily shows that there exists a neighbourhood $\mathcal{N}$ of $(0,0)$ such that

- there is an asymptotically stable slow manifold

$$\mathcal{M}_- = \{(x,y) \in \mathcal{N}: x = x_-^\star(y), y < 0\}, \qquad (2.2.14)$$

where $x_-^\star(y) = \sqrt{-y}\,[1 + \mathcal{O}_y(1)]$, and $\mathcal{O}_y(1)$ stands for a remainder $r(y)$ satisfying $\lim_{y \to 0} r(y) = 0$;

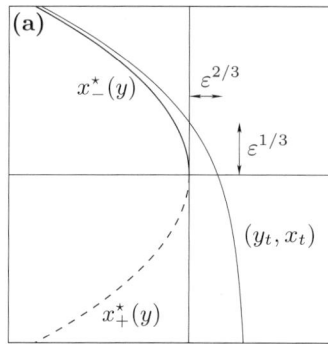 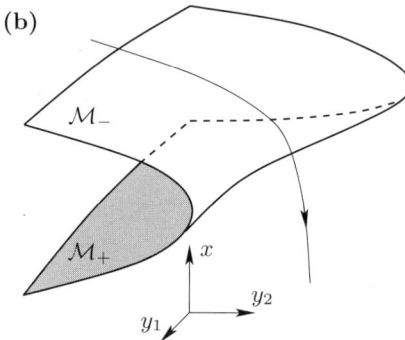

Fig. 2.4. Saddle–node bifurcation of a slow manifold. (a) Solutions track the stable branch $x_-^\star(y)$ at a distance growing up to order $\varepsilon^{1/3}$. They jump after a delay of order $\varepsilon^{2/3}$. (b) For multidimensional slow variables y, a saddle–node bifurcation corresponds to a fold in the slow manifold.

- the linearisation of f at $\mathcal{M}_-$ is given by

$$a_-^\star(y) := \partial_x f(x_-^\star(y), y) = -2\sqrt{-y}\,[1 + \mathcal{O}_y(1)]\ ; \qquad (2.2.15)$$

- there is an unstable slow manifold

$$\mathcal{M}_+ = \{(x,y) \in \mathcal{N} : x = x_+^\star(y), y < 0\}\ , \qquad (2.2.16)$$

where $x_+^\star(y) = -\sqrt{-y}\,[1 + \mathcal{O}_y(1)]$;
- there are no other slow manifolds in $\mathcal{N}$;
- $g(x,y) > 0$ in $\mathcal{N}$.

$\mathcal{M}_-$ being uniformly asymptotically stable for negative y, bounded away from 0, there is an adiabatic manifold associated with it which, in this $(1+1)$-dimensional setting, is simply a particular solution of the system. When approaching the bifurcation point, two effects make it increasingly difficult for this solution to track the slow manifold:

- The stable manifold $\mathcal{M}_-$ becomes less and less attracting,
- and $\mathcal{M}_-$ has a vertical tangent at the bifurcation point.

The interesting consequence is that trajectories no longer track the stable manifold at a distance of order ε, but of order $\varepsilon^{1/3}$. After a delay of order $\varepsilon^{2/3}$, they leave the neighbourhood $\mathcal{N}$ in the direction of negative x-values (Fig. 2.4). The following notation will be useful to formulate such scaling laws.

Notation 2.2.1 (Scaling behaviour). *For functions $x_1(t,\varepsilon)$ and $x_2(t,\varepsilon)$, defined for t in a specified interval I and $0 < \varepsilon \leqslant \varepsilon_0$, we indicate by*

$$x_1(t,\varepsilon) \asymp x_2(t,\varepsilon) \qquad (2.2.17)$$

the existence of two constants $c_\pm > 0$ such that

$$c_- x_2(t,\varepsilon) \leqslant x_1(t,\varepsilon) \leqslant c_+ x_2(t,\varepsilon) \tag{2.2.18}$$

for all $t \in I$ and $0 < \varepsilon \leqslant \varepsilon_0$.

For instance, we have $x_-^\star(y) \asymp \sqrt{-y}$ and $a_-^\star(y) \asymp -\sqrt{-y}$. The announced behaviour of the orbits is made precise in the following theorem.

Theorem 2.2.2 (Dynamic saddle–node bifurcation). *Choose an initial condition $(x_0, y_0) \in \mathcal{N}$ with constant $y_0 < 0$ and an x_0 satisfying $x_0 - x_-^\star(y_0) \asymp \varepsilon$. Then there exist constants $c_1, c_2 > 0$ such that*

$$x_t - x_-^\star(y_t) \asymp \frac{\varepsilon}{|y_t|} \quad \text{for } y_0 \leqslant y_t \leqslant -c_1 \varepsilon^{2/3}, \tag{2.2.19}$$

$$x_t \asymp \varepsilon^{1/3} \quad \text{for } -c_1 \varepsilon^{2/3} \leqslant y_t \leqslant c_2 \varepsilon^{2/3}. \tag{2.2.20}$$

Moreover, for any sufficiently small constant $L > 0$, x_t reaches $-L$ at a time $t(L)$ such that $y_{t(L)} \asymp \varepsilon^{2/3}$.

Proof. Since $\dot{y} > 0$ in $\mathcal{N}$ by assumption, we may use y instead of t as time variable, and study the equation

$$\varepsilon \frac{\mathrm{d}x}{\mathrm{d}y} = \frac{f(x,y)}{g(x,y)} =: \bar{f}(x,y). \tag{2.2.21}$$

It is straightforward to check that $\bar{f}$ satisfies the same properties (2.2.8) and (2.2.10) as f (before rescaling), and therefore, it is sufficient to study the slow–fast system (2.2.7) for $g \equiv 1$. Again we may assume by rescaling x and y that (2.2.13) holds.

- For $y \leqslant -c_1 \varepsilon^{2/3}$, where c_1 remains to be chosen, we use the change of variables $x = x_-^\star(y) + z$ and a Taylor expansion, yielding the equation

$$\varepsilon \frac{\mathrm{d}z}{\mathrm{d}y} = a_-^\star(y) z + b(z,y) - \varepsilon \frac{\mathrm{d}x_-^\star(y)}{\mathrm{d}y}, \tag{2.2.22}$$

where $|b(z,y)| \leqslant M z^2$ in $\mathcal{N}$ for some constant M. The properties of $x_-^\star(y)$ and $a_-^\star(y)$, cf. (2.2.14) and (2.2.15), imply the existence of constants $c_\pm > 0$ such that

$$\varepsilon \frac{\mathrm{d}z}{\mathrm{d}y} \leqslant -\sqrt{-y}\left[c_- - \frac{M}{\sqrt{-y}} z\right] z + \varepsilon \frac{c_+}{\sqrt{-y}}. \tag{2.2.23}$$

We now introduce the time

$$\tau = \inf\left\{t \geqslant 0 \colon (x_t, y_t) \notin \mathcal{N} \text{ or } |z_t| > \frac{c_-}{2M}\sqrt{-y_t}\right\} \in (0, \infty]. \tag{2.2.24}$$

Note that $\tau > 0$ for sufficiently small ε, since $-y_0$ is of order 1 and z_0 is of order ε by assumption. For $t \leqslant \tau$, we have

$$\varepsilon \frac{dz}{dy} \leqslant -\frac{c_-}{2}\sqrt{-y}\,z + \varepsilon \frac{c_+}{\sqrt{-y}}\,, \qquad (2.2.25)$$

and thus, solving a linear equation,

$$\begin{aligned}z_t \leqslant{}& z_0 e^{-c_-[(-y_0)^{3/2}-(-y_t)^{3/2}]/3\varepsilon} \\ &+ c_+ \int_{y_0}^{y_t} \frac{1}{\sqrt{-u}} e^{-c_-[(-u)^{3/2}-(-y_t)^{3/2}]/3\varepsilon}\,du\,,\end{aligned} \qquad (2.2.26)$$

as long as $t \leqslant \tau$ and $y_t < 0$. The integral can be evaluated by a variant of the Laplace method, see Lemma 2.2.8 below. We obtain the existence of a constant $K = K(c_-, c_+)$ such that

$$z_t \leqslant K \frac{\varepsilon}{|y_t|} \qquad (2.2.27)$$

for $t \leqslant \tau$ and $y_t < 0$. We remark in passing that the corresponding argument for the lower bound in particular shows that $z_t > 0$.

Now choosing $c_1 = (3KM/c_-)^{2/3}$, we find that for all $t \leqslant \tau$ satisfying $y_t \leqslant -c_1\varepsilon^{2/3}$, we also have $z_t \leqslant c_-\sqrt{-y_t}/3M$, so that actually $t < \tau$ whenever $y_t \leqslant -c_1\varepsilon^{2/3}$. Thus (2.2.27) holds for all t satisfying $y_t \leqslant -c_1\varepsilon^{2/3}$, which proves the upper bound in (2.2.19). The lower bound is obtained in a similar way.

- For $-c_1\varepsilon^{2/3} \leqslant y_t \leqslant c_2\varepsilon^{2/3}$, where c_2 will be chosen below, we use the fact that we rescaled x and y in such a way that

$$f(x,y) = -x^2 - y + \mathcal{O}(y) + \mathcal{O}(x^2) \qquad (2.2.28)$$

holds. The subsequent scaling $x = \varepsilon^{1/3} z$, $y = \varepsilon^{2/3} s$ thus yields the equation

$$\frac{dz}{ds} = -z^2 - s + \mathcal{O}_\varepsilon(1)\,, \qquad (2.2.29)$$

which is a small perturbation of a solvable Riccati equation. In fact, setting $z(s) = \varphi'(s)/\varphi(s)$ in (2.2.29) without the error term yields the linear second order equation $\varphi''(s) = -s\varphi(s)$, whose solution can be expressed in terms of Airy functions.

In particular, there exist constants $c_3 > c_2 > 0$ such that $z(s)$ remains positive, of order 1, for $s \leqslant c_2$, and reaches negative values of order 1 for $s = c_3$. Returning to the variables (x,y), we conclude that (2.2.20) holds, and that $x_t \asymp -\varepsilon^{1/3}$ as soon as $y_t = c_3\varepsilon^{2/3}$.

- For $y_t \geqslant c_3\varepsilon^{2/3}$, the trivial estimate $f(x,y) \leqslant -(1-\kappa)x^2$ can be employed to show that x_t reaches any negative value of order 1 after yet another time span of order $\varepsilon^{2/3}$, which concludes the proof. $\square$

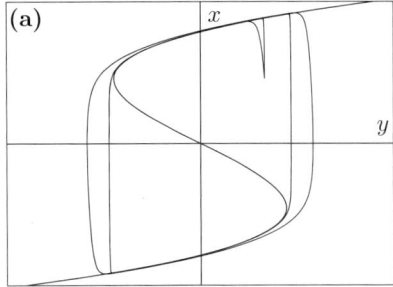 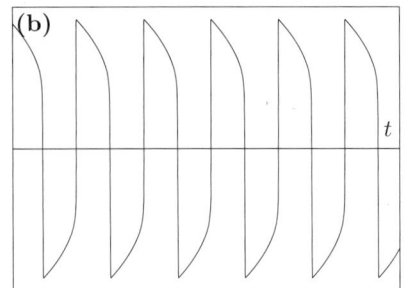

Fig. 2.5. (a) Two solutions of the van der Pol equations (2.1.13) (*light curves*) for the same initial condition $(1, \frac{1}{2})$, for $\gamma = 5$ and $\gamma = 20$. The heavy curve is the slow manifold $y = \frac{1}{3}x^3 - x$. (b) The graph of x_t (again for $\gamma = 20$) displays relaxation oscillations.

Example 2.2.3 (Van der Pol oscillator – continued). As already mentioned, the system

$$\varepsilon \dot{x} = y + x - \frac{x^3}{3}, \qquad (2.2.30)$$
$$\dot{y} = -x,$$

admits two saddle–node bifurcation points $\pm(1, -\frac{2}{3})$. Consider an orbit starting near the stable slow manifold $x = x_+^\star(y)$. This orbit will approach an ε-neighbourhood of the slow manifold in a time of order $\varepsilon|\log \varepsilon|$. Since $\dot{y} < 0$ for positive x, the bifurcation point $(1, -\frac{2}{3})$ is reached in a time of order 1.

Theorem 2.2.2, applied to the variables $(x - 1, -(y + \frac{2}{3}))$, shows that x_t "jumps" as soon as y_t has reached a value $-\frac{2}{3} - \mathcal{O}(\varepsilon^{2/3})$. Once the orbit has left a small neighbourhood of the bifurcation point, f is negative and bounded away from zero. Solutions of the associated system $x' = y + x - \frac{x^3}{3}$ enter a neighbourhood of order 1 of the slow manifold $x = x_-^\star(y)$ in a fast time of order 1, while y changes only by $\mathcal{O}(\varepsilon)$ during the same (fast) time, so that by an argument of regular perturbation theory, x_t also approaches $x_-^\star(y_t)$ in a time of order ε.

We can then apply the same analysis as before, showing that the orbit tracks the branch $x = x_-^\star(y)$ until reaching the bifurcation point $(-1, \frac{2}{3})$ and jumping back to the first branch. In fact, using a Poincaré section and the exponential contraction near adiabatic manifolds, one easily shows that orbits approach a periodic orbit. (This is known to hold for any value of the damping coefficient.) The resulting trajectory displays alternating slow and fast motions, called *relaxation oscillations* (Fig. 2.5).

Similar phenomena have been analysed for more general systems, including multidimensional slow variables. The difference is that, the Poincaré map no longer being one-dimensional, orbits do not necessarily approach a periodic one. A detailed analysis of such situations is given in [MR80] and [MKKR94].

2.2.3 Symmetric Pitchfork Bifurcation and Bifurcation Delay

We consider the reduced slow–fast system

$$\begin{aligned}\varepsilon \dot{x} &= f(x,y)\,,\\ \dot{y} &= g(x,y)\,,\end{aligned} \qquad (2.2.31)$$

for $x, y \in \mathbb{R}$, again under the assumptions

$$f(0,0) = \partial_x f(0,0) = 0 \quad \text{and} \quad g(0,0) > 0\,. \qquad (2.2.32)$$

In this section we assume that f and g are of class $\mathcal{C}^3$. If, unlike in the case of the saddle–node bifurcation, $\partial_{xx} f(0,0)$ vanishes, several new kinds of bifurcations can occur. These bifurcations are not generic, unless f and g are restricted to belong to some smaller function space, for instance functions satisfying certain symmetry conditions. We consider here the case where f is odd in x and g is even in x:

$$f(-x,y) = -f(x,y) \quad \text{and} \quad g(-x,y) = g(x,y) \qquad (2.2.33)$$

for all (x,y) in a neighbourhood $\mathcal{N}$ of $(0,0)$. We say that a *supercritical symmetric pitchfork bifurcation* occurs at $(0,0)$ if

$$\partial_{xy} f(0,0) > 0 \quad \text{and} \quad \partial_{xxx} f(0,0) < 0\,. \qquad (2.2.34)$$

The simplest example of such a system is

$$\begin{aligned}\varepsilon \dot{x} &= yx - x^3\,,\\ \dot{y} &= 1\,.\end{aligned} \qquad (2.2.35)$$

The line $x = 0$ is a slow manifold, which changes from stable to unstable as y changes from negative to positive. For positive y, there are two new branches of the slow manifold, of equation $x = \pm\sqrt{y}$, which are stable (Fig. 2.6).

Similar properties hold for general systems satisfying (2.2.32), (2.2.33) and (2.2.34). By rescaling x and y, one can always arrange for $\partial_{xy} f(0,0) = 1$ and $\partial_{xxx} f(0,0) = -6$. It is then straightforward to show that in a sufficiently small neighbourhood $\mathcal{N}$ of $(0,0)$,

- all points on the line $x = 0$ belong to a slow manifold, with stability "matrix"

$$\partial_x f(0,y) =: a(y) = y[1 + \mathcal{O}(y)]\,, \qquad (2.2.36)$$

  and thus the manifold is stable for $y < 0$ and unstable for $y > 0$, provided $\mathcal{N}$ is small enough;

- there are two other stable slow manifolds of the form $\{x = \pm x^\star(y), y > 0\}$, where $x^\star(y) = \sqrt{y}[1 + \mathcal{O}_y(1)]$; the linearisation of f at these manifolds satisfies

$$\partial_x f(\pm x^\star(y), y) =: a^\star(y) = -2y[1 + \mathcal{O}_y(1)]\,; \qquad (2.2.37)$$

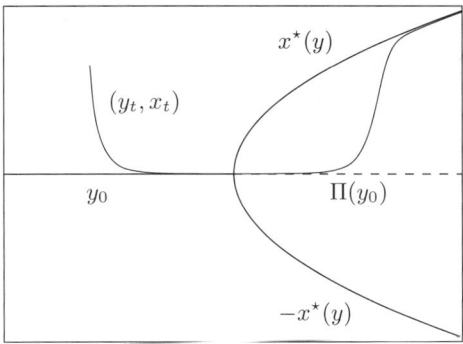

Fig. 2.6. Dynamic pitchfork bifurcation. Solutions starting in (x_0, y_0) with $x_0 > 0$ and $y_0 < 0$ track the unstable manifold $\{x = 0,\ y > 0\}$ until $y_t = \Pi(y_0)$, before jumping to the stable branch $x^\star(y)$.

- there are no other slow manifolds in $\mathcal{N}$;
- $g(x, y) > 0$ in $\mathcal{N}$.

Consider now an orbit starting in $(x_0, y_0) \in \mathcal{N}$, with $x_0 > 0$ and $y_0 < 0$. It reaches an ε-neighbourhood of the slow manifold in $x = 0$ in a time of order $\varepsilon|\log \varepsilon|$. The new feature in this situation is that $x_t \equiv 0$ is itself a particular solution of (2.2.31), and thus the slow manifold is actually an adiabatic manifold. For this reason, x_t approaches 0 exponentially closely, and when the slow manifold becomes unstable at $y = 0$, a time of order 1 is needed for x_t to move away from 0 and approach $x^\star(y_t)$ (Fig. 2.6). This phenomenon is known as *bifurcation delay*.

In order to quantify this delay, one can define, for negative y_0 such that $(0, y_0) \in \mathcal{N}$, a *bifurcation delay time*

$$\Pi(y_0) = \inf\left\{ y_1 > 0 \colon \int_{y_0}^{y_1} \frac{a(y)}{g(0, y)}\,dy > 0 \right\}. \tag{2.2.38}$$

In the particular case of the system (2.2.35), $\Pi(y_0) = -y_0$. In the general case, one easily sees that $\Pi(y_0) = -y_0 + \mathcal{O}(y_0^2)$. The following theorem states that trajectories starting in $(x_0, y_0) \in \mathcal{N}$ track the line $x = 0$ approximately until $y_t = \Pi(y_0)$.

Theorem 2.2.4 (Dynamic pitchfork bifurcation). *Let ε and $\mathcal{N}$ be sufficiently small, and choose an initial condition $(x_0, y_0) \in \mathcal{N}$ such that $x_0 > 0$ and $y_0 < 0$. Then there exist $c_1 > 0$ and*

$$\begin{aligned} y_1 &= y_0 + \mathcal{O}(\varepsilon|\log \varepsilon|), \\ y_2 &= \Pi(y_0) - \mathcal{O}(\varepsilon|\log \varepsilon|), \\ y_3 &= y_2 + \mathcal{O}(\varepsilon|\log \varepsilon|), \end{aligned} \tag{2.2.39}$$

such that

$$0 \leqslant x_t \leqslant \varepsilon \qquad \text{for } y_1 \leqslant y_t \leqslant y_2, \qquad (2.2.40)$$
$$|x_t - x^\star(y_t)| \leqslant c_1 \varepsilon \qquad \text{for } y_t \geqslant y_3, \qquad (2.2.41)$$

for all times t for which $(x_t, y_t) \in \mathcal{N}$.

Proof.

- Since $f(x, y_0) < 0$ for positive x, solutions of the associated system $x' = f(x, y_0)$ reach any neighbourhood of order 1 of $x = 0$ in a fast time of order ε. Thus, by regular perturbation theory, the orbit of the slow–fast system starting in (x_0, y_0) enters any neighbourhood of order 1 of the slow manifold $x = 0$ in a slow time of order ε. From there on, Tihonov's theorem shows that an ε-neighbourhood is reached in a time of order $\varepsilon|\log \varepsilon|$, and thus $x_{t_1} = \varepsilon$ at a time t_1 at which $y_{t_1} = y_1$.
- First note that we can write
$$f(x, y) = x\bigl[a(y) + b(x, y)x^2\bigr] \qquad (2.2.42)$$
for some bounded function b. Let $\tau = \inf\{t > t_1 : x_t > \varepsilon\}$. For $t_1 \leqslant t \leqslant \tau$, we have
$$\varepsilon \frac{\mathrm{d}x}{\mathrm{d}y} = \biggl[\frac{a(y)}{g(0, y)} + \mathcal{O}(\varepsilon^2)\biggr] x, \qquad (2.2.43)$$
and thus there is a constant $M > 0$ such that
$$\varepsilon e^{[\alpha(y_t, y_1) - M\varepsilon^2]/\varepsilon} \leqslant x_t \leqslant \varepsilon e^{[\alpha(y_t, y_1) + M\varepsilon^2]/\varepsilon}, \qquad (2.2.44)$$
where
$$\alpha(y, y_1) = \int_{y_1}^{y} \frac{a(u)}{g(0, u)} \,\mathrm{d}u. \qquad (2.2.45)$$
This shows that x_t remains smaller than ε until a time t_2 at which $\alpha(y_{t_2}, y_1) = \mathcal{O}(\varepsilon^2)$, so that $y_{t_2} = y_2 = \Pi(y_0) + \mathcal{O}(\varepsilon|\log \varepsilon|)$.
- By continuity of f, for any sufficiently small constant c_0, $f(x, y) > \frac{1}{2}xa(y_2)$ for $|x| \leqslant c_0$ and $y_2 \leqslant y \leqslant y_2 + c_0$. Thus x_t is larger than $\varepsilon e^{a(y_2)(t-t_2)/\varepsilon}$, and reaches c_0 after a time of order $\varepsilon|\log \varepsilon|$. Using the associated system and the fact that f is positive, bounded away from 0 for x in any compact interval contained in $(0, x^\star(y))$, one concludes that x_t reaches any neighbourhood of order 1 of the equilibrium branch $x^\star(y)$ after a time of order ε. From there on, Tihonov's theorem allows to conclude the proof. □

Remark 2.2.5. The proof shows that a similar bifurcation delay exists for any system for which $f(0, y)$ is identically zero. Only the behaviour after leaving the unstable equilibrium branch may differ, as it depends on the global behaviour of f.

2.2.4 How to Obtain Scaling Laws

The analysis of the dynamic saddle–node bifurcation in Section 2.2.2 revealed a nontrivial ε-dependence of solutions near the bifurcation point, involving fractional exponents as in $\varepsilon^{1/3}$ and $\varepsilon^{2/3}$. These exponents were related to the behaviour in $\sqrt{-y}$ of slow manifolds near the bifurcation point.

It is in fact possible to give a rather complete classification of the scaling laws that may occur for one-dimensional bifurcations. Such a classification involves two steps: Firstly, analyse the power-law behaviour of slow manifolds near a bifurcation point; secondly, study the equation governing the distance between solutions of the slow–fast system and the slow manifold.

The first step is best carried out with the help of the bifurcation point's *Newton polygon*. Consider again the scalar equation

$$\varepsilon \frac{\mathrm{d}x}{\mathrm{d}y} = f(x, y) \qquad (2.2.46)$$

where f is of class C^r for some $r \geqslant 2$, with all derivatives uniformly bounded in a neighbourhood $\mathcal{N}$ of the origin. We also require the bifurcation conditions

$$f(0,0) = \partial_x f(0,0) = 0 \qquad (2.2.47)$$

to hold. The function f admits a Taylor expansion in $\mathcal{N}$ of the form

$$f(x,y) = \sum_{\substack{j,k \geqslant 0 \\ j+k \leqslant r}} f_{jk} x^j y^k + R(x,y), \qquad f_{jk} = \frac{1}{j! k!} \partial_{x^j y^k} f(0,0), \qquad (2.2.48)$$

with a remainder $R(x,y) = \mathcal{O}(\|(x,y)\|^r)$, and $f_{00} = f_{10} = 0$. Assume for simplicity[6] that $f_{r0} \neq 0$ and $f_{0r} \neq 0$. Then the Newton polygon $\mathcal{P}$ is obtained by taking the convex envelope of the set of points

$$\{(j,k) \in \mathbb{N}_0^2 \colon f_{jk} \neq 0 \text{ or } j+k \geqslant r\}. \qquad (2.2.49)$$

Assume that f vanishes on a curve of equation $x = x^\star(y)$, defined for y in an interval of the form $I = (-\delta, 0]$, $[0, \delta)$ or $(-\delta, \delta)$, such that $|x^\star(y)| \asymp |y|^q$ for some $q > 0$. We will call $\{(x,y) \colon x = x^\star(y), y \in I\}$ an *equilibrium branch* with exponent q. It is well known that the Newton polygon then admits a segment of slope $-q$. For instance, in the case of the saddle–node bifurcation studied in Section 2.2.2, there are two equilibrium branches with exponent $1/2$, and $\mathcal{P}$ has two vertices $(2,0)$ and $(0,1)$, connected by a segment of slope $-1/2$.

The second step is to examine the behaviour of the distance $z = x - x^\star(y)$ between solutions of (2.2.46) and the equilibrium branch. The dynamics of z_t is governed by an equation of the form

---

[6]This assumption is made to avoid situations such as $f(x,y) = x^{5/2} - y$, in which $r = 2$ but $f_{20}(x,y) \equiv 0$, where the relation between Newton's polygon and slow manifold does not work.

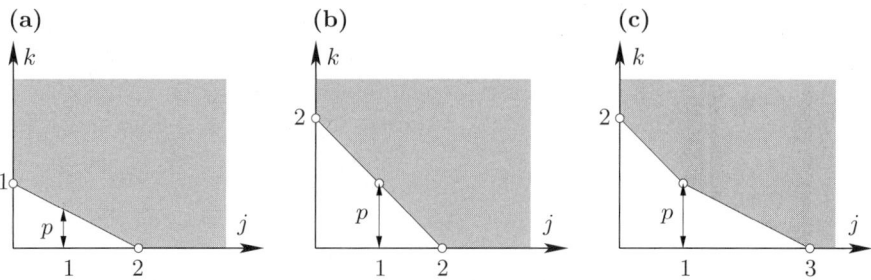

Fig. 2.7. Newton's polygon for **(a)** the saddle–node, **(b)** the transcritical and **(c)** the asymmetric pitchfork bifurcation. The saddle–node bifurcation has branches of exponent $q = 1/2$, for which $p = 1/2$. The transcritical bifurcation has branches of exponent $q = 1$, for which $p = 1$. The asymmetric pitchfork bifurcation has two types of branches, with exponents $q = 1$ and $q = 1/2$, both with $p = 1$.

$$\varepsilon \frac{\mathrm{d}z}{\mathrm{d}y} = a(y)z + b(y, z) - \varepsilon \frac{\mathrm{d}x^\star(y)}{\mathrm{d}y}, \qquad (2.2.50)$$

where $a(y) = \partial_x f(x^\star(y), y)$ is the linearisation of f at the equilibrium branch, and b is of order z^2. The behaviour of $a(y)$ near $y = 0$ can also be deduced from the Newton polygon $\mathcal{P}$. Indeed, Taylor's formula implies that

$$\partial_x f(x^\star(y), y) = \sum_{\substack{j \geq 1, k \geq 0 \\ j+k \leq r}} j f_{jk} x^\star(y)^{j-1} y^k + o(|y|^r + |y|^{rq}). \qquad (2.2.51)$$

Generically (that is, unless some unexpected cancellation occurs) we will thus have $|a(y)| \asymp |y|^p$, where

$$p = \inf\{q(j-1) + k \colon j \geq 1,\ k \geq 0,\ j+k \leq r \text{ and } f_{jk} \neq 0\}. \qquad (2.2.52)$$

Geometrically speaking, p is the ordinate at 1 of the tangent to Newton's polygon with slope $-q$ (Fig. 2.7).

Definition 2.2.6 (Tame equilibrium branch). *Let I be an interval of the form $I = (-\delta, 0]$, $[0, \delta)$ or $(-\delta, \delta)$, and assume that f vanishes on an equilibrium branch $\{(x, y) \colon x = x^\star(y), y \in I\}$ with exponent $q > 0$. The equilibrium branch is called tame if $|\partial_x f(x^\star(y), y)| \asymp |y|^p$ with p given by (2.2.52).*

Example 2.2.7.

- If $f(x, y) = -x^2 - y$, then there are two tame equilibrium branches with exponent $q = 1/2$, and such that $p = 1/2$.
- If $f(x, y) = (x-y)^2$, then there is an equilibrium branch of equation $x = y$, whose exponent is $q = 1$. This branch is not tame since $a(y) = \partial_x f(y, y)$ is identically zero, and thus $p = \infty$.

The rational numbers q and p are usually sufficient to determine the scaling behaviour of solutions near a bifurcation point. An essential rôle is played by the following variant of the Laplace method, which allows to analyse solutions of (2.2.50) when the nonlinear term $b(y,z)$ is absent.

Lemma 2.2.8. *Fix constants $y_0 < 0$, and $\zeta_0, c, p, q > 0$. Then the function*

$$\zeta(y,\varepsilon) = e^{c|y|^{p+1}/\varepsilon}\left[\varepsilon\zeta_0 e^{-c|y_0|^{p+1}/\varepsilon} + \int_{y_0}^{y} |u|^{q-1} e^{-c|u|^{p+1}/\varepsilon}\,du\right] \qquad (2.2.53)$$

satisfies

$$\zeta(y,\varepsilon) \asymp \begin{cases} \varepsilon|y|^{q-p-1} & \text{for } y_0 \leqslant y \leqslant -\varepsilon^{1/(p+1)}, \\ \varepsilon^{q/(p+1)} & \text{for } -\varepsilon^{1/(p+1)} \leqslant y \leqslant 0. \end{cases} \qquad (2.2.54)$$

Using similar ideas as in the proof Theorem 2.2.2, we obtain the following general result on the behaviour of solutions tracking a stable, tame equilibrium branch, approaching a bifurcation point. If $x^\star(y)$ is not identically zero, we may assume that it is decreasing near the bifurcation point, otherwise we change x to $-x$.

Theorem 2.2.9 (Scaling behaviour near a bifurcation point). *Assume that $x^\star : (-\delta, 0] \to \mathbb{R}_+$ describes a tame equilibrium branch of f, with exponent $q > 0$, which is stable, that is, the linearisation $a(y) = \partial_x f(x^\star(y), y)$ satisfies $a(y) \asymp -|y|^p$, where p is given by (2.2.52). Fix an initial condition $(x_0, y_0) \in \mathcal{N}$, where $y_0 \in (-\delta, 0)$ does not depend on ε and $x_0 - x^\star(y_0) \asymp \varepsilon$. Then there is a constant $c_0 > 0$ such that the solution of (2.2.46) with initial condition (x_0, y_0) satisfies*

$$x_t - x^\star(y_t) \asymp \varepsilon|y_t|^{q-p-1} \qquad \text{for } y_0 \leqslant y_t \leqslant -c_0\varepsilon^{1/(p+1)}. \qquad (2.2.55)$$

If, moreover, there is a constant $\delta_1 > 0$ such that $f(x,y) < 0$ for $-\delta \leqslant y \leqslant 0$ and $0 < x - x^\star(y) \leqslant \delta_1$, then there exists $c_1 > 0$ such that

$$x_t \asymp \varepsilon^{q/(p+1)} \qquad \text{for } -c_0\varepsilon^{1/(p+1)} \leqslant y_t \leqslant c_1\varepsilon^{1/(p+1)}. \qquad (2.2.56)$$

Proof.

- For $y_0 \leqslant y_t \leqslant -c_0\varepsilon^{1/(p+1)}$, where c_0 will be determined below, we have

$$\varepsilon\frac{dz}{dy} \leqslant -c_-|y|^p z + b(y,z) + \varepsilon c_+|y|^{q-1} \qquad (2.2.57)$$

for some constants $c_\pm > 0$. We know that $|b(y,z)| \leqslant Mz^2$ for some $M > 0$, but we will need a better bound in the case $p > q$. In fact, Taylor's formula implies that there is a $\theta \in [0,1]$ such that

$$z^{-2}b(y,z) = \frac{1}{2}\partial_{xx}f(x^\star(y) + \theta z, y),$$

$$= \frac{1}{2}\sum_{\substack{j \geqslant 2, k \geqslant 0 \\ j+k \leqslant r}} j(j-1) f_{jk}[x^\star(y) + \theta z]^{j-2} y^k + R(y), \qquad (2.2.58)$$

where $R(y) = \mathcal{O}(x^\star(y)^{r-2} + z^{r-2} + y^r)$. Using the definition (2.2.52) of p, we obtain that whenever $|z| \leqslant L|y|^q$ for some $L > 0$, there is a constant $M(L)$ such that
$$|b(y,z)| \leqslant M(L)|y|^{p-q}z^2 . \tag{2.2.59}$$
Since $M(L)$ is a nondecreasing function of L, we can choose L such that $2LM(L) \leqslant c_-$. In order to prove the upper bound in (2.2.55), we introduce
$$\tau = \inf\{t \geqslant 0 \colon |z_t| > L|y_t|^q\} . \tag{2.2.60}$$
For $t \leqslant \tau$, it easily follows from (2.2.57) and the definition of $M(L)$ that
$$\varepsilon \frac{dz}{dy} \leqslant -\frac{c_-}{2}|y|^p z + \varepsilon c_+ |y|^{q-1} , \tag{2.2.61}$$
so that, solving a linear equation and using Lemma 2.2.8,
$$z_t \leqslant K\varepsilon |y_t|^{q-p-1} \tag{2.2.62}$$
for some $K > 0$, as long as $y_t \leqslant -\varepsilon^{1/(p+1)}$. The conclusion is then obtained as in the proof of Theorem 2.2.2, taking $c_0^{p+1} = 1 \wedge (4MK/c_-)$. The lower bound is obtained in a similar way.

- For $-c_0 \varepsilon^{1/(p+1)} \leqslant y_t \leqslant c_1 \varepsilon^{1/(p+1)}$, where c_1 is determined below, we use the scaling $x = \varepsilon^{q/(p+1)}\tilde{x}$, $y = \varepsilon^{1/(p+1)}\tilde{y}$, yielding the equation
$$\frac{d\tilde{x}}{d\tilde{y}} = \varepsilon^{-(p+q)/(p+1)} f(\varepsilon^{q/(p+1)}\tilde{x}, \varepsilon^{1/(p+1)}\tilde{y}) ,$$
$$= \sum_{\substack{j,k \geqslant 0 \\ j+k \leqslant r}} f_{jk} \varepsilon^{[(j-1)q+k-p]/(p+1)} \tilde{x}^j \tilde{y}^k + \tilde{r}(x,y) , \tag{2.2.63}$$

for some remainder $\tilde{r}(x,y)$. By (2.2.52), we have $(j-1)q + k \geqslant p$ whenever $f_{jk} \neq 0$ or $j + k \geqslant r - 1$, and thus the right-hand side of (2.2.63) is of order 1 at most. In fact, we have
$$\frac{d\tilde{x}}{d\tilde{y}} = \sum f_{jk}\tilde{x}^j \tilde{y}^k + \mathcal{O}_\varepsilon(1), \tag{2.2.64}$$
where the sum is taken over all vertices of the Newton polygon belonging to a segment of slope $-q$.

If $f(x,y) < 0$ above $x^\star(y)$, x_t must decrease as long as $y_t < 0$, and it cannot cross $x^\star(y_t)$. Hence $\tilde{x} \asymp 1$, whenever $\tilde{y} = 0$. Since (2.2.64) does not depend on ε to leading order, $\tilde{x}$ remains positive and of order 1 for $\tilde{y}$ in a sufficiently small interval $[0, c_1]$. Going back to original variables, (2.2.56) is proved. □

The behaviour after y_t has reached the bifurcation value 0 depends on the number of equilibrium branches originating from the bifurcation point. In

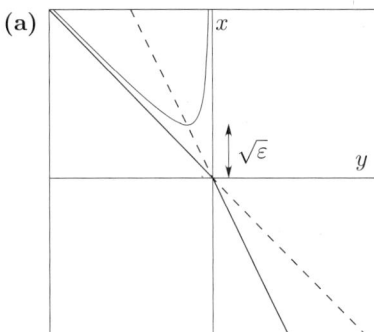 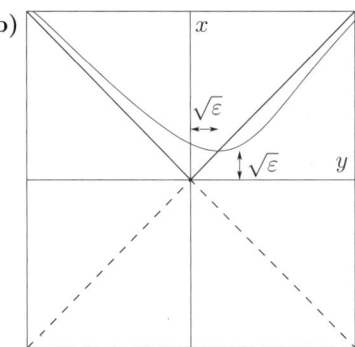

Fig. 2.8. Dynamic transcritical bifurcation. **(a)** If the unstable branch lies above the decreasing stable branch, solutions cross the unstable branch for $y \asymp -\sqrt{\varepsilon}$ and explode soon thereafter. **(b)** If the unstable branch lies below the decreasing stable branch, solutions cross the new stable branch for $y \asymp \sqrt{\varepsilon}$, before approaching it again like ε/y.

the case of the indirect saddle–node bifurcation, no equilibrium branch exists near the origin for positive y, and thus the orbits jump to some other region of phase space (or explode). If one or several stable equilibrium branches exist for $y > 0$, the behaviour of solutions can be analysed by methods similar to those of Theorem 2.2.9. Rather then attempting to give a general classification of possible behaviours, we prefer to discuss two of the most generic cases.

Example 2.2.10 (Transcritical bifurcation). Assume that $f \in \mathcal{C}^2$ satisfies, in addition to the general bifurcation condition (2.2.47), the relation $f_{01} = 0$, but with $f_{20}, f_{11}, f_{02} \neq 0$. The Newton polygon has vertices $(2, 0)$, $(1, 1)$ and $(0, 2)$, connected by two segments of slope 1 (Fig. 2.7b). We thus take $q = 1$ and look for equilibrium branches of the form

$$x = Cy(1 + \rho(y)), \qquad \rho(0) = 0. \tag{2.2.65}$$

Inserting this in the equation $f(x, y) = 0$ and taking $y \to 0$, we obtain the condition

$$f_{20}C^2 + f_{11}C + f_{02} = 0 \tag{2.2.66}$$

for C. Thus there are three cases to be considered, depending on the value of the discriminant $\Delta = f_{11}^2 - 4f_{20}f_{02}$.

1. If $\Delta > 0$, then (2.2.66) admits two solutions $C_+ > C_-$. For each of these solutions, the implicit-function theorem, applied to the pair (y, ρ), shows that there is indeed a unique equilibrium branch of the form (2.2.65). A simple computation of $\partial_x f(x, y)$ on each of these branches shows that they have opposite stability, and exchange stability as y passes through 0.
2. If $\Delta = 0$, then (2.2.66) admits one solution, but the implicit-function theorem cannot be applied. The behaviour will depend on higher order terms. The second part of Example 2.2.7 belongs into this category.

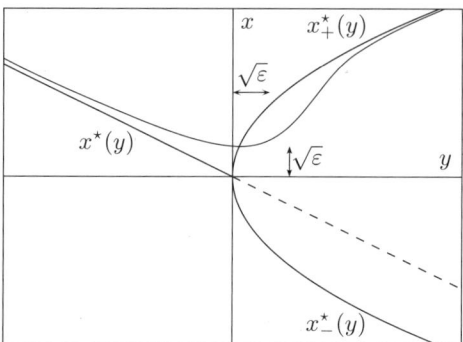

Fig. 2.9. Asymmetric dynamic pitchfork bifurcation. Solutions tracking the decreasing stable branch switch to the new branch $x_+^\star(y)$ after a time of order $\sqrt{\varepsilon}$.

3. If $\Delta < 0$, then $(0,0)$ is an isolated equilibrium point of f.

Let us consider the case $\Delta > 0$ in more detail. Assume that the stable equilibrium branch approaching the origin from the left is decreasing. The situation depends on the location of the unstable equilibrium branch, whether it lies above or below the stable one (Fig. 2.8):

- If the unstable branch lies above the stable one, which happens for $f_{20} > 0$, then Theorem 2.2.9, applied for $q = p = 1$, shows that x_t tracks the equilibrium branch at a distance of order $\varepsilon/|y_t|$ until $y = -c_0\sqrt{\varepsilon}$. Thereafter, the trajectory escapes from a neighbourhood of the bifurcation point.
- If the unstable branch lies below the stable one, which happens for $f_{20} < 0$, then the theorem can still be applied and shows, in addition, that x_t stays of order $\sqrt{\varepsilon}$ until $y = c_1\sqrt{\varepsilon}$. For later times, an analysis of the equation governing the distance between solutions and the new stable equilibrium branch shows that x_t approaches that branch like ε/y_t.

Example 2.2.11 (Asymmetric pitchfork bifurcation). Assume that $f \in \mathcal{C}^3$ satisfies $f_{01} = f_{20} = 0$, but $f_{30}, f_{11}, f_{02} \neq 0$. Then the Newton polygon has three vertices $(3,0)$, $(1,1)$ and $(0,2)$, connected by two segments of slope $1/2$ and 1 (Fig. 2.7c). A similar analysis as in the previous example shows the existence of an equilibrium branch $x^\star(y) = -(f_{02}/f_{11})y[1 + \mathcal{O}_y(1)]$, existing for both positive and negative y, and of two equilibrium branches $x_\pm^\star(y) \asymp \sqrt{|y|}$, existing either for $y < 0$ or for $y > 0$, depending on the sign of f_{11}/f_{30}.

Consider the case where the branch $x^\star(y)$ is decreasing, stable for $y < 0$ and unstable for $y > 0$, and the branches $x_\pm^\star(y)$ exist for $y > 0$. (Then the latter two branches are necessarily stable, see Fig. 2.9.) The simplest example of this kind is $f(x,y) = (x+y)(y - x^2)$.

Theorem 2.2.9, applied for $q = p = 1$, shows that x_t tracks the equilibrium branch $x^\star(y)$ at a distance of order $\varepsilon/|y_t|$ until $y = -c_0\sqrt{\varepsilon}$, and remains of order $\sqrt{\varepsilon}$ until $y = c_1\sqrt{\varepsilon}$. At this time, x_t is at a distance of order $\varepsilon^{1/4}$

from the equilibrium branch $x_+^\star(y)$. Since $f > 0$ for $x^\star(y) < x < x_+^\star(y)$, x_t increases. By the same method as in the previous example, one shows that x_t approaches $x_+^\star(y_t)$ like $\varepsilon/y_t^{3/2}$. Note in particular that, unlike in the case of the symmetric pitchfork bifurcation, there is no macroscopic delay.

2.2.5 Hopf Bifurcation and Bifurcation Delay

A *Hopf bifurcation* occurs if the Jacobian matrix $\partial_x f(\widehat{x}, \widehat{y})$ at the bifurcation point $(\widehat{x}, \widehat{y})$ has a pair of conjugate eigenvalues $\pm i\omega_0$ on the imaginary axis (with $\omega_0 \neq 0$). We assume here that all other eigenvalues of $\partial_x f(\widehat{x}, \widehat{y})$ have negative real parts. We can thus consider the reduced system

$$\begin{aligned} \varepsilon \dot{x} &= f(x, y), \\ \dot{y} &= g(x, y), \end{aligned} \qquad (2.2.67)$$

for $x \in \mathbb{R}^2$, where $f(\widehat{x}, \widehat{y}) = 0$ and $\partial_x f(\widehat{x}, \widehat{y})$ has the eigenvalues $\pm i\omega_0$.

The implicit-function theorem implies the existence, in a neighbourhood $\mathcal{D}_0 \subset \mathbb{R}^m$ of $\widehat{y}$, of a slow manifold $\{(x, y) : x = x^\star(y), y \in \mathcal{D}_0\}$, with $x^\star(\widehat{y}) = \widehat{x}$. Let us denote by $a(y) \pm i\omega(y)$ the eigenvalues of $\partial_x f(x^\star(y), y)$, where $a(0) = 0$ and $\omega(0) = \omega_0$. The associated system $x' = f(x, \lambda)$ admits $x^\star(\lambda)$ as an equilibrium (of focus type), which is stable if $a(\lambda) < 0$ and unstable if $a(\lambda) > 0$. Depending on second- and third-order terms of the Taylor expansion of f, there is (generically)

- either a stable periodic orbit near $x^\star(\lambda)$ for $a(\lambda) > 0$ and no invariant set near $x^\star(\lambda)$ for $a(\lambda) < 0$ (*supercritical* Hopf bifurcation),
- or an unstable periodic orbit near $x^\star(\lambda)$ for $a(\lambda) < 0$ and no invariant set near $x^\star(\lambda)$ for $a(\lambda) > 0$ (*subcritical* Hopf bifurcation).

We refer to [GH90, p. 152] for the precise relation between criticality and coefficients of the Taylor expansion.

The reduced system on the slow manifold is given by

$$\dot{y} = g(x^\star(y), y). \qquad (2.2.68)$$

We are interested in the following situation. Let Y_t be the solution of (2.2.68) such that $Y_{t^\star} = \widehat{y}$ for some $t^\star \in \mathbb{R}$, and assume that $a(Y_t)$ changes sign from negative to positive as t crosses $t^\star$, with positive velocity, that is, $\langle \nabla a(\widehat{y}), g(\widehat{x}, \widehat{y}) \rangle > 0$. One can then construct an affine change of variables $x = x^\star(y) + S(y)(z, \bar{z})^T$, such that the complex variable z satisfies an equation of the form

$$\varepsilon \dot{z} = [a(Y_t) + i\omega(Y_t)]z + b(z, \bar{z}, t, \varepsilon) - \varepsilon W(Y_t) \frac{dx^\star(Y_t)}{dt}, \qquad (2.2.69)$$

where $b(z, \bar{z}, t, \varepsilon) = \mathcal{O}(|z|^2 + \varepsilon |z|)$, and $W(Y_t)$ is the first line of $S(Y_t)^{-1}$.

If $x^\star(y) \equiv \widehat{x}$ identically in y, then $z_t \equiv 0$ is a particular solution of (2.2.69), and the situation is very similar to the one encountered when studying the symmetric pitchfork bifurcation in Section 2.2.3. Solutions starting near the slow manifold at a time t_0 for which $a(Y_{t_0}) < 0$ approach zero exponentially closely, and need a macroscopic *bifurcation delay* time after $t = t^\star$ before leaving a neighbourhood of order one of the origin. This delay time is given by an expression similar to (2.2.38).

The new feature of the Hopf bifurcation is that if f and g are *analytic*,[7] then a delay persists even when $x^\star(y)$ depends on y. This can be understood as a stabilising effect of fast oscillations around the slow manifold. The system (2.2.69) and its solution can be analytically continued to a complex neighbourhood of $t = t^\star$, so that we can define the analytic function

$$\Psi(t) = \int_{t^\star}^{t} \left[a(Y_s) + \mathrm{i}\omega(Y_s)\right] \mathrm{d}s \,. \tag{2.2.70}$$

For real $t_0 < t^\star$, we define

$$\Pi(t_0) = \inf\{t > t^\star \colon \operatorname{Re}\Psi(t_0) = \operatorname{Re}\Psi(t)\} \,. \tag{2.2.71}$$

Neishtadt has proved the following result [Neĭ87, Neĭ88].

Theorem 2.2.12 (Dynamic Hopf bifurcation). *There exist buffer times $t_- < t^\star$ and $t_+ = \Pi(t_-) > t^\star$ and constants $c_1, c_2 > 0$ with the following property. Any solution of (2.2.67) starting sufficiently close to $(x^\star(Y_{t_0}), Y_{t_0})$ satisfies*

$$\|(x_t, y_t) - (x^\star(Y_t), Y_t)\| \leqslant c_1 \varepsilon \tag{2.2.72}$$

for $t_0 + c_2 \varepsilon |\log \varepsilon| \leqslant t \leqslant t_1 - c_2 \varepsilon |\log \varepsilon|$, where $t_1 = \Pi(t_0) \wedge t_+$. If, moreover, $\|x_{t_0} - x^\star(Y_{t_0})\|$ is of order 1, then x_t leaves a neighbourhood of order 1 of the slow manifold at a time $t_1 + \mathcal{O}(\varepsilon |\log \varepsilon|)$.

This result means that for $t_- \leqslant t_0 \leqslant t^\star - \mathcal{O}(1)$, solutions experience a bifurcation delay until time $\Pi(t_0)$, which is similar to the delay occurring for the symmetric pitchfork bifurcation. For $t_0 \leqslant t_-$, however, this delay saturates at t_+. The computation of the buffer times $t_\pm$ is discussed in [DD91] and in [Nei95]. Roughly speaking, (t_-, t_+) are the points furthest apart on the real axis that can be connected by a path of constant $\operatorname{Re}\Psi$, satisfying certain regularity assumptions.

Example 2.2.13. Assume that $a(Y_t) = t - t^\star$ and $\omega(Y_t) = \omega_0 > 0$. Then

$$\operatorname{Re}\Psi(t) = \frac{1}{2}\left[(\operatorname{Re} t - t^\star)^2 - (\operatorname{Im} t + \omega_0)^2 + \omega_0^2\right] \tag{2.2.73}$$

is constant on hyperbolas centred in $(t^\star, -\mathrm{i}\omega_0)$. We have $\Pi(t) = 2t^\star - t$, and the buffer times are given by $t_\pm = t^\star \pm \omega_0$.

---

[7] Strictly speaking, the reduced system (2.2.67) is usually not analytic, even when the original system is, because of lack of regularity of centre manifolds. This is why the analysis in [Neĭ87, Neĭ88] does not use centre manifolds, but keeps track of all original variables. The results are, however, the same.

Note finally that analyticity is crucial for the existence of a delay if $x^\star(y)$ depends on y. Neishtadt has shown that the delay is usually destroyed even by $\mathcal{C}^\infty$-perturbations.

2.3 Periodic Orbits and Averaging

The case where the slow–fast system admits an asymptotically stable slow manifold corresponds to the simplest possible asymptotic dynamics of the associated (or fast) system: All trajectories are attracted by an equilibrium point. The reduced (slow) dynamics is then simply obtained by projecting the equations of motion on the slow manifold.

If the asymptotic dynamics of the fast system is more complicated than stationary, how should one proceed to determine the effective dynamics of the slow variables? For quite a general class of systems, one can show that the effective dynamics can be obtained by averaging the vector field $g(x, y)$ with respect to the invariant measure describing the asymptotic fast dynamics. We focus here on the case where the asymptotic fast motion is periodic.

2.3.1 Convergence towards a Stable Periodic Orbit

Assume that the associated system

$$\frac{\mathrm{d}x}{\mathrm{d}s} \equiv x' = f(x, y_0) \tag{2.3.1}$$

has, for each fixed value of y_0 in an open set $\mathcal{D}_0$, a periodic solution $\gamma^\star(s, y_0)$, with period $T(y_0)$. We further assume that this orbit is *asymptotically stable*. Recall that the stability of a periodic orbit is related to the linear system

$$\xi' = \partial_x f(\gamma^\star(s, y_0), y_0)\xi \, . \tag{2.3.2}$$

Let $U(s, s_0)$ denote the principal solution of this system, that is, such that $\xi_s = U(s, s_0)\xi_{s_0}$. The eigenvalues of the monodromy matrix $U(T(y_0), 0)$ are called *characteristic multipliers*, and their logarithms are called *characteristic exponents* or *Lyapunov exponents*. One of the multipliers is equal to 1, and the corresponding eigenvector is the tangent vector to the orbit at $s = 0$. We will require that the $n - 1$ other characteristic multipliers have a modulus strictly less than unity, which implies the orbit's asymptotic stability.

We further assume that $\gamma^\star(s, y_0)$ is continuous in both variables (note that the origin of time can be chosen arbitrarily on the orbit), and that there are positive constants T_1, T_2 such that $T_1 \leqslant T(y) \leqslant T_2$ uniformly for $y \in \mathcal{D}_0$.

We turn now to the full slow–fast system

$$\begin{aligned} \varepsilon \dot{x} &= f(x, y) \, , \\ \dot{y} &= g(x, y) \, . \end{aligned} \tag{2.3.3}$$

It seems natural that this system should admit solutions (x_t, y_t) for which x_t is close to the rapidly oscillating function $\gamma^\star(t/\varepsilon, y_t)$. The dynamics of y_t would then be governed by an equation of the form

$$\dot{y}_t \simeq g(\gamma^\star(t/\varepsilon, y_t), y_t) , \qquad (2.3.4)$$

which has a rapidly oscillating right-hand side. The general philosophy of averaging is to bring into consideration the *averaged system*

$$\dot{\bar{y}} = \bar{g}(\bar{y}) := \frac{1}{T(\bar{y})} \int_0^{T(\bar{y})} g(\gamma^\star(s, \bar{y}), \bar{y}) \, ds . \qquad (2.3.5)$$

Introducing $\Gamma^\star(\theta, y) = \gamma^\star(T(y)\theta, y)$ allows to rewrite the averaged slow vector field as

$$\bar{g}(\bar{y}) = \int_0^1 g(\Gamma^\star(\theta, \bar{y}), \bar{y}) \, d\theta . \qquad (2.3.6)$$

We assume that the solution $\bar{y}_t$ of (2.3.5) with initial condition $\bar{y}_0 = y_0$ stays in $\mathcal{D}_0$ for $0 \leqslant t \leqslant t_1$. Under these assumptions, Pontryagin and Rodygin have proved the following result [PR60].

Theorem 2.3.1 (Dynamics near a slowly varying periodic orbit). *Let x_0 be sufficiently close to $\Gamma^\star(\theta_0, y_0)$ for some θ_0. Then there exists a function Θ_t, satisfying the relation*

$$\varepsilon \dot{\Theta}_t = \frac{1}{T(\bar{y}_t)} + \mathcal{O}(\varepsilon) , \qquad (2.3.7)$$

such that the estimates

$$\begin{aligned} x_t &= \Gamma^\star(\Theta_t, \bar{y}_t) + \mathcal{O}(\varepsilon) , \\ y_t &= \bar{y}_t + \mathcal{O}(\varepsilon) \end{aligned} \qquad (2.3.8)$$

hold for $\mathcal{O}(\varepsilon|\log \varepsilon|) \leqslant t \leqslant t_1$. The error terms $\mathcal{O}(\varepsilon)$ in (2.3.7) and (2.3.8) are uniform in t on this time interval.

We shall sketch the proof in the case of a two-dimensional fast variable $x \in \mathbb{R}^2$. Let $\mathrm{n}(\theta, y)$ be the outward unit normal vector to the periodic orbit at the point $\Gamma^\star(\theta, y)$. In a neighbourhood of the periodic orbit, the dynamics can be described by coordinates (θ, r) such that

$$x = \Gamma^\star(\theta, y) + r \, \mathrm{n}(\theta, y) . \qquad (2.3.9)$$

On the one hand, we have

$$\varepsilon \dot{x} = f(\Gamma^\star(\theta, y), y) + A(\theta, y) \, \mathrm{n}(\theta, y) r + \mathcal{O}(r^2) , \qquad (2.3.10)$$

where $A(\theta, y) = \partial_x f(\Gamma^\star(\theta, y), y)$. On the other hand, we can express $\varepsilon \dot{x}$ as a function of $\dot{\theta}$ and $\dot{r}$ by differentiating (2.3.9) and using the equation for $\dot{y}$.

2.3 Periodic Orbits and Averaging

Projecting on $n(\theta, y)$ and on the unit tangent vector to the orbit yields a system of the following form, equivalent to (2.3.3):

$$\varepsilon\dot\theta = \frac{1}{T(y)} + b_\theta(\theta, r, y, \varepsilon) ,$$
$$\varepsilon\dot r = f_r(\theta, r, y, \varepsilon) , \qquad (2.3.11)$$
$$\dot y = g(\Gamma(\theta, y) + r\,n(\theta, y), y) .$$

The functions b_θ and f_r can be computed explicitly in terms of A, $\Gamma^\star$, n, and their derivatives with respect to y. They both vanish for $r = \varepsilon = 0$, and in particular the linearisation $\partial_r f_r(\theta, 0, y, 0)$ depends only on $A(\theta, y)$. In a neighbourhood of $r = 0$, $\dot\theta$ is thus positive and we can consider the equations

$$\frac{dr}{d\theta} = T(y)\frac{f_r(\theta, r, y, \varepsilon)}{1 + T(y) b_\theta(\theta, r, y, \varepsilon)} ,$$
$$\frac{dy}{d\theta} = \varepsilon T(y)\frac{g(\Gamma(\theta, y) + r\,n(\theta, y), y)}{1 + T(y) b_\theta(\theta, r, y, \varepsilon)} , \qquad (2.3.12)$$

instead of (2.3.11). Averaging the right-hand side over θ yields a system of the form

$$\frac{d\bar r}{d\theta} = T(\bar y)\bigl[\bar a(\bar y)\bar r + \mathcal{O}(\varepsilon)\bigr] ,$$
$$\frac{d\bar y}{d\theta} = \varepsilon\bigl[\bar g(\bar y) + \mathcal{O}(\bar r) + \mathcal{O}(\varepsilon)\bigr] , \qquad (2.3.13)$$

where $\bar a(\bar y) < 0$. (In fact, $\bar a(y)$ is the Lyapunov exponent of the periodic orbit $\gamma^\star(\cdot, y)$.) This is again a slow–fast system, in which θ plays the rôle of fast time. It follows thus from Tihonov's theorem that $\bar r$ approaches a slow manifold $\bar r^\star(y) = \mathcal{O}(\varepsilon)$. For such $\bar r$, we are in a situation to which the standard averaging theorem can be applied to show that $r_t - \bar r_t$ and $y_t - \bar y_t$ both remain of order ε up to "times" θ_t of order $1/\varepsilon$.

2.3.2 Invariant Manifolds

Theorem 2.3.1 is the equivalent of Tihonov's theorem for slow manifolds. In order to give a more precise description of the dynamics in a neighbourhood of the family of slowly varying periodic orbits, it is useful to have an analogue of Fenichel's theorem as well, on the existence of an invariant manifold tracking the family of periodic orbits.

To construct this invariant manifold, one can proceed as follows. Let Π be the "time" ($\theta = 1$)-Poincaré map associated with the equation (2.3.12). That is, if the trajectory with initial condition (r, y) at $\theta = 0$ passes through the point $(\hat r, \hat y)$ at $\theta = 1$, then by definition $(\hat r, \hat y) = \Pi(r, y)$. If we add ε as a dummy variable, the Poincaré map can be written in the form

$$\widehat{r} = R(r, y, \varepsilon) ,$$
$$\widehat{y} = y + \varepsilon Y(r, y, \varepsilon) , \qquad (2.3.14)$$
$$\widehat{\varepsilon} = \varepsilon .$$

Since for $\varepsilon = 0$, one recovers the dynamics of the associated system, one necessarily has $R(0, y, 0) = 0$, and $\partial_r R(0, y, 0)$ has the same eigenvalues as the monodromy matrix $U(T(y), 0)$ of the periodic orbit, except for the eigenvalue 1 associated with the variable θ.

Thus every point of the form $(0, y, 0)$ is a fixed point of the Poincaré map, and the linearisation of Π around each of these points admits 1 as an eigenvalue of multiplicity $m + 1$, while the other $n - 1$ eigenvalues are strictly smaller than 1. The centre-manifold theorem thus yields the existence of an invariant manifold of equation $r = \varepsilon \bar{r}(y, \varepsilon)$, that is,

$$\varepsilon \bar{r}\Big(y + \varepsilon Y\big(\bar{r}(y, \varepsilon), y, \varepsilon\big), \varepsilon\Big) = R\big(\varepsilon \bar{r}(y, \varepsilon), y, \varepsilon\big) . \qquad (2.3.15)$$

A perturbative calculation shows that

$$\bar{r}(y, \varepsilon) = \big[\mathbb{1} - \partial_r R(0, y, 0)\big]^{-1} \partial_\varepsilon R(0, y, 0) + \mathcal{O}(\varepsilon) . \qquad (2.3.16)$$

We can now return to the original equation (2.3.12). The set of images under the flow of the invariant manifold $r = \varepsilon \bar{r}(y, \varepsilon)$ defines a cylinder-shaped invariant object, whose parametrisation we can denote as $r = \varepsilon \bar{r}(\theta, y, \varepsilon)$. In the original x-variables, the cylinder is given by the equation

$$\bar{\Gamma}(\theta, y, \varepsilon) = \Gamma^\star(\theta, y) + \varepsilon \bar{r}(\theta, y, \varepsilon) \, \mathrm{n}(\theta, y) . \qquad (2.3.17)$$

The dynamics on the invariant cylinder is then governed by the reduced equations

$$\begin{aligned} \varepsilon \dot{\theta} &= \frac{1}{T(y)} + b_\theta(\theta, \varepsilon \bar{r}(\theta, y, \varepsilon), y, \varepsilon) , \\ \dot{y} &= g\big(\bar{\Gamma}(\theta, y, \varepsilon), y\big) . \end{aligned} \qquad (2.3.18)$$

Note that the term $b_\theta(\theta, \varepsilon \bar{r}(\theta, y, \varepsilon), y, \varepsilon)$ is at most of order ε, so that we recover the fact that $\varepsilon \dot{\theta}$ is ε-close to $1/T(y)$.

Bibliographic Comments

The fact that solutions of a slowly forced system (or, more generally, fast modes) tend to track stable equilibrium states of the corresponding frozen system is an old idea in physics, where it appears with names such as adiabatic elimination, slaving principle, or adiabatic theorem.

First mathematical results on slow–fast systems of the form considered here go back to Tihonov [Tih52] and Gradšteĭn [Gra53] for equilibria, and

Pontryagin and Rodygin [PR60] for periodic orbits. For related results based on asymptotic series for singularly perturbed systems, see also Nayfeh [Nay73], O'Malley [O'M74, O'M91], and Wasow [Was87]. Similar ideas have been used in quantum mechanics [Bog56], in particular for slowly time-dependent Hamiltonians [Nen80, Ber90, JP91].

The geometric approach to singular perturbation theory was introduced by Fenichel [Fen79]. See in particular [Jon95, JKK96] for generalisations giving conditions on $\mathcal{C}^1$-closeness between original and reduced equations.

The dynamic saddle–node bifurcation was first studied by Pontryagin [Pon57]. Later, Haberman [Hab79] addressed this problem and related ones, using matched asymptotic expansions. The theory of relaxation oscillations in several dimensions has been developed in [MR80, MKKR94]. The power-law behaviour of solutions near saddle–node bifurcation points was later rediscovered by physicists when studying bistable optical systems [JGRM90, GBS97]. See also [Arn94] for an overview.

Dynamic pitchfork and transcritical bifurcations were first studied in the generic case by Lebovitz and Schaar [LS77, LS75], as well as Haberman [Hab79]. The phenomenon of bifurcation delay for symmetric pitchfork bifurcations was observed in a laser in [ME84]. However, the same phenomenon for Hopf bifurcations was first pointed out by Shishkova [Shi73], and later analysed in detail by Neishtadt [Neĭ87, Neĭ88, Nei95]. Similar phenomena for maps undergoing period doubling bifurcations have been studied in [Bae95], while the case of a Hopf bifurcation of a periodic orbit has been considered in [NST96].

Since these early results, dynamic bifurcations have been analysed by a great variety of methods. These methods include nonstandard analysis (see, e.g., [Ben91] for a review, in particular [DD91] for bifurcation delay, and [FS03] for recent results); blow-up techniques [DR96, KS01] and a boundary function method [VBK95]. The approach based on the Newton polygon, presented in Section 2.2.4, was introduced in [Ber98, BK99].

The method of averaging was already used implicitly in celestial mechanics, where it allows to determine the secular motion of the planets' orbits, by eliminating the fast motion along the Kepler ellipses. The method was developed in particular by van der Pol, Krylov, Bogoliubov and Mitropol'skiĭ.

We did not mention another interesting characteristic of systems with slowly varying periodic orbits, the so-called geometric phase shifts. These occur, e.g., when the system's parameters are slowly modified, performing a loop in parameter space before returning to their initial value. Then the position on the orbit changes by a phase, whose first-order term in the speed of parameter variation depends only on the geometry of the loop [Ber85, KKE91]. The same phenomenon occurs in quantum mechanics (see for instance [Ber90, JKP91]).

3
One-Dimensional Slowly Time-Dependent Systems

In this chapter, we examine the effect of noise on a particular class of singularly perturbed systems, namely slowly time-dependent equations with one-dimensional state space. Such situations occur for instance for systems with slowly varying parameters, or for slowly forced systems. Doing this will allow us to develop the sample-path approach in a simpler setting, and often we will obtain sharper estimates than in the general case.

The deterministic systems that we will perturb by noise have the form

$$x' = f(x, \varepsilon s) \:. \tag{3.0.1}$$

Passing to the slow timescale $t = \varepsilon s$ yields the familiar form

$$\varepsilon \dot{x} = f(x, t) \:, \tag{3.0.2}$$

which is a particular case of the general slow–fast system (2.0.1), in which the slow vector field g is identically equal to 1, and thus $y_t = t$ for all t.

Since we assume x to be one-dimensional, we can always introduce a potential $U(x, t) = U(x, \varepsilon s)$, such that $f(x, t) = -\partial_x U(x, t)$. Example 2.1.3 shows that one can think of (3.0.1) as describing the overdamped motion of a particle in the slowly time-dependent potential $U(x, \varepsilon s)$, in the limit of strong damping (see also Section 6.1.1).

Typically, the function $f(x, t)$ vanishes on a number of curves, the *equilibrium branches*, corresponding to local maxima or minima of the potential. These branches may intersect at certain bifurcation points. Between bifurcation points, Fenichel's theorem ensures the existence of adiabatic solutions, tracking the equilibrium branches at a distance at most of order ε. Solutions starting between equilibrium branches usually reach the ε-neighbourhood of a stable branch in a time of order $\varepsilon |\log \varepsilon|$, cf. Tihonov's theorem. Combining this information with a local analysis near each bifurcation point generally allows to obtain a fairly precise picture of the deterministic dynamics.

Adding white noise to the slowly time-dependent equation (3.0.1) results in the non-autonomous Itô stochastic differential equation (SDE)

3 One-Dimensional Slowly Time-Dependent Systems

$$\mathrm{d}x_s = f(x_s, \varepsilon s)\,\mathrm{d}s + \sigma F(x_s, \varepsilon s)\,\mathrm{d}W_s\;. \tag{3.0.3}$$

Here we consider σ as another small parameter, while F is a given function of order 1. On the slow timescale, by the scaling property of the Brownian motion, this equation becomes

$$\mathrm{d}x_t = \frac{1}{\varepsilon}f(x_t, t)\,\mathrm{d}t + \frac{\sigma}{\sqrt{\varepsilon}}F(x_t, t)\,\mathrm{d}W_t\;. \tag{3.0.4}$$

Our analysis will mainly follow the same steps as in the deterministic case.

- In Section 3.1, we examine the motion near stable equilibrium branches. Starting with the equation linearised around the adiabatic solution tracking the equilibrium branch, and then incorporating nonlinear terms in a second step, we show that with high probability, paths remain concentrated in neighbourhoods of order σ of the deterministic solution. At the heart of the presented approach lies the choice of the optimal "shape" of these neighbourhoods, allowing for sharp bounds on the probability of a sample path leaving such a neighbourhood early. These concentration results also allow moments of the process to be estimated.
- In Section 3.2, we turn our attention to *unstable* equilibrium branches. In this case, in a neighbourhood of order σ of the adiabatic solution tracking the equilibrium branch, the noise term actually helps sample paths to escape from the deterministic solution. The typical time needed to leave a neighbourhood of order 1 of the unstable branch is of order $\varepsilon|\log \sigma|$.
- Section 3.3 is dedicated to a detailed study of the dynamic saddle–node bifurcation with noise, which is generic in dimension 1. Here a new phenomenon is observed, which is due to the presence of noise: For large enough σ, paths may reach and cross the unstable equilibrium branch some time before the bifurcation occurs.
- In Section 3.4, we determine the effect of noise on symmetric dynamic pitchfork bifurcations. Here the main effect of the noise term is to decrease the bifurcation delay, which is macroscopic in the absence of noise while it is of order $\sqrt{\varepsilon|\log \sigma|}$ for a large regime of non-vanishing noise intensities.
- Finally, Section 3.5 summarises results on two other types of dynamic bifurcations, namely transcritical and asymmetric pitchfork bifurcations.

Throughout this chapter, we will work in the following setting. The stochastic process $\{W_t\}_{t \geq t_0}$ is a standard one-dimensional Wiener process on some probability space $(\Omega, \mathcal{F}, \mathbb{P})$. Initial conditions x_0 are always assumed to be either deterministic, or square-integrable with respect to $\mathbb{P}$ and independent of $\{W_t\}_{t \geq t_0}$. All stochastic integrals are considered as Itô integrals. Without further mentioning, we always assume that f and F satisfy the usual Lipschitz and bounded-growth conditions, which guarantee existence and pathwise uniqueness of a strong solution $\{x_t\}_t$ of (3.0.4), cf. Theorem A.3.2. Under these conditions, there exists a continuous version of $\{x_t\}_t$. Therefore we may assume that the paths $\omega \mapsto x_t(\omega)$ are continuous for $\mathbb{P}$-almost all $\omega \in \Omega$.

We introduce the notation $\mathbb{P}^{t_0,x_0}$ for the law of the process $\{x_t\}_{t\geq t_0}$, starting in x_0 at time t_0, and use $\mathbb{E}^{t_0,x_0}$ to denote expectations with respect to $\mathbb{P}^{t_0,x_0}$. Note that the stochastic process $\{x_t\}_{t\geq t_0}$ is an inhomogeneous Markov process. We are interested in first-exit times of x_t from space–time sets. Let $\mathcal{A} \subset \mathbb{R} \times [t_0, t_1]$ be Borel-measurable. Assuming that $\mathcal{A}$ contains (x_0, t_0), we define the *first-exit time of* (x_t, t) *from* $\mathcal{A}^1$ by

$$\tau_{\mathcal{A}} = \inf\{t \in [t_0, t_1] : (x_t, t) \notin \mathcal{A}\}, \tag{3.0.5}$$

and agree to set $\tau_{\mathcal{A}}(\omega) = \infty$ for those $\omega \in \Omega$ which satisfy $(x_t(\omega), t) \in \mathcal{A}$ for all $t \in [t_0, t_1]$. Typically, we will consider sets of the form $\mathcal{A} = \{(x, t) \in \mathbb{R} \times [t_0, t_1] : g_1(t) < x < g_2(t)\}$ with continuous functions $g_1 < g_2$. Note that in this case, $\tau_{\mathcal{A}}$ is a stopping time with respect to the filtration generated by $\{x_t\}_{t\geq t_0}$.

3.1 Stable Equilibrium Branches

An overdamped particle starting near a local minimum of a static potential is attracted by this minimum. Weak noise causes the particle to fluctuate around the bottom of the well, but large excursions can only be seen on exponentially long timescales. We will now show that a similar behaviour holds for slowly moving potential wells.

We consider in this section the SDE

$$dx_t = \frac{1}{\varepsilon} f(x_t, t)\, dt + \frac{\sigma}{\sqrt{\varepsilon}} F(x_t, t)\, dW_t \tag{3.1.1}$$

in the case where f admits an asymptotically stable equilibrium branch $x^\star(t)$. This is equivalent to assuming that the potential $U(x, t)$ admits a strict local minimum at all times t. More precisely, we will require the following.

Assumption 3.1.1 (Stable case).

- Domain and differentiability: $f \in \mathcal{C}^2(\mathcal{D}, \mathbb{R})$ and $F \in \mathcal{C}^1(\mathcal{D}, \mathbb{R})$, where $\mathcal{D}$ is a domain of the form

$$\mathcal{D} = \{(x, t) : t \in I,\ d_1(t) < x < d_2(t)\}, \tag{3.1.2}$$

for an interval $I = [0, T]$ or $I = [0, \infty)$, and two continuous functions $d_1, d_2 : I \to \mathbb{R}$ such that $d_2(t) - d_1(t)$ is positive and bounded away from 0 in I. We further assume that f, F and all their partial derivatives up to order 2, respectively 1, are uniformly bounded in $\mathcal{D}$ by a constant M.

---

[1] For simplicity, we will often drop the second argument and refer to $\tau_{\mathcal{A}}$ as the first-exit time of x_t from $\mathcal{A}$.

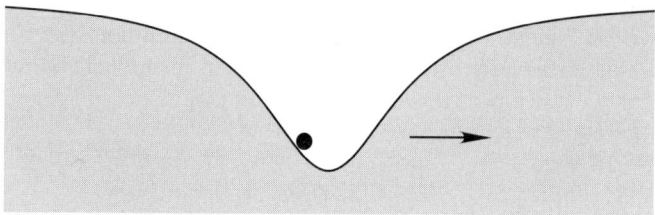

Fig. 3.1. Overdamped particle tracking a slowly moving potential well.

- Equilibrium branch: *There are a continuous function $x^\star : I \to \mathbb{R}$ and some constant $d > 0$ such that $d_1(t) + d \leqslant x^\star(t) \leqslant d_2(t) - d$, and*

$$f(x^\star(t), t) = 0 \qquad \forall t \in I . \qquad (3.1.3)$$

- Stability: *Let $a^\star(t) = \partial_x f(x^\star(t), t)$. There exists a constant $a_0^\star > 0$ such that*

$$a^\star(t) \leqslant -a_0^\star \qquad \forall t \in I . \qquad (3.1.4)$$

- Non-degeneracy of noise term: *There is a constant $F_- > 0$ such that*

$$F(x, t) \geqslant F_- \qquad \forall (x, t) \in \mathcal{D} . \qquad (3.1.5)$$

Since $\mathcal{M} = \{(x, t) \colon x = x^\star(t),\ t \in I\}$ is a uniformly asymptotically stable slow manifold of the deterministic system (cf. Definition 2.1.1), Fenichel's theorem (Theorem 2.1.8) implies the existence of an invariant manifold $\mathcal{M}_\varepsilon$ at a distance of order ε from $\mathcal{M}$. In fact, in this $(1+1)$-dimensional setting, the invariant manifold consists of a particular solution of (3.1.1) for $\sigma = 0$, of the form (see (2.1.27))

$$\bar{x}(t, \varepsilon) = x^\star(t) + \varepsilon \frac{\dot{x}^\star(t)}{a^\star(t)} + \mathcal{O}(\varepsilon^2) . \qquad (3.1.6)$$

Since $a^\star(t)$ is negative, $\bar{x}(t, \varepsilon)$ is lagging behind the equilibrium $x^\star(t)$: It is smaller than $x^\star(t)$ if $x^\star(t)$ is increasing, and larger if it is decreasing. One can associate this behaviour with the following intuitive picture (Fig. 3.1): In a slowly moving potential well, the overdamped particle settles in a position slightly behind the bottom of the well, where the force, directed towards the bottom of the well, compensates for the motion of the potential.

In the time-independent case, the theory of large deviations implies that solutions starting near the bottom $x^\star$ of a potential well tend to remain there for long time spans, a phenomenon called *metastability*. The expected time needed to reach a point x in which $V(x) = V(x^\star) + H$, often called *Kramers' time*, is of order e^{2H/σ^2}, and thus very large as soon as $|x - x^\star| \gg \sigma$.[2]

---

[2] Here we assumed that the potential well has non-vanishing curvature at the bottom. Otherwise, the condition changes to $|x - x^\star| \gg \sigma^{2/q}$, where q is the smallest integer such that $\partial_x^q V(x^\star) \neq 0$.

3.1 Stable Equilibrium Branches

Our main objective will be to characterise the noise-induced deviations of sample paths of (3.1.1) from $\bar{x}(t,\varepsilon)$ in the time-dependent case. To this end we set $x_t = \bar{x}(t,\varepsilon) + y_t$, and observe that y_t satisfies the equation

$$\begin{aligned} dy_t &= \frac{1}{\varepsilon}\big[f(\bar{x}(t,\varepsilon) + y_t, t) - f(\bar{x}(t,\varepsilon), t)\big]\,dt + \frac{\sigma}{\sqrt{\varepsilon}} F(\bar{x}(t,\varepsilon) + y_t, t)\,dW_t \\ &= \frac{1}{\varepsilon}\big[a(t,\varepsilon)y_t + b(y_t, t, \varepsilon)\big]\,dt + \frac{\sigma}{\sqrt{\varepsilon}}\big[F_0(t,\varepsilon) + F_1(y_t, t, \varepsilon)\big]\,dW_t\,, \quad (3.1.7) \end{aligned}$$

where we have set

$$\begin{aligned} a(t,\varepsilon) &= \partial_x f(\bar{x}(t,\varepsilon), t) = a^\star(t) + \mathcal{O}(\varepsilon)\,, \\ F_0(t,\varepsilon) &= F(\bar{x}(t,\varepsilon), t)\,. \end{aligned} \quad (3.1.8)$$

Note that $-a^\star(t)$ is the curvature of the potential at the bottom $x^\star(t)$ of the potential well, and $-a(t,\varepsilon)$ is the curvature at the adiabatic solution tracking the potential well. The remainders satisfy $|b(y,t,\varepsilon)| \leqslant M|y|^2$ and $|F_1(y_t,t,\varepsilon)| \leqslant M|y|$ for sufficiently small y. Henceforth, we shall suppress the ε-dependence of a, b, F_0 and F_1.

The analysis can be simplified by using new coordinates, in which the position-dependent diffusion coefficient F_1 vanishes.

Lemma 3.1.2 (Transformation of variables). Let $y_t = x_t - \bar{x}(t,\varepsilon)$. In a neighbourhood of $y = 0$, there exists a near-identity transformation $y = \tilde{y} + \mathcal{O}(\tilde{y}^2) + \mathcal{O}(\sigma^2)$, casting Equation (3.1.7) into the form

$$d\tilde{y}_t = \frac{1}{\varepsilon}\big[\tilde{a}(t)\tilde{y}_t + \tilde{b}(\tilde{y}_t, t)\big]\,dt + \frac{\sigma}{\sqrt{\varepsilon}} F_0(t)\,dW_t\,, \quad (3.1.9)$$

with $|\tilde{b}(y,t)| \leqslant \widetilde{M}(y^2 + \sigma^2)$ and $\tilde{a}(t) = a(t) + \mathcal{O}(\varepsilon) + \mathcal{O}(\sigma^2)$. If, in addition, F is twice continuously differentiable in x, with $|\partial_{xx} F|$ uniformly bounded, then one can achieve $|\tilde{b}(y,t)| \leqslant \widetilde{M} y^2$.

Proof. The function

$$h_1(y,t) = \int_0^y \frac{dz}{1 + F_1(z,t)/F_0(t)} \quad (3.1.10)$$

satisfies $h_1(y,t) = y + \mathcal{O}(y^2)$ for sufficiently small y. Itô's formula shows that $\bar{y}_t = h_1(y_t,t)$ obeys a SDE of the form

$$d\bar{y}_t = \frac{1}{\varepsilon}\big[\bar{a}(t)\bar{y}_t + \bar{b}(\bar{y}_t, t)\big]\,dt + \frac{\sigma}{\sqrt{\varepsilon}} F_0(t)\,dW_t\,, \quad (3.1.11)$$

where $\bar{a}(t) = a(t) + \mathcal{O}(\varepsilon)$, and $\bar{b}(\bar{y},t)$ contains terms of order $\bar{y}^2$ and σ^2. In case F is twice continuously differentiable with respect to x, the latter can be eliminated by a further change of variables

$$\bar{y} = \tilde{y} + \sigma^2 h_2(t) , \qquad (3.1.12)$$

provided h_2 satisfies the differential equation

$$\varepsilon \dot{h}_2(t) = \bar{a}(t) h_2(t) + \frac{1}{\sigma^2} \bar{b}(\sigma^2 h_2(t), t) . \qquad (3.1.13)$$

The right-hand side vanishes on a curve $h_2^\star(t) = -\bar{b}(0,t)/(\sigma^2 \bar{a}(t)) + \mathcal{O}(\sigma^2)$, corresponding to a uniformly asymptotically stable slow manifold if σ is sufficiently small. Thus by Tihonov's theorem (cf. Theorem 2.1.7), Equation (3.1.13) admits a bounded solution. Combining both changes of variables yields the form (3.1.9) with $\tilde{a}(t) = \bar{a}(t) + \mathcal{O}(\sigma^2)$. □

Remark 3.1.3. In fact, it is also possible to replace $F_0(t)$ in (3.1.9) by a constant, using a time-dependent rescaling of space. However, the resulting change of variables would no longer be near-identity. Carrying out the analysis for general $F_0(t)$ has the advantage of making the concentration results obtained more transparent, as the domains of concentration are little affected by the near-identity transformation.

For sufficiently small ε and σ, we may assume that $\tilde{a}(t)$ is bounded above, uniformly in I, by a negative constant $\tilde{a}_0$. In the sequel, we shall simplify the notations by dropping all tildes, which amounts to considering the original equation (3.1.7) without the position-dependent noise coefficient $F_1(y,t)$.

We will proceed in two steps. In Section 3.1.1, we analyse the linear equation obtained for vanishing nonlinear drift term b in (3.1.7). In Section 3.1.2, we examine the effect of this nonlinear term. The main result is Theorem 3.1.10, which gives rather precise estimates on the first-exit time of sample paths from a strip surrounding $\bar{x}(t,\varepsilon)$, and of width proportional to $F_0(t)^2/2|a(t)|$. Finally, Section 3.1.3 gives some consequences on the moments of y_t.

3.1.1 Linear Case

In this section, we study the non-autonomous linear SDE

$$dy_t^0 = \frac{1}{\varepsilon} a(t) y_t^0 \, dt + \frac{\sigma}{\sqrt{\varepsilon}} F_0(t) \, dW_t \qquad (3.1.14)$$

with initial condition $y_0^0 = 0$, where we assume that a and F_0 are continuously differentiable functions from I to $\mathbb{R}$, with $F_0(t)$ bounded below by $F_- > 0$, and $a(t)$ bounded above by $-a_0 < 0$. Its solution is a Gaussian process and can be represented by the Itô integrals

$$y_t^0 = \frac{\sigma}{\sqrt{\varepsilon}} \int_0^t e^{\alpha(t,s)/\varepsilon} F_0(s) \, dW_s , \qquad (3.1.15)$$

where $\alpha(t,s) = \int_s^t a(u)\,du$ is the curvature accumulated between times s and t. Thus, for each fixed time t, y_t^0 is characterised by its mean being zero and its variance given by

$$\mathrm{Var}\{y_t^0\} = \frac{\sigma^2}{\varepsilon} \int_0^t e^{2\alpha(t,s)/\varepsilon} F_0(s)^2 \, ds \,. \tag{3.1.16}$$

The variance can be computed, in principle, by evaluating two integrals. However, the expression (3.1.16) is not convenient to handle. An alternative expression is found by noting that $\mathrm{Var}\{y_t^0\} = \sigma^2 v(t)$, where $v(t)$ is a solution of the ordinary differential equation

$$\varepsilon \dot{v} = 2a(t)v + F_0(t)^2 \,, \tag{3.1.17}$$

with initial condition $v(0) = 0$. The right-hand side of (3.1.17) vanishes on the slow manifold of equation $v = v^\star(t) = F_0(t)^2/2|a(t)|$, which is uniformly asymptotically stable. We thus conclude by Tihonov's theorem that (3.1.17) admits a particular solution of the form

$$\zeta(t) = \frac{F_0(t)^2}{2|a(t)|} + \mathcal{O}(\varepsilon) \,, \tag{3.1.18}$$

where the term $\mathcal{O}(\varepsilon)$ is uniform in $t \in I$ (higher order terms in ε can be computed, provided the functions are sufficiently smooth). Note, in particular, that for sufficiently small ε, there exist constants $\zeta_+ > \zeta_- > 0$ such that

$$\zeta_- \leqslant \zeta(t) \leqslant \zeta_+ \qquad \forall t \in I \,. \tag{3.1.19}$$

The relation between $\zeta(t)$ and the variance of y_t^0 is given by

$$\mathrm{Var}\{y_t^0\} = \sigma^2 v(t) = \sigma^2 \big[\zeta(t) - \zeta(0) e^{2\alpha(t)/\varepsilon}\big] \,, \tag{3.1.20}$$

where $\zeta(0) = F_0(0)^2/2|a(0)| + \mathcal{O}(\varepsilon)$ and $\alpha(t) = \alpha(t,0) \leqslant -a_0 t$ for $t \geqslant 0$. Thus, the variance approaches $\sigma^2 \zeta(t)$ exponentially fast.

Remark 3.1.4. An expansion of $\zeta(t)$ into powers of ε can also be obtained directly from the definition (3.1.16), using successive integrations by parts.

Our aim is to show that sample paths of y_t^0 are concentrated in sets of the form

$$\mathcal{B}(h) = \big\{(y,t)\colon t \in I,\ |y| < h\sqrt{\zeta(t)}\big\} \,, \tag{3.1.21}$$

whenever we choose $h > \sigma$. At any fixed time $t \in I$, the probability that (y_t^0, t) does not belong to $\mathcal{B}(h)$ can be expressed in terms of the distribution function of the standard normal law, $\Phi(x) = (2\pi)^{-1/2} \int_{-\infty}^x e^{-u^2/2}\,du$, as

$$\mathbb{P}^{0,0}\big\{(y_t^0,t) \notin \mathcal{B}(h)\big\} = 2\Phi\bigg(-\frac{h}{\sigma}\sqrt{\frac{\zeta(t)}{v(t)}}\bigg) \leqslant 2\Phi\bigg(-\frac{h}{\sigma}\bigg) \leqslant e^{-h^2/2\sigma^2} \,. \tag{3.1.22}$$

However, the probability that the *sample path* $\{y_s\}_{0 \leq s \leq t}$ leaves $\mathcal{B}(h)$ at least once during the time interval $[0, t]$ will be slightly larger. A convenient way to write this probability uses the first-exit time

$$\tau_{\mathcal{B}(h)} = \inf\{s > 0 \colon (y_s^0, s) \notin \mathcal{B}(h)\} \tag{3.1.23}$$

of y_s^0 from $\mathcal{B}(h)$. The probability we are after is

$$\mathbb{P}^{0,0}\{\tau_{\mathcal{B}(h)} < t\} = \mathbb{P}^{0,0}\left\{\sup_{0 \leq s < t} \frac{|y_s^0|}{\sqrt{\zeta(s)}} \geq h\right\}. \tag{3.1.24}$$

Although the event $\{\tau_{\mathcal{B}(h)} < t\}$ is more likely than any of the fixed-time events $\{(y_s^0, s) \notin \mathcal{B}(h)\}$, we still expect that

$$\mathbb{P}^{0,0}\{\tau_{\mathcal{B}(h)} < t\} = C_{h/\sigma}(t, \varepsilon) e^{-h^2/2\sigma^2}, \tag{3.1.25}$$

where the prefactor $C_{h/\sigma}(t, \varepsilon)$ is increasing with time.

We will present two ways to compute the prefactor $C_{h/\sigma}(t, \varepsilon)$. The first one only gives an upper bound, but has the advantage to be easily generalised to higher dimensions. The second one allows $C_{h/\sigma}(t, \varepsilon)$ to be computed exactly to leading order in ε and σ, and will show that the upper bound derived by the first method does not overestimate the prefactor too grossly.

The first approach is based on Doob's submartingale inequality (Lemma B.1.2) and its corollary, the Bernstein-type inequality (Lemma B.1.3)

$$\mathbb{P}\left\{\sup_{0 \leq s \leq t}\left|\int_0^s \varphi(u)\,dW_u\right| \geq \delta\right\} \leq 2\exp\left\{-\frac{\delta^2}{2\int_0^t \varphi(u)^2\,du}\right\}, \tag{3.1.26}$$

stated here for stochastic integrals of Borel-measurable deterministic functions $\varphi \colon [0, t] \to \mathbb{R}$.

Proposition 3.1.5. *For any $\gamma \in (0, 1/2)$ and any $t \in I$, (3.1.25) holds with*

$$C_{h/\sigma}(t, \varepsilon) \leq 2\left\lceil\frac{|\alpha(t)|}{\varepsilon\gamma}\right\rceil \exp\left\{\gamma\frac{h^2}{\sigma^2}[1 + \mathcal{O}(\varepsilon)]\right\}. \tag{3.1.27}$$

Proof. Fix $\gamma \in (0, 1/2)$. As y_t^0 is a Gaussian process but not a martingale due to the explicit time dependence of the integrand in the representation (3.1.15), we cannot apply the Bernstein inequality (3.1.26) directly and our main task will be to approximate y_t^0 locally by Gaussian martingales. In order to do so, we introduce a partition $0 = s_0 < s_1 < \cdots < s_N = t$ of $[0, t]$, by requiring

$$-\alpha(s_{j+1}, s_j) = \varepsilon\gamma \quad \text{for} \quad 0 \leq j < N = \left\lceil\frac{|\alpha(t)|}{\varepsilon\gamma}\right\rceil. \tag{3.1.28}$$

Then, by Inequality (3.1.26) and the definition of $\zeta(t)$, for any $j \in \{1, \ldots, N\}$,

$$\begin{aligned}
&\mathbb{P}^{0,0}\{\tau_{\mathcal{B}(h)} \in [s_j, s_{j+1})\} \\
&\leqslant \mathbb{P}^{0,0}\left\{\sup_{s_j \leqslant s \leqslant s_{j+1}}\left|\int_0^s e^{-\alpha(u)/\varepsilon} F_0(u)\,dW_u\right| \geqslant \frac{\sqrt{\varepsilon}}{\sigma} \inf_{s_j \leqslant s \leqslant s_{j+1}} h\sqrt{\zeta(s)} e^{-\alpha(s)/\varepsilon}\right\} \\
&\leqslant 2\exp\left\{-\frac{h^2}{2\sigma^2} e^{2\alpha(s_{j+1},s_j)/\varepsilon} \inf_{s_j \leqslant s \leqslant s_{j+1}} \frac{\zeta(s)}{\zeta(s_{j+1})}\right\} \\
&\leqslant 2\exp\left\{-\frac{h^2}{2\sigma^2} e^{-2\gamma}[1 - \mathcal{O}(\varepsilon\gamma)]\right\}, \qquad (3.1.29)
\end{aligned}$$

where the last line is obtained from the fact that $\dot\zeta(s)$ is of order 1, which can be seen from (3.1.18) and (3.1.17). The result now follows by summing the above inequality over all intervals of the partition. □

Note that optimising over γ gives $\gamma = (\sigma^2/h^2)[1 + \mathcal{O}(\varepsilon)]^{-1}$, which is less than $1/2$ whenever $h^2 > 2\sigma^2$ as necessary for the probability (3.1.25) to be small. This choice of γ yields

$$C_{h/\sigma}(t,\varepsilon) \leqslant 2\mathrm{e}\left[\frac{|\alpha(t)|}{\varepsilon} \frac{h^2}{\sigma^2}[1 + \mathcal{O}(\varepsilon)]\right]. \qquad (3.1.30)$$

The second approach uses more elaborate results on first-passage times of Gaussian processes through a so-called *curved* boundary (see Appendix C).

Theorem 3.1.6 (Stochastic linear stable case). *There exist constants c_0, $r_0 > 0$ such that, whenever*

$$r(h/\sigma, t, \varepsilon) := \frac{\sigma}{h} + \frac{t}{\varepsilon}\mathrm{e}^{-c_0 h^2/\sigma^2} \leqslant r_0, \qquad (3.1.31)$$

then

$$\mathbb{P}^{0,0}\{\tau_{\mathcal{B}(h)} < t\} = C_{h/\sigma}(t,\varepsilon)\mathrm{e}^{-h^2/2\sigma^2}, \qquad (3.1.32)$$

with the prefactor satisfying

$$C_{h/\sigma}(t,\varepsilon) = \sqrt{\frac{2}{\pi}} \frac{|\alpha(t)|}{\varepsilon} \frac{h}{\sigma}\left[1 + \mathcal{O}\left(r(h/\sigma,t,\varepsilon) + \varepsilon + \frac{\varepsilon}{|\alpha(t)|}\log(1 + h/\sigma)\right)\right]. \qquad (3.1.33)$$

This estimate shows that as soon as we take h/σ sufficiently large, that is, when the width of the strip $\mathcal{B}(h)$ is large compared to $\sigma\sqrt{\zeta(t)}$, it is unlikely to observe any excursion of a sample path outside the strip, unless we wait for an exponentially long time. We thus recover the existence of *metastable* behaviour known in the time-independent case. The result is compatible with the classical Wentzell–Freidlin theory, which shows in particular that the first exit from the neighbourhood of a potential well typically occurs after a time exponentially large in the potential difference to be overcome. However, the expression (3.1.32) is more precise, since (3.1.33) also gives the prefactor to leading order, and the result is not only valid in the limit $\sigma \to 0$.

Note that Condition (3.1.31) amounts to requiring h/σ to be large with respect to $\log(t/\varepsilon)$, and is automatically fulfilled whenever the probability (3.1.32) is small. Fixing our choice of $h \gg \sigma$, we see that the expression (3.1.32) is useful (in the sense of the error terms in (3.1.33) being small) whenever $\log(h/\sigma) \ll t/\varepsilon \ll e^{c_0 h^2/\sigma^2}$.[3] Also note that the prefactor $C_{h/\sigma}(t,\varepsilon)$ is necessarily monotonously increasing in t.

The proof of Theorem 3.1.6 can be divided into two steps. We first introduce first-passage times at the upper and lower boundary of $\mathcal{B}(h)$, given by

$$\tau_\pm(h) = \inf\{t > 0 \colon \pm y_t^0 \geqslant h\sqrt{\zeta(t)}\}\,. \tag{3.1.34}$$

The random variables $\tau_\pm(h)$ are identically distributed, and the first-exit time from $\mathcal{B}(h)$ is given by the minimum $\tau_{\mathcal{B}(h)} = \tau_+(h) \wedge \tau_-(h)$. We start by deriving the distribution of $\tau_+(h)$.

Lemma 3.1.7. *If Condition (3.1.31) is satisfied, then the density of $\tau_+(h)$ is given by*

$$\frac{\partial}{\partial t}\mathbb{P}^{0,0}\{\tau_+(h) < t\} = \frac{1}{\sqrt{2\pi}}\frac{|a(t)|}{\varepsilon}\frac{h}{\sigma}\exp\left\{-\frac{h^2}{2\sigma^2}\frac{\zeta(t)}{v(t)}\right\}$$
$$\times \left[1 + \mathcal{O}\bigl(r(h/\sigma, t, \varepsilon) + \varepsilon + e^{-2a_0 t/\varepsilon}\bigr)\right]. \tag{3.1.35}$$

Proof. Instead of y_t^0, we consider the process

$$z_t = e^{-\alpha(t)/\varepsilon} y_t^0 = \frac{\sigma}{\sqrt{\varepsilon}} \int_0^t e^{-\alpha(s)/\varepsilon} F_0(s)\,\mathrm{d}W_s\,, \tag{3.1.36}$$

which has the advantage of being a Gaussian martingale. We denote its variance at time t by

$$\sigma^2 \tilde{v}(t) = \sigma^2 e^{-2\alpha(t)/\varepsilon} v(t)\,. \tag{3.1.37}$$

The first passage of y_t^0 at $h\sqrt{\zeta(t)}$ corresponds to the first passage of z_t at the level $d(t) = h\sqrt{\zeta(t)} e^{-\alpha(t)/\varepsilon}$, a so-called curved boundary. Some results on first-passage time densities are given in Appendix C. We plan to use Corollary C.1.6, which provides the relation

$$\frac{\partial}{\partial t}\mathbb{P}^{0,0}\{\tau_+(h) < t\} = \frac{1}{\sigma} c_0(t) e^{-d(t)^2/2\sigma^2 \tilde{v}(t)}[1 + \tilde{r}]\,, \tag{3.1.38}$$

where the exponent $d(t)^2/2\sigma^2 \tilde{v}(t)$ yields the exponent in (3.1.35). The prefactor $c_0(t)$ is given by

$$c_0(t) = \frac{\tilde{v}'(t)}{\sqrt{2\pi\tilde{v}(t)}}\left[\frac{d(t)}{\tilde{v}(t)} - \frac{d'(t)}{\tilde{v}'(t)}\right]$$
$$= \frac{h}{\sqrt{2\pi}}\frac{|a(t)|}{\varepsilon}\left[1 + \mathcal{O}(\varepsilon) + \mathcal{O}(e^{-2a_0 t/\varepsilon})\right]\,, \tag{3.1.39}$$

---

[3] Recall that by Assumption 3.1.1, the curvature $|a(t)|$ is bounded above and below for $t \in I$, so that $|\alpha(t)| \asymp t$.

and the remainder $\tilde{r}$ satisfies

$$\tilde{r} = \frac{2M_1 M_2 M_3}{\Delta} \mathcal{O}\left(\frac{4M_2\sigma}{\Delta^2} + \frac{e^{-\Delta^2/4\sigma^2}}{\sigma}t\right), \tag{3.1.40}$$

where the quantities Δ, M_1, M_2 and M_3 must be chosen such that the following conditions are met for all $t \geq 0$ and $0 \leq s \leq t$:

$$d(t)\tilde{v}'(t) - \tilde{v}(t)d'(t) \geq \Delta\tilde{v}'(t)(1 + \sqrt{\tilde{v}(t)}), \tag{3.1.41}$$

$$d(t)\tilde{v}'(t) - \tilde{v}(t)d'(t) \leq M_3(1 + \tilde{v}(t)^{3/2}), \tag{3.1.42}$$

$$1 + \tilde{v}(t) \leq M_2\tilde{v}'(t), \tag{3.1.43}$$

$$\left[\frac{d(t) - d(s)}{\tilde{v}(t) - \tilde{v}(s)} - \frac{d'(t)}{\tilde{v}'(t)}\right] \leq M_1 \frac{\sqrt{2\pi(\tilde{v}(t) - \tilde{v}(s))}}{\tilde{v}'(t)}. \tag{3.1.44}$$

A computation shows that we can choose $\Delta = \mathcal{O}(h)$, $M_1 = \mathcal{O}(h/\varepsilon)$, $M_2 = \mathcal{O}(\varepsilon)$, and $M_3 = \mathcal{O}(h/\varepsilon)$, which yields the error term in (3.1.35). □

The following result relates the distributions of the one-sided first-exit time $\tau_+(h)$ and the double-sided first-exit time $\tau_{\mathcal{B}}(h)$.

Lemma 3.1.8. *The first-exit time $\tau_{\mathcal{B}(h)}$ satisfies*

$$\mathbb{P}^{0,0}\{\tau_{\mathcal{B}(h)} < t\} = 2(1-p)\mathbb{P}^{0,0}\{\tau_+(h) < t\}, \tag{3.1.45}$$

where

$$0 \leq p \leq \sup_{0 \leq s \leq t} \mathbb{P}^{s,0}\{\tau_+(h) < t\}. \tag{3.1.46}$$

Proof. Let us write $\tau_\pm$ as a short-hand for $\tau_\pm(h)$, and recall that $\tau_{\mathcal{B}(h)} = \tau_+ \wedge \tau_-$. By symmetry of y_t^0, we have

$$\mathbb{P}^{0,0}\{\tau_+ \wedge \tau_- < t\} = 2\left[\mathbb{P}^{0,0}\{\tau_+ < t\} - \mathbb{P}^{0,0}\{\tau_- < \tau_+ < t\}\right]. \tag{3.1.47}$$

Furthermore, the Markov property implies

$$\mathbb{P}^{0,0}\{\tau_- < \tau_+ < t\} \leq \mathbb{E}^{0,0}\left\{\mathbb{1}_{\{\tau_- < t\}}\mathbb{P}^{\tau_-, -h\sqrt{\zeta(\tau_-)}}\{\tau_+ < t\}\right\}$$

$$\leq \mathbb{P}^{0,0}\{\tau_- < t\} \sup_{0 \leq s \leq t} \mathbb{P}^{s, -h\sqrt{\zeta(s)}}\{\tau_+ < t\}$$

$$\leq \mathbb{P}^{0,0}\{\tau_- < t\} \sup_{0 \leq s \leq t} \mathbb{P}^{s,0}\{\tau_+ < t\}, \tag{3.1.48}$$

where the last line follows from the fact that solutions of the same SDE, starting in different points, cannot cross. Finally, in the last line, we can replace $\mathbb{P}^{0,0}\{\tau_- < t\}$ by $\mathbb{P}^{0,0}\{\tau_+ < t\}$. □

Proof (of Theorem 3.1.6). Now the proof of (3.1.32) and (3.1.33) follows from Lemma 3.1.8 and Lemma 3.1.7, using the fact that

$$\frac{\zeta(t)}{v(t)} = \left[1 - \frac{\zeta(0)}{\zeta(t)} e^{2\alpha(t)/\varepsilon}\right]^{-1}. \tag{3.1.49}$$

Note that $\mathbb{P}^{s,0}\{\tau_+ < t\}$ is bounded by a decreasing function of s. This implies that the term p in (3.1.45) only contributes to the error terms in (3.1.33). □

Remark 3.1.9. The existence of a uniform lower bound F_- for the diffusion coefficient $F(x, t)$ is actually not necessary. It is sufficient that the lower bound ζ_- of the function $\zeta(t)$ defining the width of $\mathcal{B}(h)$ be strictly positive, which can be the case even if $F_0(t)$ vanishes at some points. Then the error terms may, however, depend on ζ_-.

3.1.2 Nonlinear Case

We return now to a description of the original equation for $y_t = x_t - \bar{x}(t,\varepsilon)$, which includes a nonlinear drift term (the noise-dependent diffusion coefficient having been eliminated by the near-identity transformation of Lemma 3.1.2):

$$\mathrm{d}y_t = \frac{1}{\varepsilon}\left[a(t)y_t + b(y_t, t)\right] \mathrm{d}t + \frac{\sigma}{\sqrt{\varepsilon}} F_0(t) \, \mathrm{d}W_t . \tag{3.1.50}$$

We may assume the existence of a constant d_0 such that the bound $|b(y,t)| \leqslant M(y^2 + \sigma^2)$ holds for $|y| \leqslant d_0$.[4]

We will use an approach which is common in the theory of dynamical systems, and treat the term b as a small perturbation of the linearised equation. Indeed, the solution of (3.1.50) can be written as the sum

$$y_t = y_t^0 + y_t^1 , \tag{3.1.51}$$

where y_t^0 is the solution (3.1.15) of the linearised equation (3.1.14), while

$$y_t^1 = \frac{1}{\varepsilon}\int_0^t e^{\alpha(t,s)/\varepsilon} b(y_s, s) \, \mathrm{d}s . \tag{3.1.52}$$

As in (3.1.21), we denote by $\mathcal{B}(h)$ the strip

$$\mathcal{B}(h) = \{(x,t): t \in I, \; |x - \bar{x}(t,\varepsilon)| < h\sqrt{\zeta(t)}\}, \tag{3.1.53}$$

centred in the deterministic solution (Fig. 3.2), whose width scales with the same function $\zeta(t)$ as before, cf. (3.1.18). If we can show that as long as the path x_t stays in $\mathcal{B}(h)$, the term y_t^1 is small compared to y_t^0 with high probability, then the distribution of the first-exit time $\tau_{\mathcal{B}(h)}$ of x_t from $\mathcal{B}(h)$ will be close to the distribution of the first-exit time for the linearised process. Implementing these ideas leads to the following result.

---

[4] The maybe unusual σ-dependence of (the bound on) b is introduced in order to handle the new drift terms arising from the transformation of variables (Lemma 3.1.2).

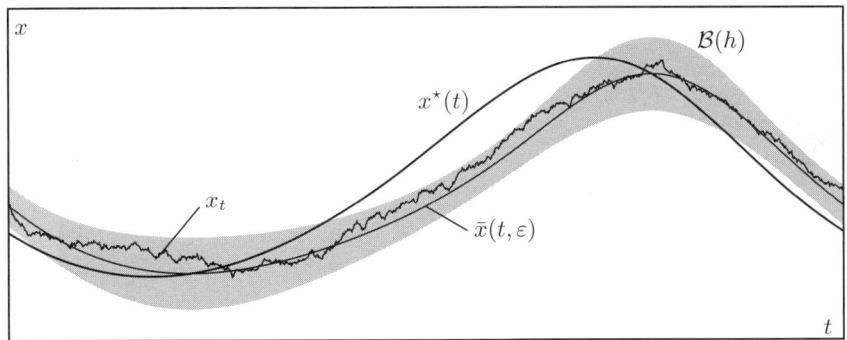

Fig. 3.2. Concentration of sample paths near a stable equilibrium branch $x^\star(t)$. The shaded set $\mathcal{B}(h)$, shown for $h = 3\sigma$, is centred in the deterministic solution $\bar{x}(t,\varepsilon)$, which tracks the equilibrium branch at a distance of order ε.

Theorem 3.1.10 (Stochastic nonlinear stable case). *There exist constants $h_0, r_0, c_0, c_1 > 0$ such that, whenever $h \leqslant h_0$ and*

$$r(h/\sigma, t, \varepsilon) := \frac{\sigma}{h} + \frac{t}{\varepsilon} e^{-c_0 h^2/\sigma^2} \leqslant r_0 \,, \tag{3.1.54}$$

then

$$C_{h/\sigma}(t,\varepsilon) e^{-\kappa_- h^2/2\sigma^2} \leqslant \mathbb{P}^{0,0}\{\tau_{\mathcal{B}(h)} < t\} \leqslant C_{h/\sigma}(t,\varepsilon) e^{-\kappa_+ h^2/2\sigma^2} \,, \tag{3.1.55}$$

where the exponents $\kappa_\pm$ are given by

$$\begin{aligned} \kappa_+ &= 1 - c_1 h \,, \\ \kappa_- &= 1 + c_1 h \,, \end{aligned} \tag{3.1.56}$$

and the prefactor satisfies

$$C_{h/\sigma}(t,\varepsilon) = \sqrt{\frac{2}{\pi}} \frac{|\alpha(t)|}{\varepsilon} \frac{h}{\sigma} \tag{3.1.57}$$
$$\times \left[1 + \mathcal{O}\!\left(r(h/\sigma,t,\varepsilon) + \varepsilon + h + \frac{\varepsilon}{|\alpha(t)|} \log(1 + h/\sigma) \right) \right].$$

Proof. As $|y_{\tau_{\mathcal{B}(h)}}| = h\sqrt{\zeta(\tau_{\mathcal{B}(h)})}$, we can write

$$\mathbb{P}^{0,0}\{\tau_{\mathcal{B}(h)} < t\} = \mathbb{P}^{0,0}\!\left\{ \sup_{0 \leqslant s \leqslant t \wedge \tau_{\mathcal{B}(h)}} \frac{|y_s|}{\sqrt{\zeta(s)}} \geqslant h \right\} \leqslant P_0 + P_1 \,, \tag{3.1.58}$$

with

$$P_0 = P_0(H_0) = \mathbb{P}^{0,0}\!\left\{ \sup_{0 \leqslant s \leqslant t} \frac{|y_s^0|}{\sqrt{\zeta(s)}} \geqslant H_0 \right\}, \tag{3.1.59}$$

$$P_1 = P_1(H_1) = \mathbb{P}^{0,0}\!\left\{ \sup_{0 \leqslant s \leqslant t \wedge \tau_{\mathcal{B}(h)}} \frac{|y_s^1|}{\sqrt{\zeta(s)}} \geqslant H_1 \right\}, \tag{3.1.60}$$

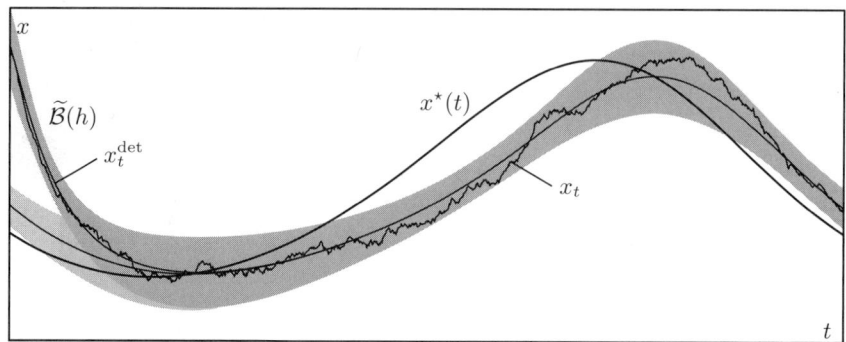

Fig. 3.3. Typical behaviour of a sample path starting at some distance from a stable equilibrium branch $x^*(t)$. The path stays in the shaded set $\widetilde{\mathcal{B}}(h)$, centred in the deterministic solution x_t^{det} starting at the same point as the sample path. The deterministic solution converges exponentially fast to the adiabatic solution $\bar{x}(t,\varepsilon)$.

for any decomposition $h = H_0 + H_1$. The probability P_0 has been estimated in Theorem 3.1.6. For P_1, we use the following deterministic bound on $|y_s^1|/\sqrt{\zeta(s)}$, valid for all $s \leqslant \tau_{\mathcal{B}(h)}$, provided $h \leqslant h_0 := d_0/\sqrt{\zeta_+}$,

$$\frac{|y_s^1|}{\sqrt{\zeta(s)}} \leqslant \frac{1}{\sqrt{\zeta(s)}} \frac{1}{\varepsilon} \int_0^s e^{\alpha(s,u)/\varepsilon} |b(y_u, u)| \, du \leqslant \frac{\zeta_+}{\sqrt{\zeta_-}} \frac{M(h^2 + \sigma^2)}{a_0}. \qquad (3.1.61)$$

Note that we have $\sigma^2 \leqslant r_0^2 h^2$ by Assumption (3.1.54). Thus, choosing $H_1 = 2M(1+r_0^2)h^2\zeta_+/(a_0\sqrt{\zeta_-})$ implies $P_1 = 0$. In addition, H_1 is significantly smaller than $H_0 = h - H_1 = h(1 - \mathcal{O}(h))$. Therefore, the contribution of the nonlinear term causes only an error term of order h in the exponent and in the prefactor obtained for P_0.

As for the lower bound, whenever $H_0 - H_1 \geqslant h$, we can bound the probability below by $P_0(H_0) - P_1(H_1)$. We thus take the same H_1 as before, and $H_0 = h + H_1$. □

The result (3.1.55) can be interpreted as follows. For a fixed choice of $h \gg \sigma$, the probability to leave $\mathcal{B}(h)$ before time t remains small as long as $|\alpha(t)|/\varepsilon \ll e^{h^2/2\sigma^2}$, that is, up to the order of Kramers' time. On the other hand, given a time t, paths are unlikely to leave $\mathcal{B}(h)$ before time t provided $h^2 \gg 2\sigma^2 \log(|\alpha(t)|/\varepsilon)$. Loosely speaking, we will say that typical sample paths are concentrated in the set $\mathcal{B}(\sigma)$.

The difference between the two exponents κ_+ and κ_- in (3.1.55) actually induces a more important inaccuracy than the error term in the prefactor. This is due to the fact that we use the same strip $\mathcal{B}(h)$ as in the linear case, although the nonlinear drift term $b(y,t)$ will typically induce an asymmetric distribution of paths around the deterministic solution.

In Theorem 3.1.10, we assumed for simplicity that x_t starts exactly on the adiabatic solution $\bar{x}(t,\varepsilon)$. What happens if the initial condition x_0 lies at a distance δ from $\bar{x}(0,\varepsilon)$? For δ which are sufficiently small, but may be of order 1, the deterministic solution x_t^{det} starting in x_0 approaches $\bar{x}(t,\varepsilon)$ exponentially fast. The linearisation $\tilde{a}(t) = \partial_x f(x_t^{\text{det}}, t)$ thus approaches $a(t)$ exponentially fast, and is bounded away from 0 if δ is small enough. One can define a neighbourhood $\widetilde{\mathcal{B}}(h)$ of x_t^{det}, using instead of $\zeta(t)$ a solution $\tilde{\zeta}(t)$ of the equation $\varepsilon \dot{z} = 2\tilde{a}(t)z + F(x_t^{\text{det}}, t)^2$; the difference $\tilde{\zeta}(t) - \zeta(t)$ decreases exponentially fast, and thus $\widetilde{\mathcal{B}}(h)$ overlaps with $\mathcal{B}(h)$ as soon as t/ε is of order $\log(\delta/h)$, and for $t/\varepsilon \gg \log(\delta/h)$, the difference between the sets becomes negligible (Fig. 3.3). The same arguments as in the case $\delta = 0$ then show that (3.1.55) still holds, with $\mathcal{B}(h)$ replaced by $\widetilde{\mathcal{B}}(h)$. In particular, the exponential decay of $\tilde{a}(t) - a(t)$ implies that $\int_0^t \tilde{a}(s)\,\mathrm{d}s = \alpha(t) + \mathcal{O}(\varepsilon)$, so that the prefactor still has the form (3.1.57).

To complete the discussion, it is useful to have some information on the motion far from equilibrium branches, that is, where the drift coefficient $f(x,t)$ is bounded away from zero. If we assume the diffusion coefficient $F(x,t)$ to be bounded away from zero as well, it is always possible to construct a change of variables which replaces F by a constant, cf. Lemma 3.1.2 and Remark 3.1.3. It is thus sufficient to consider the equation

$$\mathrm{d}x_t = \frac{1}{\varepsilon} f(x_t, t)\,\mathrm{d}t + \frac{\sigma}{\sqrt{\varepsilon}}\,\mathrm{d}W_t. \qquad (3.1.62)$$

The following result shows that sample paths starting far from equilibrium branches are likely to reach a neighbourhood of order 1 of a stable branch in a time of order ε.

Theorem 3.1.11 (Stochastic motion far from equilibria). *Assume that there are constants $\delta_0 < \delta_1$ and $0 < \rho < \delta_1 - \delta_0$ such that $f(x,t) \leqslant -f_0 < 0$ for all $x \in [\delta_0, \delta_1 + \rho]$ and $t \in [t_0, t_1]$. Let $c = 2(\delta_1 - \delta_0)/f_0$ and assume $\varepsilon \leqslant (t_1 - t_0)/c$. Then*

$$\mathbb{P}^{t_0, \delta_1}\{x_s > \delta_0 \;\forall s \in [t_0, t_0 + c\varepsilon]\} \leqslant \mathrm{e}^{-\kappa/2\sigma^2} \qquad (3.1.63)$$

holds with $\kappa = f_0 \rho^2 / 2(\delta_1 - \delta_0)$.

Proof. Define a process x_t^0 by

$$x_t^0 = \delta_1 - \frac{f_0}{\varepsilon}(t - t_0) + \frac{\sigma}{\sqrt{\varepsilon}} W_{t-t_0}, \qquad t \geqslant t_0. \qquad (3.1.64)$$

By the comparison lemma (Lemma B.3.2), if $\delta_0 \leqslant x_t^0 \leqslant \delta_1 + \rho$ for all $t \in [t_0, t_0 + c\varepsilon]$, then $x_t \leqslant x_t^0$ for those t. Consider now the decomposition

$$\mathbb{P}^{t_0, \delta_1}\{x_s > \delta_0 \;\forall s \in [t_0, t_0 + c\varepsilon]\}$$

$$\leqslant \mathbb{P}^{t_0, \delta_1}\left\{\sup_{t_0 \leqslant s \leqslant t_0 + c\varepsilon} \frac{\sigma}{\sqrt{\varepsilon}} W_{t-t_0} > \rho\right\} \qquad (3.1.65)$$

$$+ \mathbb{P}^{t_0, \delta_1}\left\{x_s > \delta_0 \;\forall s \in [t_0, t_0 + c\varepsilon],\; \sup_{t_0 \leqslant s \leqslant t_0 + c\varepsilon} \frac{\sigma}{\sqrt{\varepsilon}} W_{t-t_0} \leqslant \rho\right\}.$$

The first term on the right-hand side is bounded by $e^{-\rho^2/2c\sigma^2}$. The second term is bounded above by

$$\mathbb{P}^{t_0,\delta_1}\left\{\delta_0 < x_s^0 < \delta_1 + \rho - \frac{f_0}{\varepsilon}(s-t_0)\ \forall s \in [t_0, t_0 + c\varepsilon]\right\} = 0, \qquad (3.1.66)$$

because $\delta_1 + \rho - \frac{f_0}{\varepsilon}(s-t_0) \leqslant \delta_0$ for $s = t_0 + c\varepsilon$. □

3.1.3 Moment Estimates

The results that we have obtained on the tail probabilities of the first-exit time $\tau_{\mathcal{B}(h)}$ also yield, via standard integration-by-parts formulae, estimates of the moments of $\tau_{\mathcal{B}(h)}$ and of the deviation $|x_t - \bar{x}(t,\varepsilon)|$ of sample paths from the adiabatic solution.

The following, exponentially large lower bound on the expected first-exit time is straightforward to obtain, and is in accordance with standard large-deviation results.

Proposition 3.1.12. *Let Assumption 3.1.1 hold with $I = [0, \infty)$ as time interval. Then*

$$\mathbb{E}^{0,0}\{\tau_{\mathcal{B}(h)}\} \geqslant \frac{\varepsilon\sigma^2}{c\,h^2}e^{\kappa_+ h^2/2\sigma^2} \qquad (3.1.67)$$

holds for all $h \leqslant h_0$, where $\kappa_+ = 1 - \mathcal{O}(h)$ and c is a constant.

Proof. For any time $t \geqslant 0$, we can write

$$\mathbb{E}^{0,0}\{\tau_{\mathcal{B}(h)}\} = \int_0^\infty \mathbb{P}^{0,0}\{\tau_{\mathcal{B}(h)} \geqslant s\}\,ds$$

$$\geqslant \int_0^t \left[1 - C_{h/\sigma}(s,\varepsilon)e^{-\kappa_+ h^2/2\sigma^2}\right]ds. \qquad (3.1.68)$$

Instead of (3.1.57), we may use the estimate (3.1.30) for the prefactor $C_{h/\sigma}(s,\varepsilon)$, which is less precise but has the advantage to hold for all times s. Bounding $C_{h/\sigma}(s,\varepsilon)$ by a constant times $(s/\varepsilon)(h^2/\sigma^2)$, the result follows by optimising over t. □

In order to estimate moments of the deviation $x_t - \bar{x}(t,\varepsilon)$ of sample paths from the deterministic solution, we need to make additional assumptions on the global behaviour of f and F. For instance, in addition to Assumption 3.1.1 and the standard bounded-growth conditions, we may impose that

- f and F are defined and twice, respectively once, continuously differentiable, for all $x \in \mathbb{R}$ and $t \in I$;
- there are constants $M_0, L_0 > 0$ such that $xf(x,t) \leqslant -M_0 x^2$ for $|x| > L_0$;
- $F(x,t)$ is bounded above and below by positive constants.

The assumption of F being uniformly bounded above and below allows the transformation of Lemma 3.1.2 to be carried out globally, thereby eliminating the x-dependence of the noise coefficient. We can thus still consider the equation

$$dy_t = \frac{1}{\varepsilon}[a(t)y_t + b(y_t, t)]\,dt + \frac{\sigma}{\sqrt{\varepsilon}} F_0(t)\,dW_t \qquad (3.1.69)$$

for the deviation $y_t = x_t - \bar{x}(t,\varepsilon)$, where $b(y,t)$ is bounded by $M(y^2 + \sigma^2)$ for small $|y|$.

The assumptions on the large-x behaviour of the drift term allow the probability of sample paths making excursions of order larger than 1 to be bounded. Indeed, there are constants $L_0', M_0' > 0$ such that $y^2 a(t) + y b(y,t) \leqslant -M_0' y^2$ for $|y|$ larger than L_0' (in the sequel, we shall drop the primes). An application of the comparison lemma (Lemma B.3.2) then shows that

$$\mathbb{P}^{0,0}\left\{\sup_{0\leqslant s\leqslant t} |y_s| \geqslant L\right\} \leqslant C\left(\frac{t}{\varepsilon}+1\right) e^{-\kappa L^2/2\sigma^2} \qquad (3.1.70)$$

holds, for some constants $C, \kappa > 0$, whenever $L \geqslant L_0$.

In this setting, we obtain the following estimate on even moments of y_t; odd moments can then be deduced by an application of Schwarz' inequality.

Proposition 3.1.13. *There exist constants $c_1, c_2 > 0$ such that, for all σ satisfying $\sigma|\log \varepsilon| \leqslant c_1$ and all $k \geqslant 1$,*

$$\mathbb{E}^{0,0}\{|y_t|^{2k}\} \leqslant k!\left[1 + r(\sigma,\varepsilon,t)\right]^{2k} \sigma^{2k} \zeta(t)^k, \qquad (3.1.71)$$

with an error term satisfying $r(\sigma,\varepsilon,t) \leqslant c_2(\sigma + \varepsilon t)|\log \varepsilon|$.

Proof. We consider only the case $k = 1$, as the other cases can be treated similarly. Let $\gamma = K\sigma|\log \varepsilon|^{1/2}$, for a constant K to be chosen below. Let ζ_+ and ζ_- be upper and lower bounds on $\zeta(s)$ valid for $0 \leqslant s \leqslant t$. We may assume that $\gamma^2 \zeta_+ < h_0^2$, where h_0 is the constant in Theorem 3.1.10. Consider the event

$$A(\gamma) = \left\{\sup_{0\leqslant s \leqslant t} \frac{|y_s|}{\sqrt{\zeta(s)}} < \gamma\right\}. \qquad (3.1.72)$$

We decompose the expectation of y_t^2 as $\mathbb{E}^{0,0}\{y_t^2\} = E_1 + E_2$, where

$$E_1 = \mathbb{E}^{0,0}\{y_t^2 1_{A(\gamma)}\}, \qquad E_2 = \mathbb{E}^{0,0}\{y_t^2 1_{A(\gamma)^c}\}. \qquad (3.1.73)$$

The main contribution to the expectation comes from the term E_1. Using the representation $y_t = y_t^0 + y_t^1$, cf. (3.1.51), we can decompose it as

$$E_1 \leqslant \left(\mathbb{E}^{0,0}\{(y_t^0)^2\}^{1/2} + \mathbb{E}^{0,0}\{(y_t^1)^2 1_{A(\gamma)}\}^{1/2}\right)^2. \qquad (3.1.74)$$

We immediately have $\mathbb{E}^{0,0}\{(y_t^0)^2\} \leqslant \sigma^2 \zeta(t)$. Furthermore, using the fact that $|b(y,t)| \leqslant M(|y|^2 + \sigma^2)$, we can estimate

$$\mathbb{E}^{0,0}\left\{(y_t^1)^2 \mathbf{1}_{A(\gamma)}\right\}^{1/2} \leqslant M(\gamma^2 \zeta_+ + \sigma^2)\frac{1}{\varepsilon}\int_0^t e^{\alpha(t,s)/\varepsilon}\,ds\;. \tag{3.1.75}$$

The integral can be bounded by a constant times ε, and since $\zeta(t)$ is bounded below, we may write

$$E_1 \leqslant \sigma^2 \zeta(t)\left[1 + M_1 \sigma\left(1 + \frac{\gamma^2 \zeta_+}{\sigma^2}\right)\right]^2 \tag{3.1.76}$$

for some constant $M_1 > 0$. The term E_2 can be written, using integration by parts, as

$$E_2 = \zeta(t)\int_0^\infty 2z\,\mathbb{P}^{0,0}\left\{\sup_{0\leqslant s\leqslant t}\frac{|y_s|}{\sqrt{\zeta(s)}} \geqslant \gamma \vee z\right\} dz\;. \tag{3.1.77}$$

We split the integral at γ, h_0, and $L_0/\sqrt{\zeta_-}$, and evaluate the terms separately, with the help of Theorem 3.1.10 and (3.1.70). This yields a bound of the form

$$E_2 \leqslant \zeta(t) C\left(\frac{t}{\varepsilon} + 1\right) \tag{3.1.78}$$

$$\times\left[\left(\gamma^2 + \frac{\sigma^2}{\kappa_+}\right)e^{-\kappa_+\gamma^2/2\sigma^2} + \frac{L_0^2}{\zeta_-}e^{-\kappa_+ h_0^2/2\sigma^2} + \frac{2\sigma^2}{\kappa}e^{-\kappa L_0^2/2\sigma^2}\right].$$

Now choosing the constant K in the definition of γ sufficiently large allows E_2 to be made negligible with respect to E_1. □

3.2 Unstable Equilibrium Branches

An overdamped particle, starting near a local maximum (or barrier) of a fixed potential, moves away from the barrier with exponentially increasing speed — unless it starts exactly on the top of the barrier: Then it just sits there forever. The slightest amount of noise, however, destroys this effect, by kicking the particle to one side or the other of the barrier, thereby accelerating the first phase of its escape. We will now consider the analogous situation for slowly moving potential barriers.

We consider in this section the SDE

$$dx_t = \frac{1}{\varepsilon}f(x_t, t)\,dt + \frac{\sigma}{\sqrt{\varepsilon}}F(x_t, t)\,dW_t \tag{3.2.1}$$

in the case where f admits a uniformly hyperbolic unstable equilibrium branch $x^\star(t)$. This is equivalent to assuming that the potential $U(x,t)$ admits a strict local maximum at $x^\star(t)$ for all times t. We thus require the same hypotheses as in Section 3.1 to hold, except for the stability assumption which is replaced by its converse.

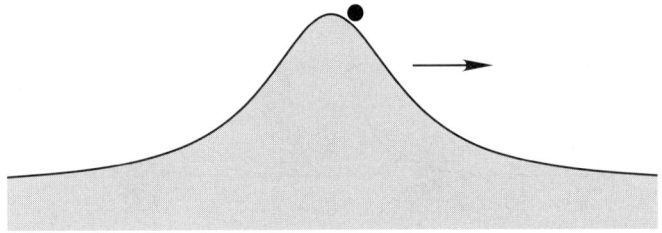

Fig. 3.4. Overdamped particle in equilibrium near a slowly moving potential barrier.

Assumption 3.2.1 (Unstable case).

- Domain and differentiability: $f \in \mathcal{C}^2(\mathcal{D}, \mathbb{R})$ and $F \in \mathcal{C}^2(\mathcal{D}, \mathbb{R})$, where $\mathcal{D}$ is a domain of the form

$$\mathcal{D} = \{(x,t): t \in I, \; d_1(t) < x < d_2(t)\}, \qquad (3.2.2)$$

for an interval $I = [0, T]$ or $I = [0, \infty)$, and two continuous functions $d_1, d_2 : I \to \mathbb{R}$ such that $d_2(t) - d_1(t)$ is positive and bounded away from 0 in I. We further assume that f, F and all their partial derivatives up to order 2, are bounded in absolute value by a constant M, uniformly in $\mathcal{D}$.[5]

- Equilibrium branch: There is a continuous function $x^\star : I \to \mathbb{R}$ such that $d_1(t) + d \leqslant x^\star(t) \leqslant d_2(t) - d$ for some $d > 0$, and

$$f(x^\star(t), t) = 0 \qquad \forall t \in I. \qquad (3.2.3)$$

- Instability: Let $a^\star(t) = \partial_x f(x^\star(t), t)$. There exists a constant $a_0^\star > 0$ such that

$$a^\star(t) \geqslant a_0^\star \qquad \forall t \in I. \qquad (3.2.4)$$

- Non-degeneracy of noise term: There is a constant $F_- > 0$ such that

$$F(x,t) \geqslant F_- \qquad \forall (x,t) \in \mathcal{D}. \qquad (3.2.5)$$

In the deterministic case $\sigma = 0$, Fenichel's theorem (cf. Theorem 2.1.8) can be applied and yields, as in the stable case, the existence of an invariant manifold given by a particular solution

$$\bar{x}(t, \varepsilon) = x^\star(t) + \varepsilon \frac{\dot{x}^\star(t)}{a^\star(t)} + \mathcal{O}(\varepsilon^2) \qquad (3.2.6)$$

of the deterministic equation. In contrast to the stable case, the adiabatic solution *precedes* the unstable equilibrium $x^\star(t)$ (Fig. 3.4). This is similar to a juggler's movements when balancing a pool cue while walking around.

---

[5] Several results remain true if F is only once continuously differentiable, but the assumption $F \in \mathcal{C}^2(\mathcal{D}, \mathbb{R})$ makes life simpler as it allows the x-dependence of F to be removed without introducing remainders of order σ^2 in the drift term.

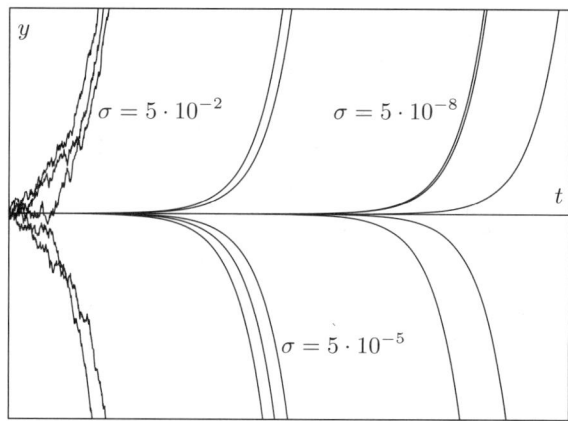

Fig. 3.5. Escape of sample paths from an unstable equilibrium branch. For each of three different noise intensities, paths are shown for five different realisations of the Brownian motion.

The invariant manifold is repelling. In fact, we can apply Tihonov's theorem (cf. Theorem 2.1.7) to the deterministic equation with reversed time, to show that solutions approach $\bar{x}(t, \varepsilon)$ exponentially fast when looking *backwards* in time. This exponential instability also implies that the time needed for neighbouring solutions to depart from $\bar{x}(t, \varepsilon)$ depends on their initial distance $y_0 = x_0 - \bar{x}(0, \varepsilon)$. When $|y_0|$ is sufficiently small, the distance $y_t = x_t - \bar{x}(t, \varepsilon)$ behaves approximately like $y_0 e^{\mathrm{const}\, t/\varepsilon}$, and thus its absolute value reaches order 1 in a time of order $\varepsilon |\log|y_0||$. However, this time can be of order 1, if $|y_0|$ is exponentially small, which is the basic mechanism responsible for bifurcation delay.

In the presence of noise, y_t obeys, as in the stable case, an equation of the form

$$dy_t = \frac{1}{\varepsilon}\bigl[a(t) y_t + b(y_t, t)\bigr] dt + \frac{\sigma}{\sqrt{\varepsilon}} F_0(t)\, dW_t, \qquad (3.2.7)$$

where, as before,

$$a(t) = \partial_x f(\bar{x}(t, \varepsilon), t) + \mathcal{O}(\sigma^2) = a^\star(t) + \mathcal{O}(\varepsilon) + \mathcal{O}(\sigma^2),$$
$$F_0(t) = F(\bar{x}(t, \varepsilon), t). \qquad (3.2.8)$$

The remainder satisfies $|b(y, t)| \leqslant M|y|^2$, for sufficiently small y. If the diffusion coefficient $F(y, t)$ is position-dependent, we transform it into the purely time-dependent coefficient $F_0(t)$, using the same arguments as in Lemma 3.1.2; this slightly affects the value of M and accounts for the σ^2-dependent term in $a(t)$.

The dynamics is dominated by the diffusion term as long as $|y| \leqslant \sigma \rho(t)$, where $\rho(t) = F_0(t)/\sqrt{2a(t)}$. For larger $|y|$, the drift term takes over and helps push sample paths away from the equilibrium branch. Our aim is to show that for $\sigma > 0$, paths will typically leave a neighbourhood of order 1 of $\bar{x}(t,\varepsilon)$ after a time of order $\varepsilon|\log\sigma|$ already, independently of the initial distance to the unstable solution (Fig. 3.5). Below we present estimates of decreasing degree of precision in three different regimes:

- In Section 3.2.1, we use results on small-ball probabilities for Brownian motion to show that for $h < \sigma$, sample paths leave a neighbourhood of size $h\rho(t)$ of $\bar{x}(t,\varepsilon)$ as soon as $\alpha(t) = \int_0^t a(s)\,\mathrm{d}s \gg (4/\pi^2)\varepsilon h^2/\sigma^2$.
- For slightly larger h, namely $\sigma < h \ll \sqrt{\sigma}$, a neighbourhood of size $h\rho(t)$ is left in a time of order $\varepsilon\log(h/\sigma)$.
- Section 3.2.2 extends the results to the escape from a neighbourhood of order δ, where δ is sufficiently small but independent of σ; this escape typically occurs after a time proportional to $\varepsilon\log(\delta/\sigma)$.

3.2.1 Diffusion-Dominated Escape

Consider first the solution y_t^0 of the linearised SDE

$$\mathrm{d}y_t^0 = \frac{1}{\varepsilon}a(t)y_t^0\,\mathrm{d}t + \frac{\sigma}{\sqrt{\varepsilon}}F_0(t)\,\mathrm{d}W_t\ , \qquad (3.2.9)$$

obtained from (3.2.7) by omitting the nonlinear drift term $b(y,t)$. As in Section 3.1.1 above, we can take advantage of the fact that y_t^0 is a Gaussian process with zero mean and variance

$$\mathrm{Var}\{y_t^0\} = \frac{\sigma^2}{\varepsilon}\int_0^t e^{2\alpha(t,s)/\varepsilon}F_0(s)^2\,\mathrm{d}s\ , \qquad (3.2.10)$$

where $\alpha(t,s) = \int_s^t a(u)\,\mathrm{d}u$. In contrast to the stable case, however, $\mathrm{Var}\{y_t^0\}$ grows exponentially fast in time. An integration by parts shows that

$$\mathrm{Var}\{y_t^0\} = \sigma^2 e^{2\alpha(t)/\varepsilon}\bigl[\rho(0)^2 - \rho(t)^2 e^{-2\alpha(t)/\varepsilon} + \mathcal{O}(\varepsilon)\bigr]\ , \qquad (3.2.11)$$

where

$$\rho(t) = \frac{F_0(t)}{\sqrt{2a(t)}}\ . \qquad (3.2.12)$$

The behaviour of the variance is thus given by

$$\mathrm{Var}\{y_t^0\} = \begin{cases} 2\sigma^2\rho(0)^2\dfrac{\alpha(t)}{\varepsilon}\bigl[1 + \mathcal{O}(t/\varepsilon) + \mathcal{O}(\varepsilon)\bigr]\ , & \text{for } t < \varepsilon\ , \\[1ex] \sigma^2\rho(0)^2 e^{2\alpha(t)/\varepsilon}\bigl[1 + \mathcal{O}(e^{-2\alpha(t)/\varepsilon}) + \mathcal{O}(\varepsilon)\bigr]\ , & \text{for } t \geqslant \varepsilon\ . \end{cases}$$
$$(3.2.13)$$

The standard deviation of y_t^0 reaches $h\rho(0)$ as soon as $\alpha(t) \simeq \varepsilon h^2/2\sigma^2$ if $h \ll \sigma$, and as soon as $\alpha(t) \simeq \varepsilon\log(h/\sigma)$ if $h \gg \sigma$.

A similar behaviour can be established for the *sample paths* of the nonlinear equation (3.2.7). We denote by $\tau_{\mathcal{S}(h)}$ the first-exit time of y_t from the strip

$$\mathcal{S}(h) = \{(y,t) \colon t \in I,\ |y| < h\rho(t)\}. \tag{3.2.14}$$

The main result of this section is the following upper bound on the probability that y_t remains in $\mathcal{S}(h)$ up to time t.

Theorem 3.2.2 (Stochastic unstable case – Diffusion-dominated escape). *Assume $|y_0| \leqslant h\rho(0)$.*

- *If $h < \sigma$, then for any time $t \in I$,*

$$\mathbb{P}^{0,y_0}\{\tau_{\mathcal{S}(h)} \geqslant t\} \leqslant \exp\left\{-\kappa(h/\sigma)\frac{\alpha(t)}{\varepsilon}\right\}, \tag{3.2.15}$$

where the exponent $\kappa(h/\sigma)$ is given by

$$\kappa(h/\sigma) = \frac{\pi^2}{4}\frac{\sigma^2}{h^2}\left[1 - \mathcal{O}\left(\frac{h}{\sigma}\sqrt{1 + \frac{\varepsilon}{\alpha(t)}}\right)\right]. \tag{3.2.16}$$

- *Let $\mu > 0$, and define $C_\mu = (2+\mu)^{-(1+\mu/2)}$. Then, for any h such that $\sigma < h < (C_\mu \sigma)^{(1+\mu)/(2+\mu)}$, and any time $t \in I$,*

$$\mathbb{P}^{0,y_0}\{\tau_{\mathcal{S}(h)} \geqslant t\} \leqslant \left(\frac{h}{\sigma}\right)^\mu \exp\left\{-\kappa_\mu \frac{\alpha(t)}{\varepsilon}\right\}, \tag{3.2.17}$$

where the exponent κ_μ is given by

$$\kappa_\mu = \frac{\mu}{1+\mu}\left[1 - \mathcal{O}\left(\varepsilon\frac{1+\mu}{\mu}\right) - \mathcal{O}\left(\frac{1}{\mu \log(1 + h/\sigma)}\right)\right]. \tag{3.2.18}$$

Relation (3.2.15) shows that most paths will already have left $\mathcal{S}(h)$ for times t such that $\alpha(t) \geqslant \varepsilon \kappa(h/\sigma)^{-1}$.

Estimate (3.2.17) allows the description to be extended to slightly larger h, up to all orders less than $\sqrt{\sigma}$. It shows that most paths will have left $\mathcal{S}(h)$ whenever $\alpha(t) \geqslant \varepsilon(1+\mu)\log(h/\sigma)$, where the admissible values for μ depend on h. For any $\nu \in (1/2, 1)$, the choice $\mu = (2\nu - 1)/(1-\nu)$, which yields an exponent κ_μ close to $2 - 1/\nu$, allows h to be taken up to order σ^ν.

Remark 3.2.3. As h approaches σ from below, Estimate (3.2.15) ceases to be applicable as the error term in (3.2.16) is no longer small. To close the gap, we will derive the additional estimate

$$\mathbb{P}^{0,y_0}\{\tau_{\mathcal{S}(h)} \geqslant t\} \leqslant \sqrt{e}\exp\left\{-\frac{\alpha(t)}{\varepsilon}\frac{1 - \mathcal{O}((h+\varepsilon)h^2/\sigma^2)}{\log(1 + (2e/\pi)h^2/\sigma^2)}\right\}, \tag{3.2.19}$$

valid for all $h \leqslant \sigma$ and all $t \in I$; see Proposition 3.2.6 below.

3.2 Unstable Equilibrium Branches

The proof of Theorem 3.2.2 is based on the following consequence of the Markov property, which we will use several times in the sequel.

Lemma 3.2.4. *Let $0 = s_0 < s_1 < \cdots < s_N = t$ be a partition of $[0, t]$. For $k = 1, \ldots, N$, let*

$$P_k = \sup_{|y| \leqslant h\rho(s_{k-1})} \mathbb{P}^{s_{k-1}, y} \left\{ \sup_{s_{k-1} \leqslant s \leqslant s_k} \frac{|y_s|}{\rho(s)} < h \right\}. \tag{3.2.20}$$

Then

$$\mathbb{P}^{0, y_0} \left\{ \sup_{0 \leqslant s \leqslant t} \frac{|y_s|}{\rho(s)} < h \right\} \leqslant \prod_{k=1}^{N} P_k \tag{3.2.21}$$

holds for any y_0 such that $|y_0| \leqslant h\rho(0)$.

Proof. Denote by A_k the event on the right-hand side of (3.2.20). As a direct consequence of the Markov property, the probability in (3.2.21) can be written as

$$\mathbb{P}^{0, y_0} \left\{ \bigcap_{k=1}^{N} A_k \right\} = \mathbb{E}^{0, y_0} \left\{ 1_{\{\bigcap_{k=1}^{N-1} A_k\}} \mathbb{P}^{s_{N-1}, y_{s_{N-1}}} \{A_N\} \right\}$$

$$\leqslant P_N \mathbb{P}^{0, y_0} \left\{ \bigcap_{k=1}^{N-1} A_k \right\} \leqslant \cdots \leqslant \prod_{k=1}^{N} P_k. \tag{3.2.22}$$

$\square$

Note that by taking the supremum over all y in (3.2.20), we overestimate the probability. A more accurate bound would be obtained by integrating over the distribution of y_{s_k} at each $k \in \{1, \ldots, N\}$.

The main point in proving Theorem 3.2.2 is to choose an appropriate partition. Its intervals should be large enough to allow the linearised process to leave $\mathcal{S}(h)$ with appreciable probability, but small enough to allow the influence of the nonlinear drift term $b(y, t)$ to be bounded.

We start by treating the case $h \ll \sigma$. The key idea is again that the linearisation y_t^0 is, at least locally, well approximated by the Gaussian martingale $e^{-\alpha(t)/\varepsilon} y_t^0$, which is equal in distribution to the time-changed Brownian motion $W_{\text{Var}(e^{-\alpha(t)/\varepsilon} y_t^0)}$. Using the Brownian scaling property and a standard estimate on small-ball probabilities for Brownian motion, cf. Appendix C.2, we obtain, for small t and h,

$$\mathbb{P}^{0,0} \{ |y_s^0| < h\rho(s) \ \forall s < t \}$$

$$\simeq \mathbb{P}^{0,0} \{ |W_{\text{Var}(e^{-\alpha(s)/\varepsilon} y_s^0)}| < h\rho(0) \ \forall s < t \}$$

$$= \mathbb{P}^{0,0} \left\{ \sqrt{\text{Var}(e^{-\alpha(t)/\varepsilon} y_t^0)} \sup_{0 \leqslant s \leqslant 1} |W_s| < h\rho(0) \right\}$$

$$\leqslant \exp\left\{ -\frac{\pi^2}{8} \frac{\text{Var}(e^{-\alpha(t)/\varepsilon} y_t^0)}{h^2 \rho(0)^2} \right\} \simeq \exp\left\{ -\frac{\pi^2}{4} \frac{\alpha(t)}{\varepsilon} \frac{\sigma^2}{h^2} \right\}. \tag{3.2.23}$$

Making the argument rigorous and applying it to each interval of the partition, yields the following result, which in turn implies (3.2.15) and (3.2.16).

Proposition 3.2.5. *Assume $h < \sigma$ and $|y_0| \leqslant h\rho(0)$. Then, for any $t \in I$,*

$$\mathbb{P}^{0,y_0}\{\tau_{\mathcal{S}(h)} \geqslant t\} \leqslant \exp\left\{-\frac{\pi^2}{4}\frac{\alpha(t)}{\varepsilon}\frac{\sigma^2}{h^2}\left[1 - \mathcal{O}\left(\frac{h}{\sigma}\sqrt{1 + \frac{\varepsilon}{\alpha(t)}}\right)\right]\right\}. \quad (3.2.24)$$

Proof.
- Let $0 = s_0 < s_1 < \cdots < s_N = t$ be a partition of $[0,t]$. We again choose this partition "equidistant with respect to α", that is, we require

$$\alpha(s_k, s_{k-1}) = \varepsilon\Delta \quad \text{for} \quad 1 \leqslant k < N = \left\lceil \frac{\alpha(t)}{\varepsilon\Delta} \right\rceil, \quad (3.2.25)$$

with $\Delta > 0$ to be chosen later. On each time interval $[s_{k-1}, s_k]$, we write the solution y_s of (3.2.7) as the sum $y_s = y_s^{k,0} + y_s^{k,1}$, where

$$y_s^{k,0} = y_{s_{k-1}} e^{\alpha(s, s_{k-1})/\varepsilon} + \frac{\sigma}{\sqrt{\varepsilon}} \int_{s_{k-1}}^{s} e^{\alpha(s,u)/\varepsilon} F_0(u)\,dW_u,$$

$$y_s^{k,1} = \frac{1}{\varepsilon} \int_{s_{k-1}}^{s} e^{\alpha(s,u)/\varepsilon} b(y_u, u)\,du. \quad (3.2.26)$$

Then the quantity P_k, defined in (3.2.20), can be decomposed as

$$P_k \leqslant \sup_{|y| \leqslant h\rho(s_{k-1})} \left[P_{k,0}(y, H_0) + P_{k,1}(y, H_1) \right], \quad (3.2.27)$$

where

$$P_{k,0}(y, H_0) = \mathbb{P}^{s_{k-1}, y}\left\{ \sup_{s_{k-1} \leqslant s < s_k} \frac{|y_s^{k,0}|}{\rho(s)} < H_0 \right\}, \quad (3.2.28)$$

$$P_{k,1}(y, H_1) = \mathbb{P}^{s_{k-1}, y}\left\{ \sup_{s_{k-1} \leqslant s < s_k} \frac{|y_s^{k,1}|}{\rho(s)} \geqslant H_1, \sup_{s_{k-1} \leqslant s < s_k} \frac{|y_s|}{\rho(s)} < h \right\},$$

provided $H_0, H_1 > 0$ satisfy $H_0 - H_1 = h$.
- Since only those sample paths y_s which do not leave $\mathcal{S}(h)$ during the time interval $[s_{k-1}, s_k)$, contribute to $P_{k,1}(y, H_1)$, the effect of the nonlinear term $b(y, t)$ can be controlled. Note that for all sufficiently small h and all $s \in [s_{k-1}, s_k)$,

$$|y_{s \wedge \tau_{\mathcal{S}(h)}}^{k,1}| \leqslant \frac{1}{\varepsilon} \int_{s_{k-1}}^{s} e^{\alpha(s,u)/\varepsilon} Mh^2 \rho(u)^2\,du$$

$$\leqslant Mh^2 \left[e^{\Delta} - 1\right] \sup_{s_{k-1} \leqslant u \leqslant s_k} \frac{\rho(u)^2}{a(u)}. \quad (3.2.29)$$

Thus, for $h \leqslant h_0$, any k and all $s \in [s_{k-1}, s_k)$, by our choice of the partition,

$$\frac{|y_{s \wedge \tau_{S(h)}}^{k,1}|}{\rho(s \wedge \tau_{S(h)})} \leqslant Mh^2[e^{\Delta} - 1][1 + \mathcal{O}(\varepsilon \Delta)] \sup_{0 \leqslant u \leqslant t} \frac{\rho(u)}{a(u)} =: \frac{\overline{H}_1}{2} . \quad (3.2.30)$$

This ensures $P_{k,1}(y, H_1) = 0$ provided we take $H_1 \geqslant \overline{H}_1$.

- To estimate the contribution of the Gaussian process $y_s^{k,0}$, we approximate it locally, that is separately on each time interval $[s_{k-1}, s_k)$, by a Gaussian martingale. The latter can be transformed into standard Brownian motion by a time change. Finally, it remains to bound the probability of Brownian motion remaining in a small ball: For $|y| \leqslant h\rho(s_{k-1})$,

$$P_{k,0}(y, H_0) \leqslant P_{k,0}(0, H_0) = \mathbb{P}^{s_{k-1}, 0}\left\{ \sup_{s_{k-1} \leqslant s < s_k} \frac{|y_s^{k,0}|}{\rho(s)} < H_0 \right\}$$

$$\leqslant \mathbb{P}^{s_{k-1}, 0}\left\{ e^{-\alpha(s, s_{k-1})/\varepsilon} |y_s^{k,0}| < H_0\rho(s) \; \forall s \in [s_{k-1}, s_k] \right\}$$

$$\leqslant \mathbb{P}^{s_{k-1}, 0}\left\{ \sup_{s_{k-1} \leqslant s \leqslant s_k} |W_{\bar{v}_k(s)/\bar{v}_k(s_k)}| < R_k(H_0, \sigma) \right\}, \quad (3.2.31)$$

where

$$R_k(H_0, \sigma) := \frac{H_0}{\sigma} \bar{v}_k(s_k)^{-1/2} \sup_{s_{k-1} \leqslant s \leqslant s_k} \rho(s) \quad (3.2.32)$$

depends on the variance $\sigma^2 \bar{v}_k(s)$ of the approximating Gaussian martingale $e^{-\alpha(s, s_{k-1})/\varepsilon} y_s^{k,0}$. This variance satisfies

$$\sigma^2 \bar{v}_k(s_k) = \frac{\sigma^2}{\varepsilon} \int_{s_{k-1}}^{s_k} e^{-2\alpha(u, s_{k-1})/\varepsilon} F_0(u)^2 \, du$$

$$\geqslant \sigma^2 \rho(s_k)^2 [1 - e^{-2\Delta}] \inf_{s_{k-1} \leqslant s \leqslant s_k} \frac{\rho(s)^2}{\rho(s_k)^2}, \quad (3.2.33)$$

and, by our choice of the partition, the infimum in this expression behaves like $1 - \mathcal{O}(\alpha(s_k, s_{k-1})) = 1 - \mathcal{O}(\varepsilon \Delta)$. Thus,

$$R_k(H_0, \sigma) \leqslant \frac{H_0}{\sigma}[1 - e^{-2\Delta}]^{-1/2}[1 + \mathcal{O}(\varepsilon \Delta)] =: R(H_0, \sigma; \Delta) . \quad (3.2.34)$$

Now Corollary C.2.2 on small-ball probabilities for Brownian motion immediately implies

$$\sup_{|y| \leqslant h\rho(s_{k-1})} P_{k,0}(y, H_0) \leqslant \frac{4}{\pi} e^{-\pi^2/8R(H_0, \sigma; \Delta)^2} . \quad (3.2.35)$$

- We fix $H_1 = \overline{H}_1$. For small values of Δ, we have $H_0 = h + H_1 = h[1 + \mathcal{O}(h\Delta(1 + \mathcal{O}(\Delta)))]$, and $1/R(H_0, \sigma; \Delta)^2 \geqslant 2\Delta(\sigma^2/h^2)[1 - \mathcal{O}(\Delta)]$ follows. Denoting the right-hand side of (3.2.35) by $q(\Delta)$ and drawing on Lemma 3.2.4, we can bound the probability of $\tau_{S(h)}$ exceeding t by $q(\Delta)^{N-1}$. We use the trivial bound 1 for P_N, since the last interval may be too small for P_N to be small. Optimising the error terms leads to the choice $\Delta = (1 + \varepsilon/\alpha(t))^{-1/2} h/\sigma$, which implies Estimate (3.2.24). □

We now turn to larger values of h. The previous proof is limited by the fact that the key estimate (3.2.35) requires $R(H_0, \sigma; \Delta)$, and thus h/σ, to be of order 1 at most. A more robust, though less precise, bound can be obtained from an endpoint estimate. At any fixed time t, the probability that the Gaussian process y_t^0 lies in a small ball of radius $h\rho(t)$, satisfies

$$\mathbb{P}^{0,0}\{|y_t^0| \leq h\rho(t)\} = 1 - 2\Phi\left(-\frac{h\rho(t)}{\sqrt{\mathrm{Var}\{y_t^0\}}}\right) \leq \frac{2h\rho(t)}{\sqrt{2\pi\,\mathrm{Var}\{y_t^0\}}}, \qquad (3.2.36)$$

where the variance $\mathrm{Var}\{y_t^0\}$ of y_t^0 is given by (3.2.11). Applying similar estimates on each interval of the partition leads to the following bounds, which complete the proof of Theorem 3.2.2 and Estimate (3.2.19) of Remark 3.2.3.

Proposition 3.2.6. *Assume $|y_0| \leq h\rho(0)$. Then, for all $h \leq \sigma$ and $t \in I$,*

$$\mathbb{P}^{0,y_0}\{\tau_{\mathcal{S}(h)} \geq t\} \leq \sqrt{e}\exp\left\{-\frac{\alpha(t)}{\varepsilon}\frac{1 - \mathcal{O}((h+\varepsilon)h^2/\sigma^2)}{\log(1 + (2e/\pi)h^2/\sigma^2)}\right\}. \qquad (3.2.37)$$

Furthermore, for any $\mu > 0$, $\sigma < h < (\sigma/(2+\mu))^{1+\mu/2})^{(1+\mu)/(2+\mu)}$ and $t \in I$,

$$\mathbb{P}^{0,y_0}\{\tau_{\mathcal{S}(h)} \geq t\} \leq \left(\frac{h}{\sigma}\right)^\mu \exp\left\{-\kappa_\mu \frac{\alpha(t)}{\varepsilon}\right\}, \qquad (3.2.38)$$

where

$$\kappa_\mu = \frac{\mu}{1+\mu}\left[1 - \mathcal{O}\left(\varepsilon^{\frac{1+\mu}{\mu}}\right) - \mathcal{O}\left(\frac{1}{\mu\log(1+h/\sigma)}\right)\right]. \qquad (3.2.39)$$

Proof. The beginning of the proof is the same as for Proposition 3.2.5. The main difference lies in the choice of the spacing $\varepsilon\Delta$ of the partition of the time interval $[0,t]$, defined implicitly by (3.2.25). While in the previous proof we were able to choose Δ small, this is not possible for larger regions. This results in the estimate (3.2.31) on $P_{k,0}(y, H_0)$ not being precise enough. Thus we start the proof by estimating $P_{k,0}(y, H_0)$ differently.

- Replacing the supremum of the (conditionally) Gaussian process $y_s^{k,0}$ by its endpoint $y_{s_k}^{k,0}$ shows, for $1 \leq k < N$,

$$P_{k,0}(y, H_0) \leq \mathbb{P}^{s_{k-1},y}\{|y_{s_k}^{k,0}| < H_0\rho(s_k)\} \qquad (3.2.40)$$

$$= \Phi\left(\frac{ye^\Delta + H_0\rho(s_k)}{\sqrt{\sigma^2 v(s_k, s_{k-1})}}\right) - \Phi\left(\frac{ye^\Delta - H_0\rho(s_k)}{\sqrt{\sigma^2 v(s_k, s_{k-1})}}\right),$$

where the (conditional) variance $\sigma^2 v(s_k, s_{k-1})$ of the endpoint $y_{s_k}^{k,0}$ satisfies

$$\sigma^2 v(s_k, s_{k-1}) = e^{2\alpha(s_k, s_{k-1})/\varepsilon}\bar{v}_k(s_k) \geq \sigma^2\rho(s_k)^2\left[e^{2\Delta} - 1\right]\inf_{s_{k-1} \leq s \leq s_k}\frac{\rho(s)^2}{\rho(s_k)^2}, \qquad (3.2.41)$$

3.2 Unstable Equilibrium Branches

cf. (3.2.33). Using the fact that in (3.2.40), the supremum over y is attained for $y = 0$,[6]

$$P_{k,0}(y, H_0) \leqslant \frac{2H_0 \rho(s_k)}{\sqrt{2\pi\sigma^2 v(s_k, s_{k-1})}} \leqslant \sqrt{\frac{2}{\pi}} \frac{H_0}{\sigma} [e^{2\Delta} - 1]^{-1/2} \sup_{s_{k-1} \leqslant s \leqslant s_k} \frac{\rho(s_k)}{\rho(s)} \quad (3.2.42)$$

follows for all y satisfying $|y| \leqslant h\rho(s_{k-1})$. Finally, by our choice of the partition, the supremum in the preceding expression can be bounded by $1 + \mathcal{O}(\alpha(s_k, s_{k-1})) = 1 + \mathcal{O}(\varepsilon\Delta)$.

- We choose $H_1 = \overline{H}_1$, where $\overline{H}_1$ is defined in (3.2.30), so that the nonlinear term becomes negligible, while $H_0 = h[1 + \mathcal{O}(h[e^\Delta - 1][1 + \mathcal{O}(\varepsilon\Delta)])]$. Then, for $1 \leqslant k < N$, P_k is bounded above, uniformly in k, by

$$q(\Delta) := \sqrt{\frac{2}{\pi}} \frac{h}{\sigma} [e^{2\Delta} - 1]^{-1/2} [1 + \mathcal{O}(\varepsilon\Delta) + \mathcal{O}(h(e^\Delta - 1)(1 + \varepsilon\Delta))] . \quad (3.2.43)$$

We use again the trivial bound 1 for P_N, since the last interval may be too small for P_N to be small. By Lemma 3.2.4, we obtain the upper bound

$$\mathbb{P}^{0,y_0} \{\tau_{\mathcal{S}(h)} \geqslant t\} \leqslant q(\Delta)^{-1} \exp\left\{ -\frac{\alpha(t) \log q(\Delta)^{-1}}{\varepsilon} \right\} . \quad (3.2.44)$$

- The remainder of the proof follows from appropriate choices of Δ. In the case $h \leqslant \sigma$, the choice

$$\Delta = \frac{1}{2} \log\left(1 + \gamma \frac{h^2}{\sigma^2}\right) \quad (3.2.45)$$

with $\gamma = 2e/\pi$ yields $q(\Delta) \simeq 1/\sqrt{e}$, and implies (3.2.37). In fact, a better bound can be found by optimising $\log q(\Delta)^{-1}/\Delta$ over γ. The optimal γ is always larger than $2e/\pi$, and tends towards $2e/\pi$ as $h^2/\sigma^2 \to 0$; for $h^2/\sigma^2 \to \pi/2$, the optimal γ diverges, and $\log q(\Delta)^{-1}/\Delta$ tends to 1. In the case $h > \sigma$, the choice

$$\Delta = \frac{1 + \mu}{2} \log\left(1 + \mu + \frac{h^2}{\sigma^2}\right) \quad (3.2.46)$$

yields

$$q(\Delta) \leqslant \sqrt{\frac{2}{\pi}} \left(\frac{h}{\sigma}\right)^{-\mu} \left(1 + \mu \frac{\sigma^2}{h^2}\right)^{-(1+\mu)/2}$$
$$\times \left[1 + \mathcal{O}(\varepsilon\Delta) + \mathcal{O}\left(h\left(1 + \mu + \frac{h^2}{\sigma^2}\right)^{(1+\mu)/2}\right)\right] . \quad (3.2.47)$$

The result (3.2.39) then follows from elementary computations. □

---

[6]Note that for $y \neq 0$, $P_{k,0}(y, H_0)$ is approximately $P_{k,0}(0, H_0)e^{-y^2/2\sigma^2 \rho(s_k)^2}$.

Remark 3.2.7. The term $\mu/(1+\mu)$ in the exponent of (3.2.39) is due to the fact that we take the supremum over all y in (3.2.27). In fact, the probability to remain in $\mathcal{S}(h)$ during $[s_{k-1}, s_k]$ decreases when $y_{s_{k-1}}$ approaches the boundary of $\mathcal{S}(h)$. A more precise estimate would be obtained by integrating over the distribution of $y_{s_{k-1}}$ at each s_{k-1}, which should decrease each P_k by a factor h/σ. However, in the sequel we will consider the escape from a larger domain, where the drift term is no longer close to its linearisation, and we will not obtain the optimal exponent anyway.

3.2.2 Drift-Dominated Escape

Theorem 3.2.2 implies that paths will typically leave a neighbourhood of order $h > \sigma$ of the unstable invariant manifold $x = \bar{x}(t,\varepsilon)$ after a time of order $\varepsilon \log(h/\sigma)$. However, the result is valid only for h of order less than $\sigma^{1/2}$. We will now extend the description to larger neighbourhoods, of the form

$$\mathcal{K}(\delta) = \{(y,t) : t \in I,\ |y| < \delta\}, \qquad (3.2.48)$$

where δ will be chosen small, but independently of σ. As soon as paths have left the small strip $\mathcal{S}(h)$, the drift term will help push the sample paths further away from $\bar{x}(t,\varepsilon)$. In fact, for δ small enough, we may assume the existence of a constant $\kappa = \kappa(\delta) > 0$, such that the drift term satisfies

$$|a(t)y + b(y,t)| \geqslant \kappa a(t)|y| \qquad (3.2.49)$$

for all $(y,t) \in \mathcal{K}(\delta)$. We set $a_\kappa(t) = \kappa a(t)$, $\alpha_\kappa(t) = \kappa \alpha(t)$, and define

$$\rho_\kappa(t) = \frac{F_0(t)}{\sqrt{2a_\kappa(t)}} = \frac{1}{\sqrt{\kappa}} \rho(t). \qquad (3.2.50)$$

We also introduce the shorthands

$$\underline{\rho}_\kappa(t) = \inf_{0 \leqslant s \leqslant t} \rho_\kappa(s), \qquad \bar{\rho}_\kappa(t) = \sup_{0 \leqslant s \leqslant t} \rho_\kappa(s). \qquad (3.2.51)$$

The following theorem is our main result on the first-exit time $\tau_{\mathcal{K}(\delta)}$ of y_t from $\mathcal{K}(\delta)$.

Theorem 3.2.8 (Stochastic unstable case – Drift-dominated escape).
Fix δ, κ and μ in such a way that (3.2.49) holds and $\mu > \kappa/(1-\kappa)$. Then, for any $T > 0$, there exist $\varepsilon_0 > 0$ and $L(T,\mu) > 0$ such that, for all $\varepsilon \leqslant \varepsilon_0$, all

$$\sigma \leqslant \frac{L(T,\mu)}{|\log \varepsilon|^{1+\mu/2}}, \qquad (3.2.52)$$

all $t \in (0,T]$, and for any initial condition $(y_0, 0) \in \mathcal{K}(\delta)$, the bound

$$\mathbb{P}^{0,y_0}\{\tau_{\mathcal{K}(\delta)} \geqslant t\} \leqslant C(t) \frac{\delta}{\sigma} |\log \varepsilon|^{\mu/2} \left(1 + \frac{\alpha_\kappa(t)}{\varepsilon}\right) \frac{e^{-\alpha_\kappa(t)/\varepsilon}}{\sqrt{1 - e^{-2\alpha_\kappa(t)/\varepsilon}}} \qquad (3.2.53)$$

holds, with
$$C(t) = \sqrt{\frac{2}{\pi} \frac{1}{\rho_\kappa(t)}} \left(\frac{2\bar\rho_\kappa(t)}{\sqrt{\kappa}\rho_\kappa(t)} \right)^\mu . \qquad (3.2.54)$$

The probability in (3.2.53) becomes small as soon as
$$\kappa\alpha(t) = \alpha_\kappa(t) \gg \varepsilon \log\left(\frac{\delta}{\sigma}\right) + \varepsilon\frac{\mu}{2}\log|\log\varepsilon| , \qquad (3.2.55)$$

where the second term on the right-hand side hardly matters because of Condition (3.2.52). The optimal choice of κ and μ depends on the particular situation. Generically, (3.2.49) is satisfied for some $\kappa \in (0,1)$, and thus it is best to choose the smallest possible μ. In certain cases, however, $\kappa = 1$ may be a possible choice, but the need to take a finite μ imposes a choice of $\kappa < 1$ for the theorem to be applicable.

The estimate (3.2.53) holds for t of order 1 only. To extend it to even larger values of t, we can use the Markov property to restart after a finite time T. This gives the following bound.

Corollary 3.2.9. *Fix a T of order 1. Then, under the assumptions of Theorem 3.2.8, for any initial condition $(y_0, 0) \in \mathcal{K}(\delta)$ and for all $t > T$*

$$\mathbb{P}^{0,y_0}\{\tau_{\mathcal{K}(\delta)} \geq t\} \leq \exp\left\{ -\frac{\alpha_\kappa(t)}{\varepsilon}[1 - \mathcal{O}(\varepsilon|\log\varepsilon|) - \mathcal{O}(\varepsilon|\log\sigma|)] \right\}, \qquad (3.2.56)$$

where the $\mathcal{O}(\cdot)$-terms depend on μ, T, $C(T)$ and δ.

The main idea of the proof of Theorem 3.2.8 is to compare solutions of the original equation for y_t with those of the linear one
$$dy_t^\kappa = \frac{1}{\varepsilon} a_\kappa(t) y_t^\kappa \, dt + \frac{\sigma}{\sqrt{\varepsilon}} F_0(t) \, dW_t . \qquad (3.2.57)$$

Given an initial time s and an initial condition y_0, the solution of (3.2.57) is a Gaussian process of variance
$$\sigma^2 v_\kappa(t,s) = \frac{\sigma^2}{\varepsilon} \int_s^t e^{2\alpha_\kappa(t,u)/\varepsilon} F_0(u)^2 \, du . \qquad (3.2.58)$$

A trivial bound for the function $v_\kappa(t,s)$, is given by
$$\underline{\rho}_\kappa(t)^2 e^{2\alpha_\kappa(t,s)/\varepsilon}[1 - e^{-2\alpha_\kappa(t,s)/\varepsilon}] \leq v_\kappa(t,s) \leq \bar\rho_\kappa(t)^2 e^{2\alpha_\kappa(t,s)/\varepsilon} . \qquad (3.2.59)$$

Assume now that y_t has left the small strip $\mathcal{S}(h)$ through its upper boundary at time $\tau_{\mathcal{S}(h)}$. As long as y_t remains positive, the comparison lemma B.3.2 shows that y_t stays above the solution y_t^κ of the linear equation (3.2.57), starting at the same point. In particular, if y_t^κ leaves $\mathcal{K}(\delta)$ without returning

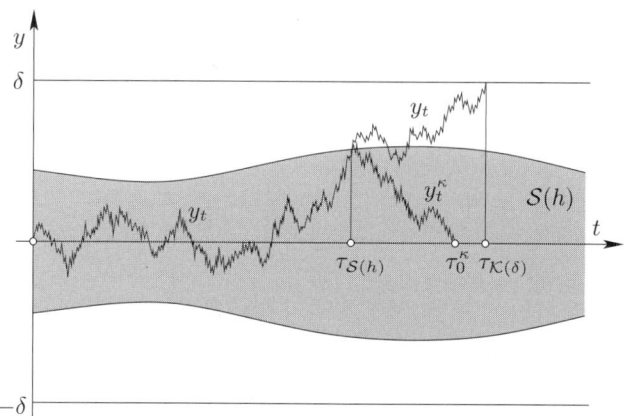

Fig. 3.6. Assume the path y_t leaves the diffusion-dominated region $\mathcal{S}(h)$ through its upper boundary at time $\tau_{\mathcal{S}(h)}$. For larger times, it is bounded below by the solution y_t^κ of the linear equation (3.2.57), starting on the boundary at time $\tau_{\mathcal{S}(h)}$, as long as y_t does not return to zero.

to zero, then y_t must have left $\mathcal{K}(\delta)$ already at an earlier time (Fig. 3.6). We will express this with the help of the stopping time

$$\tau_0^\kappa = \inf\{u : y_u^\kappa \leq 0\}, \tag{3.2.60}$$

which satisfies the following relation.

Lemma 3.2.10. *Assume the original and linear processes y_t and y_t^κ both start in some $y_0 \in (0, \delta)$ at time s. Then, for all times $t > s$,*

$$\mathbb{P}^{s,y_0}\{\tau_{\mathcal{K}(\delta)} \geq t, \tau_0^\kappa \geq t\} \leq \frac{\delta}{\sqrt{2\pi\sigma^2 v_\kappa(t,s)}}. \tag{3.2.61}$$

Proof. By the choice of κ, the comparison lemma B.3.2 implies $y_t \geq y_t^\kappa$ for all $t \leq \tau_{\mathcal{K}(\delta)} \wedge \inf\{u > s : y_u > 0\}$. As a consequence, y_t cannot reach 0 before y_t^κ does. Therefore,

$$\mathbb{P}^{s,y_0}\{\tau_{\mathcal{K}(\delta)} \geq t, \tau_0^\kappa \geq t\} \leq \mathbb{P}^{s,y_0}\{0 < y_u^\kappa < \delta \,\forall u \in [s,t]\}, \tag{3.2.62}$$

which is bounded above by $\mathbb{P}^{s,y_0}\{y_t^\kappa \in [0,\delta]\}$. Now, the result follows from the fact that y_t^κ is a Gaussian random variable with variance $\sigma^2 v_\kappa(t,s)$. □

Next we bound the probability that y_t^κ returns to zero before time t. The following lemma actually provides more precise information on the law of τ_0^κ. We will apply it for initial conditions satisfying $y_0 \geq h\rho(s)$.

Lemma 3.2.11. *Let y_t^κ start in $y_0 \geq \sqrt{2}\sigma\bar{\rho}_\kappa(t)$ at initial time s. Then, the probability of reaching zero before time $t > s$, satisfies the bound*

3.2 Unstable Equilibrium Branches

$$\mathbb{P}^{s,y_0}\{\tau_0^\kappa < t\} \leqslant \exp\left\{-\frac{y_0^2 e^{2\alpha_\kappa(t,s)/\varepsilon}}{2\sigma^2 v_\kappa(t,s)}\right\}. \tag{3.2.63}$$

Moreover, the density $\psi^\kappa(t;s,y_0) = \frac{\partial}{\partial t}\mathbb{P}^{s,y_0}\{\tau_0^\kappa < t\}$ satisfies

$$\psi^\kappa(t;s,y_0) \leqslant \sqrt{\frac{2}{\pi}}\,\frac{y_0}{\sigma}\,\frac{a_\kappa(t)}{\varepsilon}\,\frac{e^{-\alpha_\kappa(t,s)/\varepsilon}}{\sqrt{v_\kappa(t,s)}}\,e^{-y_0^2/2\sigma^2 \bar{\rho}_\kappa(t)^2}. \tag{3.2.64}$$

Proof. Let us introduce the shorthand

$$\Xi = \frac{y_0^2 e^{2\alpha_\kappa(t,s)/\varepsilon}}{2\sigma^2 v_\kappa(t,s)} \geqslant \frac{y_0^2}{2\sigma^2}\,\frac{1}{\bar{\rho}_\kappa(t)^2} \geqslant 1. \tag{3.2.65}$$

The reflection principle (Lemma B.4.1) implies that

$$\mathbb{P}^{s,y_0}\{\tau_0^\kappa < t\} = 2\mathbb{P}^{s,y_0}\{y_t^\kappa < 0\} = 2\Phi\left(-\frac{y_0 e^{\alpha_\kappa(t,s)/\varepsilon}}{\sigma\sqrt{v_\kappa(t,s)}}\right) \leqslant e^{-\Xi}, \tag{3.2.66}$$

which yields (3.2.63).

Using the fact that $v_\kappa(t,s)$ satisfies the differential equation

$$\frac{\partial}{\partial t}v_\kappa(t,s) = \frac{1}{\varepsilon}\bigl[2a_\kappa(t)v_\kappa(t,s) + F_0(t)^2\bigr], \tag{3.2.67}$$

facilitates taking the derivative of (3.2.66), and yields

$$\begin{aligned}\psi^\kappa(t;s,y_0) &= \frac{2}{\sqrt{2\pi}}\,e^{-\Xi}\,\frac{\partial}{\partial t}\frac{(-y_0)e^{\alpha_\kappa(t,s)/\varepsilon}}{\sigma\sqrt{v_\kappa(t,s)}} \\ &= \sqrt{\frac{2}{\pi}}\,\frac{1}{y_0}\,\frac{\sigma}{\varepsilon}\,\frac{F_0(t)^2}{\sqrt{v_\kappa(t,s)}}\,e^{-\alpha_\kappa(t,s)/\varepsilon}\,\Xi e^{-\Xi}.\end{aligned} \tag{3.2.68}$$

The bound (3.2.64) then follows from the fact that the function $\Xi \mapsto \Xi e^{-\Xi}$ is decreasing for $\Xi \geqslant 1$. □

These results already yield a bound for the probability we are interested in, namely

$$\mathbb{P}^{s,h\rho(s)}\{\tau_{\mathcal{K}(\delta)} \geqslant t\} \leqslant \frac{\delta}{\sqrt{2\pi}\sigma\bar{\rho}_\kappa(t)}\,\frac{e^{-\alpha_\kappa(t,s)/\varepsilon}}{\sqrt{1-e^{-2\alpha_\kappa(t,s)/\varepsilon}}} + \exp\left\{-\frac{h^2}{2\sigma^2}\,\frac{\rho(s)^2}{\bar{\rho}_\kappa(t)^2}\right\}. \tag{3.2.69}$$

The first term decreases in the expected way, but the second is at best $e^{-h^2/2\sigma^2} \geqslant e^{-1/2\sigma}$, as we cannot choose h of larger order than $\sigma^{1/2}$. In order to improve this bound, we have to take into account the fact that even when returning to zero, paths will again be expelled of $\mathcal{S}(h)$, cf. Theorem 3.2.2; this substantially decreases $\mathbb{P}^{s,h\rho(s)}\{\tau_{\mathcal{K}(\delta)} \geqslant t\}$.

82   3 One-Dimensional Slowly Time-Dependent Systems

Proof (of Theorem 3.2.8).

- We first introduce some notations. We want to choose h slightly larger than σ; so we set

$$h = \gamma\sigma\sqrt{|\log\varepsilon|}, \qquad (3.2.70)$$

where a suitable γ of order 1 will be chosen below. One easily checks that for such a choice, the condition on h in Theorem 3.2.2 follows from Condition (3.2.52) on σ. The various exit probabilities, used in the sequel, are denoted by

$$\Phi_t(s,y) = \mathbb{P}^{s,y}\{\tau_{\mathcal{K}(\delta)} \geqslant t\}, \qquad (3.2.71)$$

$$q_t(s) = \sup_{|y| \leqslant h\rho(s)} \Phi_t(s,y), \qquad (3.2.72)$$

$$Q_t(s) = \sup_{h\rho(s) \leqslant |y| \leqslant \delta} \Phi_t(s,y). \qquad (3.2.73)$$

When applying Lemma 3.2.10, we will use the shorthand

$$g(t,s) = \frac{e^{-\alpha_\kappa(t,s)/\varepsilon}}{\sqrt{1 - e^{-2\alpha_\kappa(t,s)/\varepsilon}}}. \qquad (3.2.74)$$

We will start by estimating $q_t(s)$ as a function of (a bound on) Q_t. This bound is then used to find a self-consistent equation for Q_t, which is solved by iterations.

- Without loss of generality, we may assume that y_t starts in $y \geqslant 0$. Consider first the case $0 \leqslant y \leqslant h\rho(s)$. The strong Markov property allows us to decompose

$$\Phi_t(s,y) = \mathbb{P}^{s,y}\{\tau_{\mathcal{S}(h)} \geqslant t\} + \mathbb{P}^{s,y}\{\tau_{\mathcal{K}(\delta)} \geqslant t, \tau_{\mathcal{S}(h)} < t\}$$

$$\leqslant \left(\frac{h}{\sigma}\right)^\mu e^{-\alpha_\kappa(t,s)/\varepsilon} + \mathbb{E}^{s,y}\{1_{\{\tau_{\mathcal{S}(h)} < t\}} Q_t(\tau_{\mathcal{S}(h)})\}, \qquad (3.2.75)$$

where we used Theorem 3.2.2 on the diffusion-dominated escape to estimate the first term on the right-hand side.[7] Let $\overline{Q}_t(u)$ be an upper bound on $Q_t(u)$ which is continuously differentiable and non-decreasing as a function of u, and takes values in $[0,1]$. Using a suitable variant of the integration-by-parts formula (see Appendix B), the second term on the right-hand side is seen to be bounded by

$$\left(\frac{h}{\sigma}\right)^\mu e^{-\alpha_\kappa(t,s)/\varepsilon} + \left(\frac{h}{\sigma}\right)^\mu \int_s^t \overline{Q}_t(u) \frac{a_\kappa(u)}{\varepsilon} e^{-\alpha_\kappa(u,s)/\varepsilon}\,du. \qquad (3.2.76)$$

We thus obtain the bound

$$q_t(s) \leqslant \left(\frac{h}{\sigma}\right)^\mu \left[2e^{-\alpha_\kappa(t,s)/\varepsilon} + \int_s^t \overline{Q}_t(u) \frac{a_\kappa(u)}{\varepsilon} e^{-\alpha_\kappa(u,s)/\varepsilon}\,du\right]. \qquad (3.2.77)$$

---

[7] Recall that we have chosen μ and κ in such a way that κ_μ, as defined in Theorem 3.2.2, is larger than κ for sufficiently small ε.

3.2 Unstable Equilibrium Branches

- We turn now to the case $h\rho(s) \leqslant y \leqslant \delta$. Here we can decompose

$$\Phi_t(s,y) = \mathbb{P}^{s,y}\{\tau_{\mathcal{K}(\delta)} \geqslant t, \tau_0^\kappa \geqslant t\} + \mathbb{P}^{s,y}\{\tau_{\mathcal{K}(\delta)} \geqslant t, \tau_0^\kappa < t\}, \quad (3.2.78)$$

where Lemma 3.2.10 provides a bound for the first term on the right-hand side. Again using the strong Markov property, the second term can be bounded by

$$\mathbb{P}^{s,y}\{\tau_{\mathcal{K}(\delta)} \geqslant t, \tau_0^\kappa < t\} = \mathbb{E}^{s,y}\{1_{\{\tau_0^\kappa < t\}} \mathbb{P}^{\tau_0^\kappa, y_{\tau_0^\kappa}}\{\tau_{\mathcal{K}(\delta)} \geqslant t\}\}$$
$$\leqslant \int_s^t [q_t(u) + \overline{Q}_t(u)] \psi^\kappa(u; s, y)\, du, \quad (3.2.79)$$

where the density $\psi^\kappa(u; s, y)$ of τ_0^κ can be estimated by Lemma 3.2.11.

- In order to proceed, we need to estimate some integrals involving $g(t, u)$ and $g(u, s)$. The first estimate is

$$\frac{1}{\varepsilon} \int_s^t a_\kappa(u) e^{-\alpha_\kappa(u,s)/\varepsilon} g(u, s)\, du \leqslant \frac{1}{\varepsilon} \int_s^t a_\kappa(u) g(u, s)\, du \leqslant \frac{\pi}{2}. \quad (3.2.80)$$

Let $\phi = e^{-\alpha_\kappa(t,s)/\varepsilon}$. Then, using the change of variable $e^{-2\alpha_\kappa(u,s)/\varepsilon} = x(1-\phi^2) + \phi^2$, we can compute the integral

$$\frac{1}{\varepsilon} \int_s^t a_\kappa(u) e^{-\alpha_\kappa(u,s)/\varepsilon} g(t, u) g(u, s)\, du = \frac{\phi}{2} \int_0^1 \frac{dx}{\sqrt{x(1-x)}} = \frac{\pi}{2}\phi. \quad (3.2.81)$$

A third integral can be estimated with the help of the change of variables $x^2 = 1 - e^{-2\alpha_\kappa(t,u)/\varepsilon}$:

$$\frac{1}{\varepsilon} \int_s^t a_\kappa(u) e^{-\alpha_\kappa(u,s)/\varepsilon} g(t, u)\, du \leqslant \phi \int_0^{\sqrt{1-\phi^2}} \frac{dx}{1 - x^2} \quad (3.2.82)$$

$$= \frac{\phi}{2} \log \frac{1 + \sqrt{1-\phi^2}}{1 - \sqrt{1-\phi^2}} \leqslant \phi \log \frac{2}{\phi} \leqslant \left(1 + \frac{\alpha_\kappa(t,s)}{\varepsilon}\right) e^{-\alpha_\kappa(t,s)/\varepsilon}.$$

- We now replace $q_t(u)$ in (3.2.79) by its upper bound (3.2.77), and use Lemma 3.2.11 to estimate the density $\psi^\kappa(u; s, y)$. With the help of (3.2.80) to bound the integrals, one arrives at

$$Q_t(s) \leqslant C_0 g(t, s) + c e^{-\alpha_\kappa(t,s)/\varepsilon}$$
$$+ c \int_s^t \frac{a_\kappa(u)}{\varepsilon} \overline{Q}_t(u) e^{-\alpha_\kappa(u,s)/\varepsilon} \left[\frac{1}{2} + \frac{1}{\pi} g(u, s)\right] du, \quad (3.2.83)$$

where

$$c := c(t, s) := \sqrt{2\pi\kappa} \left(\frac{h}{\sigma}\right)^{1+\mu} \frac{\bar{\rho}_\kappa(t)}{\underline{\rho}_\kappa(t)} e^{-\kappa h^2 \underline{\rho}_\kappa(t)^2 / 2\sigma^2 \bar{\rho}_\kappa(t)^2},$$

$$C_0 := C_0(t) := \frac{1}{\sqrt{2\pi}} \frac{\delta}{\sigma} \frac{1}{\underline{\rho}_\kappa(t)}. \quad (3.2.84)$$

Taking σ sufficiently small, we may assume that $C_0 \geq 5$. Furthermore, the choice

$$\gamma = \frac{2}{\sqrt{\kappa}} \frac{\bar{\rho}_\kappa(t)}{\rho_\kappa(t)} \tag{3.2.85}$$

guarantees that $c(1 + \alpha_\kappa(t,s)/\varepsilon) \leq 2/5$ for $0 \leq s < t \leq T$ and sufficiently small ε (recall that smaller ε implies larger h by our choice of h).

- We construct by iteration two sequences $(a_n)_{n \geq 1}$ and $(b_n)_{n \geq 1}$ such that

$$Q_t(u) \leq C_0 g(t,u) + a_n e^{-\alpha_\kappa(t,u)/\varepsilon} + b_n \qquad \forall u \in [s,t] \tag{3.2.86}$$

holds for all $n \geq 1$. Using $\overline{Q}_t(u) = 1$ as a trivial bound in (3.2.83) and applying (3.2.80) again, we see that (3.2.86) is satisfied with

$$a_1 = c \quad \text{and} \quad b_1 = c. \tag{3.2.87}$$

Replacing $\overline{Q}_t(u)$ by the right-hand side of (3.2.86) shows by induction that the same bound holds for $n \mapsto n+1$ if

$$a_{n+1} = c\left[1 + \frac{C_0}{2}\left(2 + \frac{\alpha_\kappa(t,s)}{\varepsilon}\right) + \frac{a_n}{2}\left(1 + \frac{\alpha_\kappa(t,s)}{\varepsilon}\right)\right], \tag{3.2.88}$$

$$b_{n+1} = cb_n. \tag{3.2.89}$$

Since $\lambda := c(1 + \alpha_\kappa(t,s)/\varepsilon)/2 \leq 1/5$, we obtain

$$\lim_{n \to \infty} a_n = \frac{1}{1-\lambda}\left[\lambda C_0 + c\left(1 + \frac{C_0}{2}\right)\right] \leq \frac{1}{2}(1 + C_0), \tag{3.2.90}$$

$$\lim_{n \to \infty} b_n = 0, \tag{3.2.91}$$

and thus

$$Q_t(s) \leq C_0 g(t,s) + \frac{1}{2}(1 + C_0) e^{-\alpha_\kappa(t,s)/\varepsilon} \leq \frac{1}{2}(1 + 3C_0) g(t,s). \tag{3.2.92}$$

Combining this estimate with the bound (3.2.77) for $q_t(s)$ finally yields the result, by taking the maximum of the bounds on $Q_t(s)$ and $q_t(s)$, using $C_0 \geq 5$ and the definitions of h and γ. □

3.3 Saddle–Node Bifurcation

Up to now, we have considered the dynamics near uniformly hyperbolic, stable or unstable equilibrium branches. We turn now to new situations, arising when hyperbolicity is lost as one approaches a bifurcation point. In the picture of an overdamped particle in a potential, this situation arises, for instance, when a potential well becomes increasingly flat, until it disappears altogether in a saddle–node bifurcation (Fig. 3.7).

3.3 Saddle–Node Bifurcation

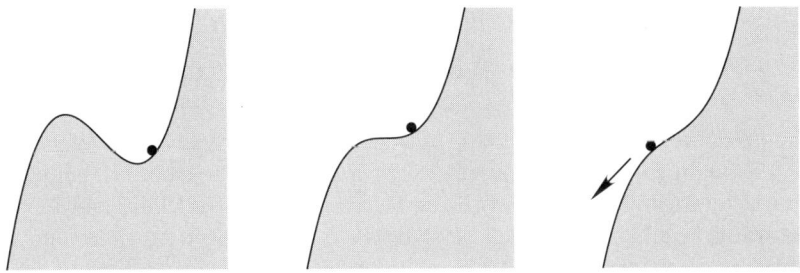

Fig. 3.7. Overdamped particle in a potential undergoing a saddle–node bifurcation.

We consider in this section the SDE

$$\mathrm{d}x_t = \frac{1}{\varepsilon} f(x_t, t) \,\mathrm{d}t + \frac{\sigma}{\sqrt{\varepsilon}} F(x_t, t) \,\mathrm{d}W_t \qquad (3.3.1)$$

in the case where f undergoes an indirect saddle–node bifurcation at $(x, t) = (0, 0)$. Precisely, we will require the following.

Assumption 3.3.1 (Saddle–node bifurcation).

- Domain and differentiability: $f \in \mathcal{C}^2(\mathcal{D}, \mathbb{R})$ and $F \in \mathcal{C}^1(\mathcal{D}, \mathbb{R})$, where $\mathcal{D}$ is a domain of the form

$$\mathcal{D} = \{(x, t) : |t| \leq T, \, |x| \leq d\} \,, \qquad (3.3.2)$$

with $T, d > 0$. We further assume that f, F and all their partial derivatives up to order 2, respectively 1, are uniformly bounded in $\mathcal{D}$ by a constant M.
- Saddle–node bifurcation: The point $(0, 0)$ is a bifurcation point,

$$f(0, 0) = \partial_x f(0, 0) = 0 \,, \qquad (3.3.3)$$

satisfying the conditions

$$\partial_t f(0, 0) < 0 \qquad \text{and} \qquad \partial_{xx} f(0, 0) < 0 \,. \qquad (3.3.4)$$

- Non-degeneracy of noise term: There is a constant $F_- > 0$ such that

$$F(x, t) \geq F_- \qquad \forall (x, t) \in \mathcal{D} \,. \qquad (3.3.5)$$

We can always scale x and t in such a way that $\partial_t f(0, 0) = -1$ and $\partial_{xx} f(0, 0) = -2$, like in the simplest example $f(x, t) = -t - x^2$. As mentioned in Section 2.2.2, by decreasing T and d if necessary, we can guarantee that f vanishes in $\mathcal{D}$ if and only if (x, t) belongs to one of the two equilibrium branches $x_+^\star : [-T, 0] \to [-d, 0]$ or $x_-^\star : [-T, 0] \to [0, d]$, which satisfy

$$x_\pm^\star(t) = \mp \sqrt{|t|} \left[1 + \mathcal{O}_t(1)\right] \,. \qquad (3.3.6)$$

Here the signs $\pm$ refer to the linearisation of f, which obeys the relations

$$a_\pm^\star(t) := \partial_x f(x_\pm^\star(t), t) = \pm\sqrt{|t|}\bigl[1 + \mathcal{O}_t(1)\bigr] . \qquad (3.3.7)$$

Thus the equilibrium branch $x_-^\star(t)$ is stable, while $x_+^\star(t)$ is unstable.

The main result of Section 2.2.2, Theorem 2.2.2, describes the behaviour of a particular solution $\bar{x}(t, \varepsilon)$ of the deterministic equation. This solution tracks the stable equilibrium branch at a distance increasing with time, and reacts to its disappearance only after a delay of order $\varepsilon^{2/3}$. More precisely,

$$\bar{x}(t, \varepsilon) - x_-^\star(t) \asymp \frac{\varepsilon}{|t|} \qquad \text{for } -T \leqslant t \leqslant -c_0 \varepsilon^{2/3}, \qquad (3.3.8)$$

$$\bar{x}(t, \varepsilon) \asymp \varepsilon^{1/3} \qquad \text{for } -c_0 \varepsilon^{2/3} \leqslant t \leqslant t_1 := c_1 \varepsilon^{2/3} , \qquad (3.3.9)$$

for some positive constants c_0, c_1. Moreover, $\bar{x}(t, \varepsilon)$ reaches $-d$ at a time of order $\varepsilon^{2/3}$.

Let us now consider the effect of noise on the dynamics. Since we work in a small neighbourhood of the bifurcation point, it is convenient to use coordinates in which the diffusion coefficient is constant.

Lemma 3.3.2. *For sufficiently small d and T, there exists a change of variables $\tilde{x} = x/F(0, t) + \mathcal{O}(x^2) + \mathcal{O}(\sigma^2) + \mathcal{O}(\varepsilon)$ transforming Equation (3.3.1) into*

$$d\tilde{x}_t = \frac{1}{\varepsilon} \tilde{f}(\tilde{x}_t, t)\,dt + \frac{\sigma}{\sqrt{\varepsilon}}\,dW_t , \qquad (3.3.10)$$

where $\tilde{f}$ satisfies the same assumptions as f.

Proof. Define the function

$$h(x, t) = \int_0^x \frac{dz}{F(z, t)} = \frac{x}{F(0, t)} + \mathcal{O}(x^2) . \qquad (3.3.11)$$

Then Itô's formula shows that $\bar{x}_t = h(x_t, t)$ satisfies an equation of the form (3.3.10) with a drift term given by

$$\tilde{f}(h(x, t), t) = \frac{f(x, t)}{F(x, t)} + \mathcal{O}(\sigma^2) + \mathcal{O}(\varepsilon) . \qquad (3.3.12)$$

The corresponding deterministic system admits a saddle–node bifurcation point in a neighbourhood of order $\sigma^2 + \varepsilon$ of the origin, as can be checked by a direct computation, using the implicit-function theorem. It can also be seen as a consequence of the structural stability of the saddle–node bifurcation. It then suffices to translate the origin to the new bifurcation point, and rescale space and time, in order to obtain a drift term with the same properties as f. □

In the sequel, we will simplify the notation by considering Equation (3.3.10) without the tildes.[8] For negative t, bounded away from 0, the results of Section 3.1 show that sample paths are concentrated in a neighbourhood of order σ of $\bar{x}(t,\varepsilon)$. The main obstacle to extending this result up to $t = 0$ is that the uniform stability of $x_-^\star(t)$ is violated, since the curvature at the bottom of the potential well tends to zero. As a consequence, the spreading of sample paths will increase. For strong enough noise, paths are even likely to reach and overcome the potential barrier, located in $x_+^\star(t)$, some time before the barrier vanishes.

We will divide the analysis into two steps. In Section 3.3.1, we extend as far as possible the domain of validity of Theorem 3.1.10 on the concentration of sample paths near the deterministic solution $\bar{x}(t,\varepsilon)$. In Sections 3.3.2 and 3.3.3, we analyse the subsequent behaviour, when sample paths either follow $\bar{x}(t,\varepsilon)$ in its jump towards negative x, or overcome the potential barrier already for negative t.

3.3.1 Before the Jump

The stochastic motion in a vicinity of the deterministic solution $\bar{x}(t,\varepsilon)$ can again be described by introducing the deviation $y_t = x_t - \bar{x}(t,\varepsilon)$. The relevant quantity is the curvature $a(t) = \partial_x f(\bar{x}(t,\varepsilon),t)$ of the potential at $\bar{x}(t,\varepsilon)$.

As a direct consequence of (3.3.8), (3.3.9) and Taylor's formula, we have

$$a(t) \asymp -\bigl(|t|^{1/2} \vee \varepsilon^{1/3}\bigr) \qquad \text{for } -T \leq t \leq t_1 = c_1 \varepsilon^{2/3}, \qquad (3.3.13)$$

indicating that the curvature of the potential at $\bar{x}(t,\varepsilon)$ stays bounded away from zero up to time t_1, although the potential well disappears already at time 0. As in Section 3.1.1, we introduce a neighbourhood $\mathcal{B}(h)$ of the deterministic solution $\bar{x}(t,\varepsilon)$, defined by

$$\mathcal{B}(h) = \bigl\{ (x,t) \colon -T \leq t \leq t_1,\ |x - \bar{x}(t,\varepsilon)| < h\sqrt{\zeta(t)} \bigr\}. \qquad (3.3.14)$$

The function $\zeta(t)$, which is related to the variance of the linearisation of (3.3.10), is defined by the integral

$$\zeta(t) = \zeta(-T) e^{2\alpha(t,-T)/\varepsilon} + \frac{1}{\varepsilon} \int_{-T}^{t} e^{2\alpha(t,s)/\varepsilon}\, ds, \qquad (3.3.15)$$

where $\alpha(t,s) = \int_s^t a(u)\, du$ and $\zeta(-T) = 1/2|a(-T)| + \mathcal{O}(\varepsilon)$.[9] For negative t, bounded away from zero, we have $\zeta(t) = 1/2|a(t)| + \mathcal{O}(\varepsilon)$. The behaviour of

---

[8] Mathematical textbooks must be littered with dropped tildes, much as Space in Earth's neighbourhood is littered with satellite debris.

[9] The initial value $\zeta(-T)$ can be chosen in such a way as to match a previously defined concentration domain of paths, if the stable equilibrium branch $x_-^\star(t)$ already exists for $t \leq T$.

$\zeta(t)$ as t approaches zero can be determined by an integration by parts, and an application of Lemma 2.2.8. The final result is

$$\zeta(t) - \frac{1}{2|a(t)|} \asymp \frac{\varepsilon}{t^2} \qquad \text{for } -T \leqslant t \leqslant -c_0\varepsilon^{2/3}, \qquad (3.3.16)$$

$$\zeta(t) \asymp \varepsilon^{-1/3} \qquad \text{for } -c_0\varepsilon^{2/3} \leqslant t \leqslant t_1 = c_1\varepsilon^{2/3}. \qquad (3.3.17)$$

It follows that the relations

$$\zeta(t) \asymp \frac{1}{|t|^{1/2} \vee \varepsilon^{1/3}} \asymp \frac{1}{|a(t)|} \qquad (3.3.18)$$

hold for all $t \in [-T, t_1]$. The set $\mathcal{B}(h)$ thus becomes wider as t approaches the bifurcation value 0, but stays of constant order in a neighbourhood of order $\varepsilon^{2/3}$ of $t = 0$. We also introduce

$$\hat{\zeta}(t) = \sup_{-T \leqslant s \leqslant t} \zeta(s), \qquad (3.3.19)$$

which satisfies $\hat{\zeta}(t) \asymp \zeta(t)$ for $t \leqslant t_1$. Theorem 3.1.10 admits the following extension.

Theorem 3.3.3 (Stochastic saddle–node bifurcation – stable phase).
Let $t_0 \in [-T, 0)$ be fixed, and $x_0 = \bar{x}(t_0, \varepsilon)$. There exist constants $h_0, r_0, c, c_2 > 0$ such that, for all $h \leqslant h_0 \hat{\zeta}(t)^{-3/2}$ and $t_0 \leqslant t \leqslant t_1 = c_1\varepsilon^{2/3}$ satisfying

$$r(h/\sigma, \varepsilon) := \frac{\sigma}{h} + \frac{1}{\varepsilon}e^{-ch^2/\sigma^2} \leqslant r_0, \qquad (3.3.20)$$

one has

$$C_{h/\sigma}(t, \varepsilon) e^{-\kappa_- h^2/2\sigma^2} \leqslant \mathbb{P}^{t_0, x_0}\{\tau_{\mathcal{B}(h)} < t\} \leqslant C_{h/\sigma}(t, \varepsilon) e^{-\kappa_+ h^2/2\sigma^2}, \qquad (3.3.21)$$

where the exponents $\kappa_\pm$ are given by

$$\begin{aligned} \kappa_+ &= 1 - c_2 h \hat{\zeta}(t)^{3/2}, \\ \kappa_- &= 1 + c_2 h \hat{\zeta}(t)^{3/2}, \end{aligned} \qquad (3.3.22)$$

and the prefactor satisfies

$$C_{h/\sigma}(t, \varepsilon) = \sqrt{\frac{2}{\pi} \frac{|\alpha(t, t_0)|}{\varepsilon} \frac{h}{\sigma}} \qquad (3.3.23)$$

$$\times \left[1 + \mathcal{O}\left(\frac{\varepsilon[|\log \varepsilon| + \log(1 + h/\sigma)]}{|\alpha(t, t_0)|} + r(h/\sigma, \varepsilon) + h\hat{\zeta}(t)^{3/2} \right) \right].$$

Proof. The proof is essentially the same as the proof of Theorem 3.1.10, the only difference lying in more careful estimates of the effect of the maximal value of $\zeta(t)$ and the minimal value of $|a(t)|$. The proof of Lemma 3.1.7 remains unchanged, except for the error terms of $c_0(t)$ in (3.1.39). In fact,

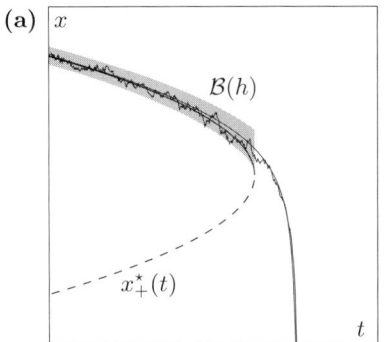 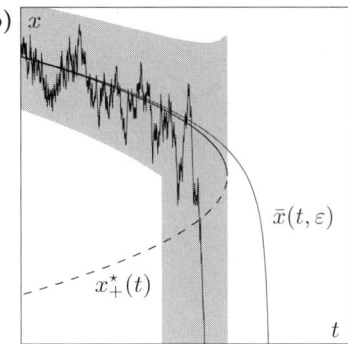

Fig. 3.8. Behaviour of sample paths near a dynamic saddle–node bifurcation. **(a)** In the weak-noise regime $\sigma < \sqrt{\varepsilon}$, paths remain concentrated in the shaded set $\mathcal{B}(h)$ centred in the deterministic solution; **(b)** in the strong-noise regime $\sigma > \sqrt{\varepsilon}$, paths are likely to overcome the unstable solution some time before the bifurcation.

$$c_0(t) = \frac{h}{\sqrt{2\pi}} \frac{1}{2\varepsilon\zeta(t)} \left[1 + \mathcal{O}\left(\frac{\zeta(0)}{\zeta(t)} e^{-2|\alpha(t,t_0)|/\varepsilon} \right) \right], \qquad (3.3.24)$$

where (3.3.16) yields $1/\zeta(t) = |2a(t)|[1 + \mathcal{O}(\varepsilon|t|^{-3/2})] = |2a(t)| + \mathcal{O}(\varepsilon|t|^{-1})$. The error term $\mathcal{O}(\varepsilon|t|^{-1})$ is responsible for the error term $\varepsilon|\log \varepsilon|$ in (3.3.23). The remainder of the proof is similar, except that more careful estimates of the terms occurring in (3.1.41)–(3.1.44) yield $M_1 = \mathcal{O}(h/\varepsilon\zeta(t))$, $M_2 = \mathcal{O}(\varepsilon\zeta(t))$ and $M_3 = \mathcal{O}(h/\varepsilon)$. The integral in (3.1.61), which deals with the contribution of the nonlinear drift term, is bounded by

$$\frac{1}{\sqrt{\zeta(s)}} \frac{Mh^2\hat{\zeta}(s)}{\inf_{t_0 \leqslant u \leqslant s}|a(u)|} \leqslant \mathrm{const}\, h^2 \hat{\zeta}(t)^{3/2}\, . \qquad (3.3.25)$$

Thus the result follows by taking H_1 of order $h^2\hat{\zeta}(t)^{3/2}$. □

Theorem 3.3.3 shows that paths are concentrated in a neighbourhood of order $\sigma\sqrt{\hat{\zeta}(t)} \asymp \sigma/(|t|^{1/4} \vee \varepsilon^{1/6})$ of $\bar{x}(t,\varepsilon)$, as long as $\hat{\zeta}(t)^{3/2}$ is small compared to σ^{-1}. There are thus two regimes to be considered:

- *Regime I:* $\sigma < \sigma_c = \sqrt{\varepsilon}$. In this case, the theorem can be applied up to time $t_1 := c_1 \varepsilon^{2/3}$, when the spreading of paths reaches order $\sigma/\varepsilon^{1/6}$ (Fig. 3.8a). In particular, the probability that paths have reached the t-axis before time t_1 is of order $C_{h_0 \sigma_c/\sigma}(t_1,\varepsilon) e^{-\mathrm{const}\,\varepsilon/\sigma^2}$.
- *Regime II:* $\sigma \geqslant \sigma_c = \sqrt{\varepsilon}$. In this case, the theorem can only be applied up to time $-\sigma^{4/3}$, when the spreading of paths reaches order $\sigma^{2/3}$, so that paths may reach the t-axis with appreciable probability (Fig. 3.8b).

In Regime I, which we shall call *weak-noise regime*, the behaviour of typical sample paths does not differ much from the behaviour of the deterministic solution up to time t_1, and we will show in Section 3.3.3 that typical paths reach $x = -d$ about at the same time as $\bar{x}(t,\varepsilon)$.

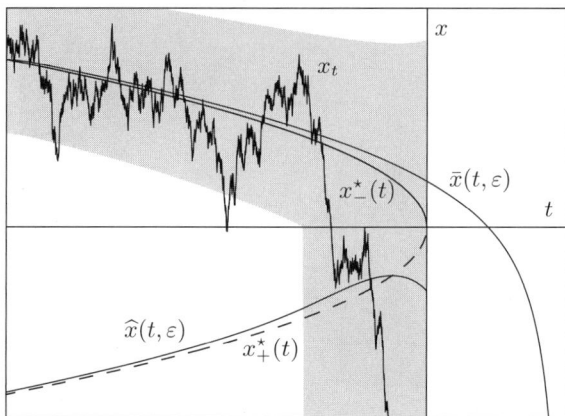

Fig. 3.9. Behaviour of sample paths near a dynamic saddle–node bifurcation in the strong-noise regime. Note the repeated attempts to reach the adiabatic solution tracking the potential barrier.

In Regime II, which we shall call *strong-noise regime*, the situation is different. Taking the largest allowed value $h_{\max} \asymp |t|^{3/4}$ of h in Theorem 3.3.3 shows that for $t \ll -\sigma^{4/3}$, the probability to cross the saddle before time t is bounded by

$$C_{h_{\max}/\sigma}(t,\varepsilon) e^{-\text{const}\,|t|^{3/2}/\sigma^2}. \tag{3.3.26}$$

The result does not exclude (but does not prove either) that paths reach the unstable equilibrium branch $x_+^\star(t)$ already at times of order $-\sigma^{4/3}$. We will prove in the following section that this is indeed the case, and bound the probability of the unlikely event that paths do *not* cross the potential barrier.

3.3.2 Strong-Noise Regime

We assume in this subsection that $\sigma > \sqrt{\varepsilon}$. Up to times of order $-\sigma^{4/3}$, Theorem 3.3.3 shows that sample paths are concentrated near the adiabatic solution $\bar{x}(t,\varepsilon)$ tracking the stable potential well at $x_-^\star(t)$, although their spreading increases as the potential well becomes flatter with increasing t. We will now show that, as time increases, it quickly becomes very unlikely not to reach and overcome the potential barrier at $x_+^\star(t)$, so that most paths reach negative values of absolute order 1 some time *before* the potential well disappears at time 0.

The analysis will be simplified by taking d and T small enough for the relation

$$\partial_{xx} f(x,t) \leqslant 0 \tag{3.3.27}$$

to hold for all $(x,t) \in \mathcal{D}$, which is possible by Assumption 3.3.1.

An important rôle is played by an adiabatic solution $\widehat{x}(t,\varepsilon)$ tracking the potential barrier at $x_+^\star(t)$ (Fig. 3.9). This solution is not uniquely defined, but

3.3 Saddle–Node Bifurcation

we may select it, for instance, by imposing that $\widehat{x}(0,\varepsilon) = -\varepsilon^{1/3}$. Reversing the direction of time, and using the results of Section 2.2.4, it is not difficult to show that

$$\widehat{x}(t,\varepsilon) \asymp -\varepsilon^{1/3} \qquad \text{for } -c_0\varepsilon^{2/3} \leqslant t \leqslant 0 , \qquad (3.3.28)$$

$$\widehat{x}(t,\varepsilon) - x_+^\star(t) \asymp \frac{\varepsilon}{|t|} \qquad \text{for } -T \leqslant t \leqslant -c_0\varepsilon^{2/3} , \qquad (3.3.29)$$

where it may be necessary to increase c_0. This implies that the linearisation of f at $\widehat{x}(t,\varepsilon)$ satisfies

$$\widehat{a}(t) := \partial_x f(\widehat{x}(t,\varepsilon),t) \asymp \sqrt{|t|} \vee \varepsilon^{1/3} \asymp |a(t)| \asymp \frac{1}{\zeta(t)} . \qquad (3.3.30)$$

The least favourable situation for overcoming the potential barrier occurs when a sample path starts, at time $-\sigma^{4/3}$, near the upper boundary of $\mathcal{B}(h)$. In order to reach $-d$, it has to cross first $\bar{x}(t,\varepsilon)$, and then $\widehat{x}(t,\varepsilon)$. The basic idea is to consider that paths make a certain number of excursions, during which they attempt to overcome the potential barrier. If each excursion has a probability of success of, say, $1/3$, then the probability *not* to overcome the barrier during a given time span, containing N attempts, will be of order $(2/3)^N = \exp\{-N\log(3/2)\}$. Making these ideas precise leads to the following result.

Theorem 3.3.4 (Stochastic saddle–node bifurcation – the strong-noise regime). *Let $t_0 \in [-T,0)$ be fixed, and $x_0 = \bar{x}(t_0,\varepsilon)$. There exist constants $r_0, c, c_3 > 0$ such that, whenever $t \in [t_2, 0] = [-c_3\sigma^{4/3}, 0]$ and h satisfy the conditions*

$$\bar{x}(s,\varepsilon) + h\sqrt{\zeta(s)} \leqslant d \qquad \forall s \in [t_0, t] , \qquad (3.3.31)$$

and

$$r(h/\sigma,\varepsilon) := \frac{\sigma}{h} + \frac{1}{\varepsilon} e^{-ch^2/\sigma^2} \leqslant r_0 , \qquad (3.3.32)$$

there is a constant $\kappa > 0$ such that

$$\mathbb{P}^{t_0,x_0}\{x_s > -d \; \forall s \in [t_0, t]\} \leqslant \frac{3}{2} \exp\left\{-\kappa \frac{|\alpha(t,t_2)|}{\varepsilon(|\log\sigma| \vee \log h/\sigma)}\right\} + C_{h/\sigma}(t,\varepsilon) e^{-h^2/2\sigma^2} , \qquad (3.3.33)$$

where

$$C_{h/\sigma}(t,\varepsilon) = \sqrt{\frac{2}{\pi}} \frac{|\alpha(t,t_0)|}{\varepsilon} \frac{h}{\sigma} \qquad (3.3.34)$$

$$\times \left[1 + \mathcal{O}\left(\frac{\varepsilon[|\log\varepsilon| + \log(1+h/\sigma)]}{|\alpha(t,t_0)|} + r(h/\sigma,\varepsilon)\right)\right] .$$

The two terms on the right-hand side of (3.3.33) bound, respectively, the probability that x_s does not reach $-d$ before time t, while staying below $\bar{x}(s,\varepsilon) + h\sqrt{\zeta(s)}$, and the probability that x_s crosses the level $\bar{x}(s,\varepsilon) + h\sqrt{\zeta(s)}$ before time t.

The term $|\alpha(t,t_2)|$ in the exponent behaves like $\sigma^{2/3}(t-t_2)$, and is thus of order σ^2 as soon as $t-t_2$ reaches order $\sigma^{4/3}$. Generically, d is of order 1, and the condition on h is less restrictive for t of order $-\sigma^{4/3}$, where it imposes $h = \mathcal{O}(\sigma^{1/3})$. The probability not to reach the level $-d$ before time t is thus bounded by

$$\frac{3}{2}\exp\left\{-\mathrm{const}\,\frac{\sigma^{2/3}(t-t_2)}{\varepsilon|\log\sigma|}\right\} + \frac{\mathrm{const}}{\varepsilon}e^{-\mathrm{const}/\sigma^{4/3}}, \qquad (3.3.35)$$

which becomes small as soon as $t - t_2 \gg \varepsilon|\log\sigma|/\sigma^{2/3}$. In particular, the saddle is crossed before the bifurcation occurs with a probability larger than

$$1 - \mathcal{O}\left(\exp\left\{-\mathrm{const}\left[\frac{\sigma^2}{\varepsilon|\log\sigma|} \wedge \frac{1}{\sigma^{4/3}}\right]\right\}\right), \qquad (3.3.36)$$

which is close to 1 as soon as the noise intensity satisfies $\sigma \gg \sqrt{\varepsilon|\log\sigma|}$. In special cases, it may be possible to choose arbitrarily large d, which allows for larger values of h, and thus a smaller second term in the exponent.

The main part of the proof of Theorem 3.3.4 is contained in the following estimate.

Proposition 3.3.5. *Under the assumptions of Theorem 3.3.4,*

$$\mathbb{P}^{t_0,x_0}\left\{-d < x_s < \bar{x}(s,\varepsilon) + h\sqrt{\zeta(s)}\;\forall s \in [t_0,t]\right\}$$
$$\leqslant \frac{3}{2}\exp\left\{-\kappa\frac{|\alpha(t,t_2)|}{\varepsilon(|\log\sigma|\vee\log h/\sigma)}\right\}. \qquad (3.3.37)$$

Proof.

- We introduce again a partition $t_0 = s_0 < s_1 < \cdots < s_N = t$ of $[t_0,t]$. It is defined by $s_1 = t_2 = -c_3\sigma^{4/3}$, where c_3 will be chosen later, and

$$\alpha(s_k, s_{k-1}) = \varrho\varepsilon \qquad \text{for } 1 < k < N = \left\lceil\frac{|\alpha(t,t_2)|}{\varrho\varepsilon}\right\rceil + 1. \qquad (3.3.38)$$

The constant ϱ will also be determined later. The same argument as in Lemma 3.2.4 shows that the probability in (3.3.37) can be bounded by the product $\prod_{k=1}^{N} P_k$, where

$$P_k = \sup_{x \in I_{k-1}} \mathbb{P}^{s_{k-1},x}\left\{-d < x_s < \bar{x}(s,\varepsilon) + h\sqrt{\zeta(s)}\;\forall s \in [s_{k-1},s_k]\right\},$$
$$(3.3.39)$$

with $I_{k-1} = [-d, \bar{x}(s_{k-1}, \varepsilon) + h\sqrt{\zeta(s_{k-1})}]$. We use the trivial bound 1 for P_1 (since $-d$ is unlikely to be reached before time s_1) and P_N (since the last time interval may be too short). Our plan is to show that for an appropriate choice of ϱ and c_3, $P_k \leqslant 2/3$ for $2 \leqslant k \leqslant N-1$, and thus the probability in (3.3.37) is bounded by

$$\left(\frac{2}{3}\right)^{N-1} = \frac{3}{2}\exp\left\{-N\log\frac{3}{2}\right\} \leqslant \frac{3}{2}\exp\left\{-\frac{|\alpha(t,t_2)|}{\varrho\varepsilon}\log\frac{3}{2}\right\}. \quad (3.3.40)$$

- In order to estimate P_k, we should take into account the fact that, in the worst case, x_t has to cross $\bar{x}(t, \varepsilon)$ and $\hat{x}(t, \varepsilon)$ before reaching $-d$. The dynamics is thus divided into three steps, which turn out to require comparable times. Hence we further subdivide $[s_{k-1}, s_k]$ into three intervals, delimited by times $\tilde{s}_{k,1} < \tilde{s}_{k,2}$ such that

$$|\alpha(s_k, \tilde{s}_{k,2})| = |\alpha(\tilde{s}_{k,2}, \tilde{s}_{k,1})| = |\alpha(\tilde{s}_{k,1}, s_{k-1})| = \frac{1}{3}\varrho\varepsilon. \quad (3.3.41)$$

To ease notation, we introduce the stopping times

$$\begin{aligned}
\tau_k^+ &= \inf\{s \in [s_{k-1}, s_k] : x_s > \bar{x}(s, \varepsilon) + h\sqrt{\zeta(s)}\}, \\
\tau_{k,1} &= \inf\{s \in [s_{k-1}, s_k] : x_s < \bar{x}(s, \varepsilon)\}, \\
\tau_{k,2} &= \inf\{s \in [\tau_{k,1}, s_k] : x_s < \hat{x}(s, \varepsilon)\}, \\
\tau_{k,3} &= \inf\{s \in [\tau_{k,2}, s_k] : x_s < -d\},
\end{aligned} \quad (3.3.42)$$

corresponding to the times at which x_s crosses the various relevant levels. By convention, these stopping times are infinite if the set on the right-hand side is empty. Using the Markov property, we can decompose P_k as

$$\begin{aligned}
P_k &= \sup_{x \in I_{k-1}} \mathbb{P}^{s_{k-1}, x}\{\tau_{k,3} \wedge \tau_k^+ > s_k\} \\
&\leqslant \sup_{x \in I_{k-1}} \mathbb{P}^{s_{k-1}, x}\{\tau_{k,1} \wedge \tau_k^+ > \tilde{s}_{k,1}\} \quad (3.3.43) \\
&\quad + \sup_{x \in I_{k-1}} \mathbb{E}^{s_{k-1}, x}\{1_{\{\tau_{k,1} \leqslant \tilde{s}_{k,1}\}} \mathbb{P}^{\tau_{k,1}, x_{\tau_{k,1}}}\{\tau_{k,3} \wedge \tau_k^+ > s_k\}\}.
\end{aligned}$$

The probability appearing in the second term on the right-hand side can be further decomposed as

$$\begin{aligned}
\mathbb{P}^{\tau_{k,1}, x_{\tau_{k,1}}}&\{\tau_{k,3} \wedge \tau_k^+ > s_k\} \\
&\leqslant \mathbb{P}^{\tau_{k,1}, x_{\tau_{k,1}}}\{\tau_{k,2} \wedge \tau_k^+ > \tilde{s}_{k,2}\} \quad (3.3.44) \\
&\quad + \mathbb{E}^{\tau_{k,1}, x_{\tau_{k,1}}}\{1_{\{\tau_{k,2} \leqslant \tilde{s}_{k,2}\}} \mathbb{P}^{\tau_{k,2}, x_{\tau_{k,2}}}\{\tau_{k,3} \wedge \tau_k^+ > s_k\}\}.
\end{aligned}$$

We thus have to estimate the probabilities of three events, of the form $\{\tau_{k,j} \wedge \tau_k^+ > \tilde{s}_{k,j}\}$, each time assuming the worst values for initial time and position.

- The first phase of dynamics is described by the variable $y_s = x_s - \bar{x}(s,\varepsilon)$. We may assume $y_{s_{k-1}} =: y > 0$, since otherwise the first term on the right-hand side of (3.3.43) is zero. The nonlinear part of the drift term is negative by our assumption (3.3.27) on $\partial_{xx} f$, so that y_s is bounded above by

$$y_s^0 := y e^{\alpha(s,s_{k-1})/\varepsilon} + \frac{\sigma}{\sqrt{\varepsilon}} \int_{s_{k-1}}^{s} e^{\alpha(s,u)/\varepsilon} \, dW_u \qquad (3.3.45)$$

for all $s \in [s_k, \tilde{s}_{k,1}]$. The expectation and variance of the Gaussian random variable $y_{\tilde{s}_{k,1}}^0$ satisfy the following relations:

$$\mathbb{E}\{y_{\tilde{s}_{k,1}}^0\} = y e^{-\varrho/3}, \qquad (3.3.46)$$

$$\mathrm{Var}\{y_{\tilde{s}_{k,1}}^0\} = \frac{\sigma^2}{\varepsilon} \int_{s_{k-1}}^{\tilde{s}_{k,1}} e^{2\alpha(\tilde{s}_{k,1},u)/\varepsilon} \, du \geqslant \inf_{u \in [s_{k-1}, \tilde{s}_{k,1}]} \frac{\sigma^2 [1 - e^{-2\varrho/3}]}{|2a(u)|}.$$

We may assume that $|a(u)|$ is decreasing, so that the infimum is reached for $u = s_{k-1}$. By the reflection principle,

$$\mathbb{P}^{s_{k-1},y}\{\tau_{k,1} > \tilde{s}_{k,1}\} = 1 - 2\mathbb{P}^{s_{k-1},y}\{y_{\tilde{s}_{k,1}}^0 \leqslant 0\}$$

$$= 1 - 2\Phi\left(-\frac{\mathbb{E}\{y_{\tilde{s}_{k,1}}^0\}}{\sqrt{\mathrm{Var}\{y_{\tilde{s}_{k,1}}^0\}}}\right) \leqslant \frac{2\mathbb{E}\{y_{\tilde{s}_{k,1}}^0\}}{\sqrt{2\pi \, \mathrm{Var}\{y_{\tilde{s}_{k,1}}^0\}}}$$

$$\leqslant \sqrt{\frac{2}{\pi}} \frac{e^{-\varrho/3} \sqrt{\zeta(s_{k-1})}}{\sigma \sqrt{1 - e^{-2\varrho/3}}} \sqrt{2|a(s_{k-1})|}. \qquad (3.3.47)$$

- To describe the second phase of motion, we use as variable the deviation $z_s = x_s - \hat{x}(s,\varepsilon)$ from the deterministic solution tracking the unstable equilibrium. It satisfies a similar equation as y_s, but with $\hat{a}(s)$ instead of $a(s)$. Relation (3.3.30) implies the existence of a constant $L > 0$ such that $\hat{a}(t) \geqslant |a(t)|/L$. For similar reasons as before,

$$z_s \leqslant z_s^0 = z_{\tau_{k,1}} e^{\hat{\alpha}(s,\tau_{k,1})/\varepsilon} + \frac{\sigma}{\sqrt{\varepsilon}} \int_{\tau_{k,1}}^{s} e^{\hat{\alpha}(s,u)/\varepsilon} \, dW_u, \qquad (3.3.48)$$

where $\hat{\alpha}(s,u) = \int_u^s \hat{a}(v) \, dv \geqslant |\alpha(s,u)|/L$. If $|y_{\tau_{k,1}}| < |a(\tau_{k,1})|$, then (3.3.29) and (3.3.30) imply that $z_{\tau_{k,1}}$ is bounded by a constant times $|a(\tau_{k,1})|$. Using again the reflection principle, we obtain that

$$\mathbb{P}^{\tau_{k,1}, z_{\tau_{k,1}}}\{\tau_{k,2} > \tilde{s}_{k,2}\} \leqslant \sqrt{\frac{2}{\pi}} \frac{z_{\tau_{k,1}}}{\sigma \sqrt{1 - e^{-2\varrho/3L}}} \sqrt{2\hat{a}(s_{k-1})}, \qquad (3.3.49)$$

uniformly in $\tau_{k,1} \leqslant \tilde{s}_{k,1}$.
- In order to describe the last phase of motion, we use the simple endpoint estimate

$$\mathbb{P}^{\tau_{k,2}, \hat{x}(\tau_{k,2},\varepsilon)}\{\tau_{k,3} > s_k\} \leqslant \mathbb{P}^{\tau_{k,2}, 0}\{\tilde{z}_{s_k}^0 > -d\}. \qquad (3.3.50)$$

An estimation of the corresponding Gaussian integral yields the bound

$$\mathbb{P}^{\tau_k,2,0}\{\tilde{z}^0_{s_k} > -d\} \leqslant \frac{1}{2} + \frac{1}{\sqrt{2\pi}} \frac{d e^{-\varrho/3L}}{\sigma\sqrt{1 - e^{-2\varrho/3L}}} \sqrt{2\hat{a}(s_{k-1})}. \qquad (3.3.51)$$

- To conclude the proof, we choose the parameters c_3 and ϱ in such a way that $P_k \leqslant 2/3$. It follows from the estimates (3.3.47), (3.3.49) and (3.3.51) that

$$P_k \leqslant \frac{1}{2} + \frac{C_0}{\sigma} \frac{\sqrt{\hat{a}(s_{k-1})}}{\sqrt{1 - e^{-2\varrho/3L}}} \left[(h\sqrt{\zeta(s_{k-1})} + d)e^{-\varrho/3L} + \hat{a}(s_{k-1}) \right] \quad (3.3.52)$$

for a constant C_0. The properties of $\hat{a}(t)$ and $\zeta(t)$ imply the existence of another constant C_1 such that

$$P_k \leqslant \frac{1}{2} + C_1 \left[\left(\frac{h}{\sigma} + c_3^{1/2} \frac{d}{\sigma^{2/3}} \right) e^{-\varrho/3L} + c_3^{3/2} \right]. \qquad (3.3.53)$$

We thus choose first c_3 in such a way that $c_3^{3/2} \leqslant 1/(12C_1)$, and then ϱ in such a way that the term in brackets is also smaller than $1/(12C_1)$. This requires ϱ to be of order $\log(h/\sigma) + |\log \sigma|$, and gives $P_k \leqslant 2/3$. $\qquad \square$

In order to complete the proof of Theorem 3.3.4, we need to estimate the probability that x_t exceeds $\bar{x}(t,\varepsilon) + h\sqrt{\zeta(t)}$, i.e., that it makes a large excursion in the "wrong" direction, away from the potential barrier.

Proposition 3.3.6. *Under the assumptions of Theorem 3.3.4,*

$$\mathbb{P}^{t_0,x_0}\left\{ \sup_{t_0 \leqslant s \leqslant t} \frac{x_s - \bar{x}(s,\varepsilon)}{\sqrt{\zeta(s)}} \geqslant h, \inf_{t_0 \leqslant s \leqslant t} x_s > -d \right\} \leqslant C_{h/\sigma}(t,\varepsilon) e^{-h^2/2\sigma^2}, \qquad (3.3.54)$$

where

$$C_{h/\sigma}(t,\varepsilon) = \sqrt{\frac{2}{\pi} \frac{|\alpha(t,t_0)|}{\varepsilon}} \frac{h}{\sigma} \qquad (3.3.55)$$

$$\times \left[1 + \mathcal{O}\left(\frac{\varepsilon[|\log \varepsilon| + \log(1 + h/\sigma)]}{|\alpha(t,t_0)|} + r(h/\sigma,\varepsilon) \right) \right].$$

Proof. As a consequence of Lemma 3.3.2 and the fact that the nonlinear part of the drift term is negative, we deduce from the comparison lemma (Lemma B.3.2) that $y_s = x_s - \bar{x}(s,\varepsilon)$ is bounded above by

$$y_s^0 = \frac{\sigma}{\sqrt{\varepsilon}} \int_{t_0}^{s} e^{\alpha(s,u)/\varepsilon} \, dW_u, \qquad (3.3.56)$$

as long as x_s has not left the domain $\mathcal{D}$. This shows in particular that the probability in (3.3.54) is bounded above by

$$\mathbb{P}^{t_0,0}\left\{\sup_{t_0\leqslant s\leqslant t}\frac{y_s^0}{\sqrt{\zeta(s)}}>h\right\}. \qquad (3.3.57)$$

The same arguments as in the first part of the proof of Theorem 3.3.3, but without the nonlinear term, show that this probability can in turn be bounded above by $C_{h/\sigma}(t,\varepsilon)e^{-h^2/2\sigma^2}$. This concludes the proof of the proposition, and thus also the proof of the theorem. □

3.3.3 Weak-Noise Regime

In this section, we complete the discussion of the dynamic saddle–node bifurcation with noise by discussing the behaviour of paths after the bifurcation, when $\sigma < \sqrt{\varepsilon}$. Theorem 3.3.3 implies that at time $t_1 = c_1\varepsilon^{2/3}$, the paths are concentrated in a neighbourhood of order $\sigma\varepsilon^{-1/6}$ of the deterministic solution $\bar{x}(t,\varepsilon) \asymp \varepsilon^{1/3}$. In particular,

$$\mathbb{P}^{x_0,t_0}\{x_{t_1}\leqslant 0\}\leqslant C_{\sqrt{\varepsilon}/\sigma}(t_1,\varepsilon)e^{-\mathrm{const}\,\varepsilon/\sigma^2}. \qquad (3.3.58)$$

We also know that the deterministic solution reaches the lower level $x = -d$ in a time of order $\varepsilon^{2/3}$. The description of the stochastic dynamics in this case is somewhat complicated by the fact that a region of instability is crossed, where paths repel each other. The following estimate is a rather rough bound, which suffices, however, to show that solutions of the stochastic differential equation are unlikely to jump much later than in the deterministic case.

Theorem 3.3.7 (Stochastic saddle–node bifurcation – the weak-noise regime). *There is a constant $\kappa > 0$ such that for all $x_1 \leqslant d$ and $t_1 \leqslant t \leqslant T$,*

$$\mathbb{P}^{t_1,x_1}\{x_s > -d \;\forall s \in [t_1,t]\} \leqslant \exp\left\{-\frac{\varepsilon}{\sigma^2}\left[\frac{\kappa}{|\log\varepsilon|}\frac{t-t_1}{\varepsilon^{2/3}}-1\right]\right\}. \qquad (3.3.59)$$

Proof. It follows from the assumptions on f that

$$f(x,t) \leqslant -\mathrm{const}\,(x^2 + \varepsilon^{2/3}) \qquad (3.3.60)$$

for all $t \geqslant t_1$. It is thus possible to compare the solutions of the original equation with those of an autonomous one. The probability we are looking for can be estimated in a similar way as in Proposition 3.3.5 by introducing a partition of $[t_1,t]$. In this case, it turns out that when taking the intervals of the partition of constant length $\gamma\varepsilon^{2/3}$, a γ of order $|\log\varepsilon|$ is sufficient to make each P_k smaller than $e^{-\mathcal{O}(\varepsilon/\sigma^2)}$. The probability in (3.3.59) is then bounded by the product of all P_k. □

The probability that a path has not reached the lower level $-d$ by time t becomes small as soon as $t - t_1 \gg \varepsilon^{2/3}|\log\varepsilon|$, that is, shortly after the deterministic solution has reached $-d$.

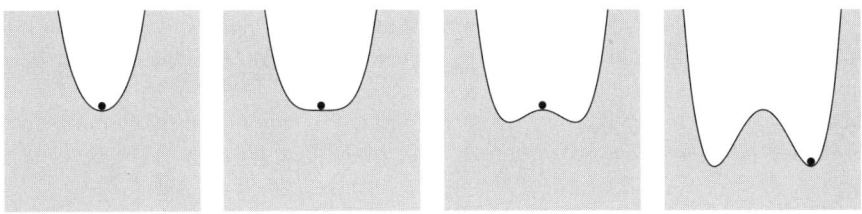

Fig. 3.10. Overdamped particle in a potential undergoing a pitchfork bifurcation.

3.4 Symmetric Pitchfork Bifurcation

Another interesting situation arises when a single-well potential transforms into a double-well potential as time increases (Fig. 3.10). Then the overdamped particle has the choice between falling into one of the new wells, or remaining, in unstable equilibrium, on the top of the barrier.

We consider in this section the SDE

$$\mathrm{d}x_t = \frac{1}{\varepsilon} f(x_t, t) \, \mathrm{d}t + \frac{\sigma}{\sqrt{\varepsilon}} F(x_t, t) \, \mathrm{d}W_t \qquad (3.4.1)$$

in the case where f undergoes a subcritical symmetric pitchfork bifurcation at $(x, t) = (0, 0)$. The precise assumptions are the following.

Assumption 3.4.1 (Symmetric pitchfork bifurcation).

- Domain and differentiability: $f \in \mathcal{C}^3(\mathcal{D}, \mathbb{R})$ and $F \in \mathcal{C}^2(\mathcal{D}, \mathbb{R})$, where $\mathcal{D}$ is a domain of the form

$$\mathcal{D} = \{(x, t) \colon |t| \leqslant T, \ |x| \leqslant d\}, \qquad (3.4.2)$$

with $T, d > 0$. We further assume that f, F and all their partial derivatives up to order 3, respectively 2, are uniformly bounded in $\mathcal{D}$ by a constant M.
- Symmetry: For all $(x, t) \in \mathcal{D}$,

$$f(x, t) = -f(-x, t) \qquad \text{and} \qquad F(x, t) = F(-x, t) . \qquad (3.4.3)$$

- Pitchfork bifurcation: The point $(0, 0)$ is a bifurcation point,

$$f(0, 0) = \partial_x f(0, 0) = 0 , \qquad (3.4.4)$$

satisfying the conditions

$$\partial_{tx} f(0, 0) > 0 \qquad \text{and} \qquad \partial_{xxx} f(0, 0) < 0 . \qquad (3.4.5)$$

- Non-degeneracy of noise term: There is a constant $F_- > 0$ such that

$$F(x, t) \geqslant F_- \qquad \forall (x, t) \in \mathcal{D} . \qquad (3.4.6)$$

We can scale x and t in such a way that $\partial_{tx}f(0,0) = 1$ and $\partial_{xxx}f(0,0) = -6$, like in the simplest example $f(x,t) = tx - x^3$. As mentioned in Section 2.2.3, by decreasing T and d if necessary, we can guarantee that f vanishes in $\mathcal{D}$ if and only if $x = 0$ or (x,t) belongs to one of the two equilibrium branches $x = \pm x^\star(t)$, $t > 0$, where $x^\star : [0,T] \to [0,d]$ satisfies

$$x^\star(t) = \sqrt{t}\bigl[1 + \mathcal{O}_t(1)\bigr] . \tag{3.4.7}$$

The linearisation of f at $x = 0$ is given by

$$a(t) := \partial_x f(0,t) = t\bigl[1 + \mathcal{O}(t)\bigr] , \tag{3.4.8}$$

while the linearisation at $x = \pm x^\star(t)$ satisfies

$$a^\star(t) := \partial_x f(\pm x^\star(t), t) = -2t\bigl[1 + \mathcal{O}_t(1)\bigr] . \tag{3.4.9}$$

Taking T small enough, we can guarantee that $a^\star(t) < 0$ for $0 < t \leqslant T$, while $a(t)$ has the same sign as t. In other words, the equilibrium branch $x = 0$ changes from stable to unstable as t becomes positive, while the branches $\pm x^\star(t)$ are always stable.

The main result of Section 2.2.3, Theorem 2.2.4, states that any solution starting at time $t_0 < 0$ sufficiently close to 0 tracks the particular solution $x_t \equiv 0$ at least up to the bifurcation delay time

$$\Pi(t_0) = \inf\{t > 0 \colon \alpha(t, t_0) > 0\} , \tag{3.4.10}$$

where $\alpha(t,s) = \int_s^t a(u)\,du$. If x_{t_0} is of order one, then x_t jumps, near time $\Pi(t_0)$, from the saddle at $x = 0$ to a neighbourhood of order ε of $x^\star(t)$, in a time of order $\varepsilon|\log \varepsilon|$.

Let us now consider the effect of noise on the dynamics. We first note that the x-dependence of the noise coefficient $F(x,t)$ can be eliminated by the same transformation as in Lemma 3.3.2. The symmetries of f and F imply that the new drift term still vanishes for $x = 0$, so that no terms of order σ^2 need to be eliminated. We may thus consider the equation

$$\mathrm{d}x_t = \frac{1}{\varepsilon}f(x_t, t)\,\mathrm{d}t + \frac{\sigma}{\sqrt{\varepsilon}}\,\mathrm{d}W_t \tag{3.4.11}$$

instead of (3.4.1).

The results of Section 3.1 can be applied to show that, for negative times, bounded away from zero, sample paths are concentrated in a neighbourhood of order σ of the deterministic solution with the same initial condition. As t approaches 0, the bottom of the potential well becomes increasingly flat, allowing paths to spread more and more. In Section 3.4.1, we extend the domain of validity of Theorem 3.1.10, to show that paths remain concentrated near $x = 0$ up to a certain time depending on the noise intensity (typically, that time is of order $\sqrt{\varepsilon|\log \sigma|}$).

Sections 3.4.1 and 3.4.2 describe the subsequent behaviour of paths, from the moment they leave a neighbourhood of the saddle until they approach one of the potential wells at $x^\star(t)$ or $-x^\star(t)$.

3.4.1 Before the Bifurcation

Let us consider a general initial condition $(x_0, t_0) \in \mathcal{D}$ with $t_0 < 0$. The deterministic solution x_t^{det} starting in x_0 at time t_0 approaches 0 exponentially fast, and remains close to 0 almost up to time $\Pi(t_0)$. The SDE linearised around x_t^{det} has the form

$$dy_t^0 = \frac{1}{\varepsilon}\bar{a}(t) y_t^0 \, dt + \frac{\sigma}{\sqrt{\varepsilon}} dW_t , \qquad (3.4.12)$$

where

$$\bar{a}(t) := \partial_x f(x_t^{\text{det}}, t) = a(t) + \mathcal{O}(x_0 e^{\alpha(t,t_0)/\varepsilon}) . \qquad (3.4.13)$$

This implies in particular that, for $t_0 \leqslant t \leqslant \Pi(t_0)$,

$$\bar{\alpha}(t, t_0) := \int_{t_0}^{t} \bar{a}(s) \, ds = \alpha(t, t_0) + \mathcal{O}(x_0 \varepsilon) . \qquad (3.4.14)$$

In analogy with previous sections, we define a neighbourhood of x_t^{det} by

$$\mathcal{B}(h) = \{(x, t) \in \mathcal{D}: t_0 \leqslant t \leqslant T, \, |x - x_t^{\text{det}}| < h\sqrt{\zeta(t)}\} , \qquad (3.4.15)$$

where

$$\zeta(t) = \zeta(t_0) e^{2\bar{\alpha}(t,t_0)/\varepsilon} + \frac{1}{\varepsilon} \int_{t_0}^{t} e^{2\bar{\alpha}(t,s)/\varepsilon} \, ds \qquad (3.4.16)$$

and $\zeta(t_0) = 1/|2\bar{a}(t_0)| + \mathcal{O}(\varepsilon)$. The behaviour of $\zeta(t)$ can be analysed with the help of Lemma 2.2.8 and is summarised as follows.

Lemma 3.4.2. *There exists a constant $c_0 > 0$ such that*

$$\begin{aligned}
\zeta(t) - \frac{1}{|2\bar{a}(t)|} &\asymp \frac{\varepsilon}{|t|^3} & & \text{for } t_0 \leqslant t \leqslant -c_0\sqrt{\varepsilon} , \\
\zeta(t) &\asymp \frac{1}{\sqrt{\varepsilon}} & & \text{for } |t| \leqslant c_0\sqrt{\varepsilon} , \qquad (3.4.17) \\
\zeta(t) &\asymp \frac{1}{\sqrt{\varepsilon}} e^{2\bar{\alpha}(t)/\varepsilon} & & \text{for } c_0\sqrt{\varepsilon} \leqslant t \leqslant \Pi(t_0) ,
\end{aligned}$$

where $\bar{\alpha}(t) = \bar{\alpha}(t, 0)$.

The set $\mathcal{B}(h)$ thus slowly widens as the potential well becomes flatter when t approaches 0. It has a width of order $h\varepsilon^{-1/4}$ for $|t| \leqslant c_0\sqrt{\varepsilon}$, and for $t > \sqrt{\varepsilon}$, its width grows exponentially as $h\varepsilon^{-1/4} e^{\bar{\alpha}(t)/\varepsilon}$. Note that $\bar{\alpha}(t) \asymp t^2$ for $t > 0$. We set again

$$\hat{\zeta}(t) = \sup_{t_0 \leqslant s \leqslant t} \zeta(s) \asymp \zeta(t) . \qquad (3.4.18)$$

The generalisation of Theorem 3.1.10 takes the following form.

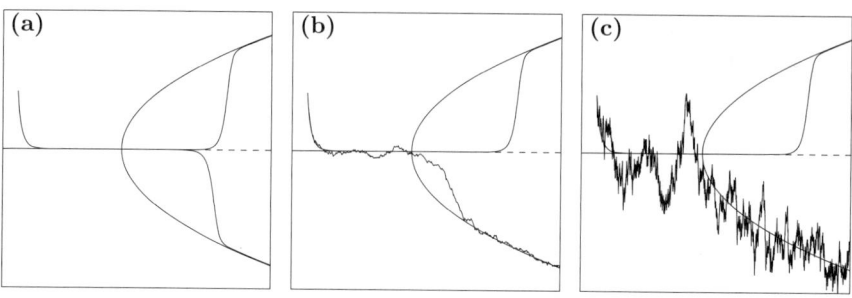

Fig. 3.11. Sample paths of a stochastic pitchfork bifurcation, in the case $f(x,t) = tx - x^3$. The adiabatic parameter is $\varepsilon = 0.02$, and the noise intensities correspond to the three qualitatively different regimes: **(a)** $\sigma = 0.00001$ (Regime I), **(b)** $\sigma = 0.02$ (Regime II), **(c)** $\sigma = 0.2$ (Regime III). In each case, the deterministic solution is shown for comparison.

Theorem 3.4.3 (Stochastic pitchfork bifurcation – stable phase). *There exist constants $h_0, r_0, c, c_2 > 0$ such that, for all $h \leqslant h_0 \hat{\zeta}(t)^{-1}$ satisfying*

$$r(h/\sigma, \varepsilon) := \frac{\sigma}{h} + \frac{1}{\varepsilon} e^{-ch^2/\sigma^2} \leqslant r_0, \qquad (3.4.19)$$

one has

$$C_{h/\sigma}(t,\varepsilon) e^{-\kappa_- h^2/2\sigma^2} \leqslant \mathbb{P}^{t_0, x_0}\{\tau_{\mathcal{B}(h)} < t\} \leqslant C_{h/\sigma}(t,\varepsilon) e^{-\kappa_+ h^2/2\sigma^2}, \qquad (3.4.20)$$

where the exponents $\kappa_\pm$ are given by

$$\begin{aligned} \kappa_+ &= 1 - c_2 h^2 \hat{\zeta}(t)^2, \\ \kappa_- &= 1 + c_2 h^2 \hat{\zeta}(t)^2, \end{aligned} \qquad (3.4.21)$$

and the prefactor satisfies

$$C_{h/\sigma}(t,\varepsilon) = \sqrt{\frac{2}{\pi}} \frac{|\bar{a}(t \wedge 0, t_0)|}{\varepsilon} \frac{h}{\sigma} \qquad (3.4.22)$$
$$\times \left[1 + \mathcal{O}\left(\frac{\varepsilon[|\log \varepsilon| + \log(1 + h/\sigma)]}{|\bar{a}(t \wedge 0, t_0)|} + r(h/\sigma, \varepsilon) + h^2 \hat{\zeta}(t)^2 \right) \right].$$

Proof. The proof follows closely the proof of Theorem 3.3.3. The different error terms are a consequence of the different behaviour of $\zeta(t)$, and the fact that $f(x,t) = a(t)x + \mathcal{O}(x^3)$ because of our symmetry assumptions. The term $\bar{a}(t \wedge 0, t_0)$ in the prefactor is due to the fact that the first-passage density decreases exponentially fast for $t > c_0 \sqrt{\varepsilon}$ because of the behaviour of $\zeta(t)$, compare (3.3.24). □

Theorem 3.4.3 shows that paths are concentrated in a neighbourhood of order $\sigma \sqrt{\zeta(t)}$ of x_t^{det} as long as $\hat{\zeta}(t)$ is small compared to σ^{-1}. One can distinguish between three regimes.

- *Regime I:* $\sigma \leqslant e^{-K/\varepsilon}$ for some $K > 0$. In this case, paths are concentrated around x_t^{det} as long as $2\bar{\alpha}(t) \ll K$, and even longer if $2\bar{\alpha}(\Pi(t_0)) = 2|\bar{\alpha}(t_0)| \ll K$. With high probability, one observes a bifurcation delay of order 1 as in the deterministic case (Fig. 3.11a).
- *Regime II:* $\sigma = \varepsilon^{p/2}$ for some $p > 1$. Paths are concentrated around x_t^{det} as long as $4\bar{\alpha}(t) < (p-1)\varepsilon|\log \varepsilon|$, that is, up to a time of order $\sqrt{(p-1)\varepsilon|\log \varepsilon|}$. This time can be considered as a microscopic bifurcation delay (Fig. 3.11b).
- *Regime III:* $\sigma \geqslant \sqrt{\varepsilon}$. The theorem can be applied up to a time of order $-\sigma$, when the spreading of paths reaches order $\sqrt{\sigma}$. The potential being quartic near $t = 0$, the spreading remains of order $\sqrt{\sigma}$ up to times of order σ after the bifurcation (Fig. 3.11c).

In the sequel, we will mainly consider Regime II. We will show that after a time of order $\sqrt{\varepsilon|\log \sigma|}$, paths not only leave a neighbourhood of $x = 0$, but also reach one of the two new potential wells at $\pm x^\star(t)$, in a time of the same order.

3.4.2 Leaving the Unstable Branch

We consider now the dynamics in Regime II, that is, when $e^{-K/\varepsilon} \ll \sigma < \sqrt{\varepsilon}$. Theorem 3.4.3 shows that paths stay in $\mathcal{B}(h)$ up to times of order $\sqrt{\varepsilon|\log \varepsilon|}$, and that at time $\sqrt{\varepsilon}$, their typical spreading has order $\sigma\varepsilon^{-1/4} < \varepsilon^{1/4}$.

We start by characterising the first-exit time from the diffusion-dominated strip

$$\mathcal{S}(h) = \{(x,t) \in \mathcal{D}: t \geqslant \sqrt{\varepsilon},\ |x| < h\rho(t)\}, \qquad (3.4.23)$$

where, as in Section 3.2.1,

$$\rho(t) = \frac{1}{\sqrt{2a(t)}}. \qquad (3.4.24)$$

Since $a(t) \asymp t$, $\rho(t)$ is decreasing for sufficiently small t.

The following estimate is proved in exactly the same way as Proposition 3.2.6, taking into account the symmetries of f, which imply a smaller nonlinear term.

Theorem 3.4.4 (Stochastic pitchfork bifurcation – diffusion-dominated escape). *Let $\mu > 0$, and define $C_\mu = (2+\mu)^{-(1+\mu/2)}$. Then, for any initial condition $(x_0, t_0) \in \mathcal{S}(h)$, any time $t \in [t_0, T]$, and any h such that $\sigma < h < (t_0^2 C_\mu \sigma^{1+\mu})^{1/(3+\mu)}$,*

$$\mathbb{P}^{t_0, y_0}\{\tau_{\mathcal{S}(h)} \geqslant t\} \leqslant \left(\frac{h}{\sigma}\right)^\mu \exp\left\{-\kappa_\mu \frac{\alpha(t, t_0)}{\varepsilon}\right\}, \qquad (3.4.25)$$

where the exponent κ_μ is given by

$$\kappa_\mu = \frac{\mu}{1+\mu}\left[1 - \mathcal{O}\left(\varepsilon\frac{1+\mu}{\mu}\right) - \mathcal{O}\left(\frac{1}{\mu \log(1+h/\sigma)}\right)\right]. \qquad (3.4.26)$$

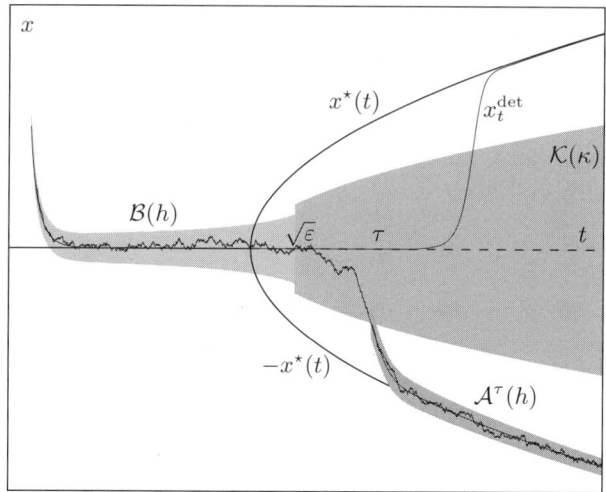

Fig. 3.12. Sample path of a stochastic pitchfork bifurcation in Regime II. The path is likely to remain in the set $\mathcal{B}(h)$ until time $\sqrt{\varepsilon}$. It typically leaves the set $\mathcal{K}(\kappa)$ after at a random time $\tau = \tau_{\mathcal{K}(\kappa)}$ of order $\sqrt{\varepsilon|\log\sigma|}$, after which it concentrates in a neighbourhood $\mathcal{A}^\tau(h)$ of another deterministic solution.

This result shows that sample paths typically leave $\mathcal{S}(h)$ as soon as $\alpha(t, t_0) \gg \varepsilon \log(h/\sigma)$. Since $\alpha(t, t_0)$ grows quadratically with t, the typical exit time is of order $\sqrt{\varepsilon \log(h/\sigma)}$. Once paths have left the small neighbourhood $\mathcal{S}(h)$ of the saddle, the drift term becomes appreciable, and helps push paths away from the saddle. In order to describe this process, we introduce a family of larger regions

$$\mathcal{K}(\kappa) = \left\{ (x, t) \in \mathcal{D} : t \geqslant \sqrt{\varepsilon},\ \frac{f(x, t)}{x} \geqslant \kappa a(t) x^2 \right\}, \qquad (3.4.27)$$

where $\kappa \in (0, 1)$. We set $a_\kappa(t) = \kappa a(t)$ and $\alpha_\kappa(t, s) = \kappa \alpha(t, s)$. The assumptions on f imply that the upper and lower boundaries of $\mathcal{K}(\kappa)$ are two curves of equation $x = \pm\tilde{x}(t)$, where

$$\tilde{x}(t) = \sqrt{(1 - \kappa) t}\, [1 + \mathcal{O}(t)]. \qquad (3.4.28)$$

Using the same line of thought as in Section 3.2.2, we obtain the following estimate on the law of the first-exit time $\tau_{\mathcal{K}(\kappa)}$.

Theorem 3.4.5 (Stochastic pitchfork bifurcation – drift-dominated escape). *Assume that $\sigma|\log\sigma|^{3/2} \leqslant \sqrt{\varepsilon}$. For all $\kappa \in (0, 1)$ and all initial conditions $(x_0, t_0) \in \mathcal{K}(\kappa)$,*

$$\mathbb{P}^{t_0, x_0}\{\tau_{\mathcal{K}(\kappa)} \geqslant t\} \leqslant C(t) \frac{|\log\sigma|}{\sigma} \left(1 + \frac{\alpha_\kappa(t, t_0)}{\varepsilon}\right) \frac{e^{-\alpha_\kappa(t, t_0)/\varepsilon}}{\sqrt{1 - e^{-2\alpha_\kappa(t, t_0)/\varepsilon}}}, \qquad (3.4.29)$$

where
$$C(t) = \frac{2}{\sqrt{\pi}}\sqrt{a_\kappa(t)}\tilde{x}(t) \ . \tag{3.4.30}$$

Proof. The proof is basically the same as the proof of Theorem 3.2.8. A slight simplification arises from the fact that $\rho(t)$ is decreasing. Since we have to assume $\sigma \leqslant \sqrt{\varepsilon}$ anyway, we choose $h = 2\sigma\sqrt{|\log\sigma|}$ and $\mu = 2$, which yields the slightly different prefactor. □

We conclude that paths typically leave $\mathcal{K}(\kappa)$ as soon as $\alpha_\kappa(t, t_0) \gg \varepsilon|\log\sigma|$. Since $\alpha_\kappa(t, t_0)$ is close to $\kappa t^2/2$ for small t, this implies
$$\mathbb{P}^{t_0,x_0}\{\tau_{\mathcal{K}(\kappa)} \geqslant \sqrt{2c\varepsilon|\log\sigma|}\} = \mathcal{O}(\sigma^{c\kappa-1}|\log\sigma|) \ , \tag{3.4.31}$$
which becomes small as soon as $c > 1/\kappa$. In the regime $\mathrm{e}^{-K/\varepsilon} \ll \sigma < \sqrt{\varepsilon}$, the typical bifurcation delay is thus of order $\sqrt{\varepsilon|\log\sigma|}$ (Fig. 3.12).

3.4.3 Reaching a Stable Branch

To complete the analysis of Regime II (that is, $\mathrm{e}^{-1/K\varepsilon} \ll \sigma \leqslant \sqrt{\varepsilon}$), it remains to show that paths approach one of the two stable equilibrium branches $\pm x^\star(t)$ after they have left a neighbourhood $\mathcal{K}(\kappa)$ of the saddle. This can simply be done by comparing the dynamics of sample paths leaving $\mathcal{K}(\kappa)$, say, through its upper boundary at time $\tau = \tau_{\mathcal{K}(\kappa)} \geqslant \sqrt{\varepsilon}$, with the deterministic solution $x_t^{\mathrm{det},\tau}$ starting at the same point.

The behaviour of $x_t^{\mathrm{det},\tau}$ can be analysed with the methods of Section 2.2.4. We restrict the discussion to values of κ in $(1/2, 2/3)$. The condition $\kappa < 2/3$ guarantees that the potential is convex outside $\mathcal{K}(\kappa)$, so that deterministic solutions attract each other. The condition $\kappa > 1/2$ allows excluding the possibility that $x_t^{\mathrm{det},\tau}$ reenters $\mathcal{K}(\kappa)$. We summarise the results in the following proposition.

Proposition 3.4.6. *Assume that* $\kappa \in (1/2, 2/3)$. *For sufficiently small* T, *there exists* $\eta = 2 - 3\kappa - \mathcal{O}_T(1)$, *such that the solution* $x_t^{\mathrm{det},\tau}$ *of the deterministic equation* $\varepsilon\dot{x} = f(x,t)$ *starting at time* $\tau \geqslant \sqrt{\varepsilon}$ *in* $x = \tilde{x}(\tau)$ *satisfies*
$$0 \leqslant x^\star(t) - x_t^{\mathrm{det},\tau} \leqslant \mathrm{const}\left[\frac{\varepsilon}{t^{3/2}} + (x^\star(\tau) - \tilde{x}(\tau))\mathrm{e}^{-\eta\alpha(t,\tau)/\varepsilon}\right] \tag{3.4.32}$$
for $\tau \leqslant t \leqslant T$. *Furthermore, the curvature* $a^\tau(t) = \partial_x f(x_t^{\mathrm{det},\tau}, t)$ *of the potential at* $x_t^{\mathrm{det},\tau}$ *satisfies*
$$a^\star(t) \leqslant a^\tau(t) \leqslant a^\star(t) + \mathrm{const}\left[\frac{\varepsilon}{t} + t\mathrm{e}^{-\eta\alpha(t,\tau)/\varepsilon}\right] \ . \tag{3.4.33}$$

Finally, if $\alpha^\tau(t,s) = \int_s^t a^\tau(u)\,\mathrm{d}u$, *the function*

$$\zeta^\tau(t) = \frac{1}{2|a^\tau(\tau)|} e^{2\alpha^\tau(t,\tau)/\varepsilon} + \frac{1}{\varepsilon} \int_{t_0}^{t} e^{2\alpha^\tau(t,s)/\varepsilon} \, ds \,, \tag{3.4.34}$$

satisfies

$$\left| \zeta^\tau(t) - \frac{1}{2|a^\tau(t)|} \right| \leqslant \mathrm{const} \left[\frac{\varepsilon}{t^3} + \frac{1}{t} e^{-\eta\alpha(t,\tau)/\varepsilon} \right]. \tag{3.4.35}$$

The bounds (3.4.32) imply that as soon as $t - \tau \gg \varepsilon/\tau$, the particular solution $x_t^{\mathrm{det},\tau}$ approaches the bottom of the potential well like $\varepsilon/t^{3/2}$. In fact, all solutions $x_t^{\mathrm{det},\tau}$ starting on the upper boundary of $\mathcal{K}(\kappa)$ converge exponentially fast to the same particular solution, approaching $x^\star(t)$ like $\varepsilon/t^{3/2}$. The bounds (3.4.33) show that $a^\tau(t) \asymp t$. Thus it follows from (3.4.35) that $\zeta^\tau(t)$ decreases like $1/t$.

In analogy with the stable case, we introduce the set

$$\mathcal{A}^\tau(h) = \left\{ (x,t) \in \mathcal{D} : t \geqslant \tau, \ |x - x_t^{\mathrm{det},\tau}| \leqslant h\sqrt{\zeta^\tau(t)} \right\}, \tag{3.4.36}$$

which is centred at the deterministic solution $x_t^{\mathrm{det},\tau}$, and has a width decreasing like $h/\sqrt{t}$. The following analogue of Theorem 3.4.3 shows that sample paths leaving $\mathcal{K}(\kappa)$ at time τ are unlikely to leave the set $\mathcal{A}^\tau(h)$ if $h \gg \sigma$. The proof uses a slightly more precise estimate as mentioned in Remark C.1.5 of Appendix C.1.

Theorem 3.4.7 (Stochastic pitchfork bifurcation – approaching $x^\star(t)$).
There exist constants $h_0, r_0, c, c_2 > 0$ such that, for all $h \leqslant h_0 \tau$ satisfying

$$r(h/\sigma, \varepsilon) := \frac{\sigma}{h} + \frac{1}{\varepsilon} e^{-ch^2/\sigma^2} \leqslant r_0 \,, \tag{3.4.37}$$

on the set $\{\tau < T\}$, one has

$$C_{h/\sigma}(t,\varepsilon) e^{-\kappa_- h^2/2\sigma^2} \leqslant \mathbb{P}^{\tau, \tilde{x}(\tau)} \{ \tau_{\mathcal{A}^\tau(h)} < t \} \leqslant C_{h/\sigma}(t,\varepsilon) e^{-\kappa_+ h^2/2\sigma^2} \,, \tag{3.4.38}$$

where the exponents $\kappa_\pm$ are given by

$$\begin{aligned} \kappa_+ &= 1 - c_2 h/\tau \,, \\ \kappa_- &= 1 + c_2 h/\tau \,, \end{aligned} \tag{3.4.39}$$

and the (random) prefactor satisfies

$$C_{h/\sigma}(t,\varepsilon) = \sqrt{\frac{2}{\pi}} \frac{|\alpha^\tau(t,\tau)|}{\varepsilon} \frac{h}{\sigma} \tag{3.4.40}$$

$$\times \left[1 + \mathcal{O}\left(\frac{\varepsilon[|\log \tau| + \log(1 + h/\sigma)]}{|\alpha^\tau(t,\tau)|} + r(h/\sigma, \varepsilon) + \frac{h}{\tau} \right) \right].$$

Note that several quantities in (3.4.38) depend on the random first-exit time $\tau = \tau_{\mathcal{K}(\kappa)}$. However, the result is understood to be valid on the set $\{\sqrt{\varepsilon} \leqslant \tau \leqslant T\}$, and this restriction on τ allows to bound all random quantities by deterministic constants.

Since the deterministic solutions starting on the boundary of $\mathcal{K}(\kappa)$ approach each other exponentially fast, the sets $\mathcal{A}^\tau(h)$ start overlapping as soon as $\eta\alpha(t,\tau) \gg \varepsilon$. Together with the results of the previous section, this implies that a time of order $\sqrt{\varepsilon}|\log \sigma|$ is sufficient for most sample paths to concentrate near the bottom of a potential well.

An interesting related question is whether paths, starting at some $t_0 < 0$ in a given point x_0, say with $x_0 > 0$, will reach the equilibrium branch $x^\star(t)$ or $-x^\star(t)$. A simple answer can be given by comparing x_t with the solutions of the linear equation

$$\mathrm{d}x_t^0 = \frac{1}{\varepsilon}a(t)x_t^0\,\mathrm{d}t + \frac{\sigma}{\sqrt{\varepsilon}}\,\mathrm{d}W_t\;. \tag{3.4.41}$$

Since $f(x,t) \leqslant a(t)x$ for sufficiently small t and positive x, x_t is bounded above by x_t^0 as long as $x_t \geqslant 0$. The reflection principle thus implies

$$\mathbb{P}^{t_0,x_0}\{x_t \geqslant 0\} = 1 - \frac{1}{2}\mathbb{P}^{t_0,x_0}\{\exists s \in [t_0,t): x_s = 0\}$$

$$\leqslant 1 - \frac{1}{2}\mathbb{P}^{t_0,x_0}\{\exists s \in [t_0,t): x_s^0 = 0\}$$

$$= \mathbb{P}^{t_0,x_0}\{x_t^0 \geqslant 0\}\;. \tag{3.4.42}$$

Performing a Gaussian integral and using Lemma 3.4.2, one obtains

$$\frac{1}{2} \leqslant \mathbb{P}^{t_0,x_0}\{x_t \geqslant 0\} \leqslant \frac{1}{2} + \mathcal{O}\left(\frac{\varepsilon^{1/4}}{\sigma}\mathrm{e}^{-|\alpha(0,t_0)|/\varepsilon}\right) \tag{3.4.43}$$

for all $t \geqslant 0$. Since $|\alpha(0,t_0)|$ is of order t_0^2, paths reach either potential well with almost the same probability, provided σ is not exponentially small.

In a sense, Regime II is the most favourable for an experimental determination of the bifurcation diagram, by slowly sweeping the control parameter. Indeed, exponentially small noise intensities lead to an undesired bifurcation delay, which fails to reveal part of the stable bifurcation branches. On the other hand, noise intensities larger than $\sqrt{\varepsilon}$ yield a large spreading of paths around the bifurcation diagram. The relation between bifurcation delay and noise intensity can also be used to determine the noise level experimentally, by performing experiments with different sweeping rates ε.

3.5 Other One-Dimensional Bifurcations

3.5.1 Transcritical Bifurcation

The behaviour of sample paths when approaching a transcritical or a saddle–node bifurcation point is quite similar. Depending on the noise intensity, paths either remain close to the stable equilibrium branch, or they cross the unstable branch and escape. The weak-noise behaviour *after* passing the bifurcation point is, by contrast, very different for the transcritical and the saddle–node bifurcation. It moreover depends, as in the deterministic case discussed in Example 2.2.10, on the relative location of equilibrium branches (see Fig. 2.8).

For simplicity, we discuss here the particular case of the SDE

$$\mathrm{d}x_t = \frac{1}{\varepsilon}(t^2 - x_t^2)\,\mathrm{d}t + \frac{\sigma}{\sqrt{\varepsilon}}\,\mathrm{d}W_t \;. \tag{3.5.1}$$

The deterministic system has a stable equilibrium branch $x_-^\star(t) = |t|$ and an unstable branch $x_+^\star(t) = -|t|$, touching at $t = 0$. The linearisations of the drift term $f(x,t) = t^2 - x^2$ at these branches are $a_\pm^\star(t) = \pm 2|t|$.

We have seen that the adiabatic solution $\bar{x}(t,\varepsilon)$ associated with the stable branch satisfies, for some constant c_0,

$$\bar{x}(t,\varepsilon) - x_-^\star(t) \asymp \frac{\varepsilon}{|t|} \qquad \text{for } t \leqslant -c_0\sqrt{\varepsilon} \;, \tag{3.5.2}$$

$$\bar{x}(t,\varepsilon) \asymp \sqrt{\varepsilon} \qquad \text{for } -c_0\sqrt{\varepsilon} \leqslant t \leqslant c_0\sqrt{\varepsilon} \;, \tag{3.5.3}$$

$$\bar{x}(t,\varepsilon) - x_-^\star(t) \asymp -\frac{\varepsilon}{|t|} \qquad \text{for } t \geqslant c_0\sqrt{\varepsilon} \;. \tag{3.5.4}$$

The linearisation $a(t) = \partial_x f(\bar{x}(t,\varepsilon),t) = -2\bar{x}(t,\varepsilon)$ of the drift term at $\bar{x}(t,\varepsilon)$ thus scales as $a(t) \asymp -(|t| \vee \sqrt{\varepsilon})$ for all t. As usual, $\alpha(t,s)$ denotes the integral of $a(u)$ from s to t. Fixing some initial time $t_0 = -T < 0$, we can define $\zeta(t)$ exactly as in (3.3.15) and (3.4.16); it satisfies $\zeta(t) \asymp |a(t)|^{-1}$. The set

$$\mathcal{B}(h) = \bigl\{(x,t)\colon t \geqslant t_0,\ |x - \bar{x}(t,\varepsilon)| < h\sqrt{\zeta(t)}\bigr\} \tag{3.5.5}$$

thus has a maximal width of order $h\varepsilon^{-1/4}$, attained for $t \in [-c_0\sqrt{\varepsilon}, c_0\sqrt{\varepsilon}]$. The function $\hat{\zeta}(t) = \sup_{t_0 \leqslant s \leqslant t} \zeta(s)$ grows like $|t|^{-1}$ for $t \leqslant -c_0\sqrt{\varepsilon}$, and stays of constant order $1/\sqrt{\varepsilon}$ for subsequent times.

The analogue of Theorem 3.3.3 is the following.

Theorem 3.5.1 (Stochastic transcritical bifurcation – stable phase).
Let $x_0 = \bar{x}(t_0,\varepsilon)$. There exist constants $h_0, r_0, c, c_2 > 0$ such that, for all $h \leqslant h_0 \hat{\zeta}(t)^{-3/2}$ and $t \geqslant t_0$ satisfying

$$r(h/\sigma, t, \varepsilon) := \frac{\sigma}{h} + \frac{(t-t_0)}{\varepsilon}\mathrm{e}^{-ch^2/\sigma^2} \leqslant r_0 \;, \tag{3.5.6}$$

one has

$$C_{h/\sigma}(t,\varepsilon)\mathrm{e}^{-\kappa_- h^2/2\sigma^2} \leqslant \mathbb{P}^{t_0, x_0}\bigl\{\tau_{\mathcal{B}(h)} < t\bigr\} \leqslant C_{h/\sigma}(t,\varepsilon)\mathrm{e}^{-\kappa_+ h^2/2\sigma^2} \;, \tag{3.5.7}$$

where the exponents $\kappa_\pm$ are given by

$$\begin{aligned}\kappa_+ &= 1 - c_2 h \hat{\zeta}(t)^{3/2} \;, \\ \kappa_- &= 1 + c_2 h \hat{\zeta}(t)^{3/2} \;,\end{aligned} \tag{3.5.8}$$

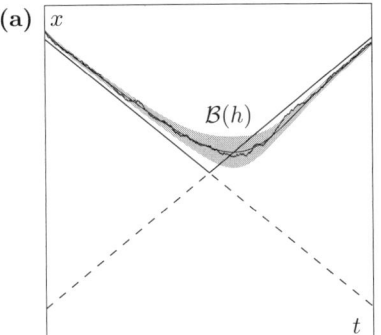 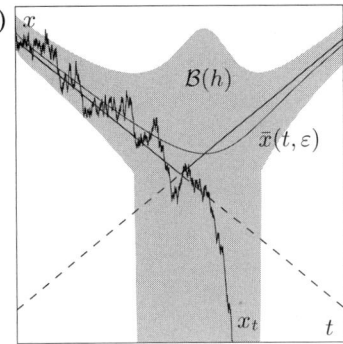

Fig. 3.13. Behaviour of sample paths near a dynamic transcritical bifurcation. (a) In the weak-noise regime $\sigma < \varepsilon^{3/4}$, paths remain concentrated in the shaded set $\mathcal{B}(h)$ centred in the deterministic solution; (b) in the strong-noise regime $\sigma > \varepsilon^{3/4}$, paths are likely to overcome the unstable solution some time before the bifurcation.

and the prefactor satisfies

$$C_{h/\sigma}(t,\varepsilon) = \sqrt{\frac{2}{\pi}} \frac{|\alpha(t,t_0)|}{\varepsilon} \frac{h}{\sigma} \qquad (3.5.9)$$
$$\times \left[1 + \mathcal{O}\left(\frac{\varepsilon[|\log \varepsilon| + \log(1 + h/\sigma)]}{|\alpha(t,t_0)|} + r(h/\sigma, t, \varepsilon) + h\hat{\zeta}(t)^{3/2}\right)\right].$$

Theorem 3.5.1 shows that paths are concentrated in a neighbourhood of order $\sigma\sqrt{\zeta(t)} \asymp \sigma/(\sqrt{|t|} \vee \varepsilon^{1/4})$ of $\bar{x}(t,\varepsilon)$, as long as $\hat{\zeta}(t)^{3/2}$ is small compared to σ^{-1}. There are thus two regimes to be considered:

- *Regime I*: $\sigma < \sigma_{\mathrm{c}} = \varepsilon^{3/4}$. In this case, the theorem can be applied for all (finite) times t, and shows that paths are likely to remain in $\mathcal{B}(h)$ if $\sigma \ll h \leqslant \varepsilon^{3/4}$. The maximal spreading of typical paths has order $\sigma\varepsilon^{-1/4}$ (Fig. 3.13a). In particular, the probability that paths reach the t-axis is bounded by $C_{h_0\sigma_{\mathrm{c}}/\sigma}(t,\varepsilon)\mathrm{e}^{-\mathrm{const}\,\varepsilon^{3/2}/\sigma^2}$.
- *Regime II*: $\sigma \geqslant \sigma_{\mathrm{c}} = \varepsilon^{3/4}$. In this case, the theorem can only be applied up to time $-\sigma^{2/3}$, when the spreading of paths reaches order $\sigma^{2/3}$, so that paths may reach the t-axis with appreciable probability (Fig. 3.13b).

The following result, which is proved exactly in the same way as Theorem 3.3.4, shows that in Regime II, paths are indeed likely to cross the unstable equilibrium branch.

Theorem 3.5.2 (Stochastic transcritical bifurcation – the strong-noise regime). *Assume that $\sigma \geqslant \varepsilon^{3/4}$ and fix $-d < 0$. Let $x_0 = \bar{x}(t_0, \varepsilon)$. There exist constants $c_3, \kappa > 0$ such that, for all $t \in [t_1, t_2] = [-c_3\sigma^{2/3}, c_3\sigma^{2/3}]$,*

$$\mathbb{P}^{t_0,x_0}\{x_s > -d \;\forall s \in [t_0, t]\} \leqslant \frac{3}{2} \exp\left\{-\kappa \frac{|\alpha(t,t_1)|}{\varepsilon(|\log \sigma| + |\log \varepsilon|)}\right\}. \qquad (3.5.10)$$

Proof. One starts by establishing the same upper bound as in Theorem 3.3.4, see (3.3.33), which contains a second term bounding the probability of excursions away from the unstable branch. However, owing to the particular nature of the drift term in (3.5.1), the parameter h can be chosen arbitrarily large. The choice $h = \sigma^{5/3}/\sqrt{\varepsilon}$ makes the second term negligible. Its only effect is then the term $|\log \varepsilon|$ in the denominator of the exponent. □

The term $|\alpha(t, t_1)|$ in the exponent behaves like $\sigma^{2/3}(t - t_1)$, and is thus of order $\sigma^{4/3}$ as soon as $t - t_1$ reaches order $\sigma^{2/3}$. The probability not to cross the unstable solution is thus of order

$$\frac{3}{2} \exp\left\{-\text{const}\, \frac{\sigma^{4/3}}{\varepsilon(|\log \sigma| + |\log \varepsilon|)}\right\}. \qquad (3.5.11)$$

Two important facts can be noted:

1. In the strong-noise regime $\sigma \geqslant \varepsilon^{3/4}$, transitions are only likely in a time window of width of order $\sigma^{2/3}$ around the bifurcation point. Indeed, paths which have not reached the t-axis by time $c_3 \sigma^{2/3}$ are most likely attracted again by the stable branch, and concentrate in $\mathcal{B}(\sigma)$.
2. This is a first example where sufficiently strong noise induces, with probability close to 1, a radically different behaviour than in the deterministic case. In combination with global mechanisms (for drift terms behaving only locally as in (3.5.1)), this phenomenon can lead to organised noise-induced behaviours such as stochastic resonance.

3.5.2 Asymmetric Pitchfork Bifurcation

In this last section, we briefly discuss the effect of noise on asymmetric dynamic pitchfork bifurcations. Consider for simplicity the particular case

$$\mathrm{d}x_t = \frac{1}{\varepsilon}(x_t + t)(t - x_t^2)\,\mathrm{d}t + \frac{\sigma}{\sqrt{\varepsilon}}\,\mathrm{d}W_t. \qquad (3.5.12)$$

The equilibrium branch $x^\star(t) = -t$ is stable for $t < 0$ and unstable for $t > 0$ (and $t < 1$). As t becomes positive, two stable equilibrium branches $x_\pm^\star(t) = \pm\sqrt{t}$ appear in a pitchfork bifurcation. The associated potential transforms from a single-well to a double-well potential, the only difference with the symmetric case being the fact that $x^\star(t)$ depends on time.

As we have seen in Example 2.2.11, in the deterministic case $\sigma = 0$, (3.5.12) admits an adiabatic solution $\bar{x}(t, \varepsilon)$, tracking $x^\star(t)$ at a distance of order $\varepsilon/|t|$ for $t \leqslant -\sqrt{\varepsilon}$. For $|t| \leqslant \sqrt{\varepsilon}$, $\bar{x}(t, \varepsilon)$ remains positive and of order $\sqrt{\varepsilon}$, while for $t \geqslant \sqrt{\varepsilon}$ it approaches the upper equilibrium branch at $x_+^\star(t)$ like $\varepsilon/|t|^{3/2}$ (see Fig. 2.9). One can also show the existence of a deterministic solution $\widehat{x}(t, \varepsilon)$, tracking the saddle at $x^\star(t)$ for $t > 0$. It separates the basins of attraction of the two potential wells, and satisfies $\widehat{x}(0, \varepsilon) \asymp -\sqrt{\varepsilon}$ (Fig. 3.14).

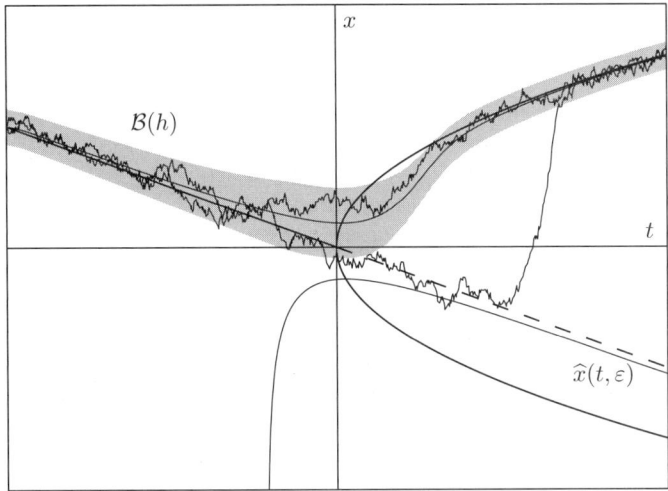

Fig. 3.14. A typical and a less typical sample path for the asymmetric dynamic pitchfork bifurcation.

Let $-a(t)$ denote the curvature of the potential at $\bar{x}(t,\varepsilon)$. One easily shows that there is a constant c_0 such that $a(t) \asymp -|t|$ for $|t| \geqslant c_0\sqrt{\varepsilon}$ (and t not approaching 1). For $|t| \leqslant c_0\sqrt{\varepsilon}$, $|a(t)|$ is at most of order $\sqrt{\varepsilon}$, and it may briefly become positive for some positive times. This slight peculiarity, however, happens on too short a time interval to have visible effects on the dynamics. Indeed, one shows that the function $\zeta(t)$ defined exactly as in (3.3.15) and (3.4.16), still satisfies $\zeta(t) \asymp 1/(|t| \vee \sqrt{\varepsilon})$. The set

$$\mathcal{B}(h) = \{(x,t) \colon t \geqslant t_0,\ |x - \bar{x}(t,\varepsilon)| < h\sqrt{\zeta(t)}\} \quad (3.5.13)$$

thus has a maximal width of order $h\varepsilon^{-1/4}$, attained for times of order $\sqrt{\varepsilon}$.

Again, one obtains two regimes:

- If $\sigma < \sigma_c = \varepsilon^{3/4}$, sample paths are likely to remain in $\mathcal{B}(h)$ for all times, and thus to end up tracking the bottom of the right-hand potential well.
- If $\sigma \geqslant \sigma_c = \varepsilon^{3/4}$, sample paths have an appreciable probability of reaching the adiabatic solution $\widehat{x}(t,\varepsilon)$ tracking the saddle. They are thus also likely to visit the left-hand potential well.

The subsequent dynamics for $\sigma \geqslant \sigma_c$ is more difficult to analyse. As long as the potential barrier is sufficiently low, sample paths may jump back and forth between the two potential wells. It is not straightforward to obtain the probability to end up tracking the bottom of the left-hand rather than the right-hand well. We expect, however, that if the depths of the two wells are different, sample paths are more likely to settle in the deeper one of the two wells (which is the right-hand well for the drift term in (3.5.12)).

Bibliographic Comments

In the time-independent case, the fact that large excursions away from stable equilibria have exponentially small probability has been known to physicists for a long time, see [Arr89, Eyr35, Kra40]. Rigorous proofs of exponential estimates (in the general multidimensional case) have first been obtained by Wentzell and Freidlin [VF69, VF70, FW98]. More precise information on the law of first-exit times from the neighbourhood of stable equilibrium points has been obtained by Day [Day83].

In the time-dependent setting, first-passage times of Gaussian processes have been studied by many authors (see, for instance, [Fer82, Dur85, Ler86, Dur92]). More details can be found in the bibliographic comments at the end of Appendix C. The approach used in Proposition 3.1.5 and Theorem 3.1.10 was introduced in [BG02d], while the improved bounds of Theorem 3.1.6 are based on estimates on first-passage densities obtained in [BG04].

Results on escape times in the time-independent unstable case are related to estimates on small-ball probabilities, first studied by Chung [Chu48] for random walks, and extended to Brownian motion by Ciesielski and Taylor [CT62]. The technique employed here for the nonlinear time-dependent case was first developed in [BG02d].

Static bifurcations with noise have been studied by Arnold, Crauel, Flandoli and others from the point of view of random dynamical systems [CF94, Arn98]. In particular, the stochastic pitchfork bifurcation has been considered in [CF98].

The results on the dynamic saddle–node bifurcation presented here are a particular case of those in [BG02b], which build on certain results in [BG02e].

Investigations of the dynamic pitchfork bifurcation with noise go back to the mid-eighties. Experimental and numerical results showing that the random bifurcation delay is of order $\sqrt{\varepsilon|\log \sigma|}$ were obtained, for instance, in [ME84, TSM88, SMM89, SHA91]. Approximate analytical results are given in [JL98, Kus99]. The results presented here are from [BG02d].

4

Stochastic Resonance

In some of the examples studied in the previous chapter, the noise term was able to induce transitions between equilibrium branches, with appreciable probability, even some time before a bifurcation occurred. In fact, the existence of a bifurcation is not even necessary — an avoided bifurcation, in which equilibrium branches come close without actually touching, can be sufficient to produce such transitions. In this chapter, we apply the methods just developed to a much studied phenomenon involving noise-induced transitions, known as *stochastic resonance* (SR).

SR occurs when a multistable system, subject to noise, displays an organised behaviour which is qualitatively different from its behaviour in the absence of noise. A typical example is a bistable system driven by weak periodic forcing, which, in combination with noise, induces close-to-periodic large-amplitude oscillations between the stable equilibria. For instance, the SDE

$$\mathrm{d}x_s = \left[x_s - x_s^3 + A\cos\varepsilon s\right]\mathrm{d}s + \sigma\,\mathrm{d}W_s \qquad (4.0.1)$$

describes the dynamics of an overdamped particle in a symmetric double-well potential $U(x) = \frac{1}{4}x^4 - \frac{1}{2}x^2$, subject to periodic driving $A\cos\varepsilon s$ and additive noise. The potential $U(x)$ has local minima in $x = \pm 1$. In the absence of noise, and when the amplitude A of the forcing lies below a threshold, solutions of (4.0.1) always remain in the same potential well. The noise term may cause the particle to switch potential wells. Roughly speaking, the expression stochastic resonance refers to the fact that these interwell transitions are not uniformly distributed in time, but show instead some trace of the periodic forcing.

One of the difficulties in quantitative studies of SR arises from the fact that several parameters are involved. Even in the simple example (4.0.1), there are already three of them: noise intensity σ, amplitude A and frequency ε of the forcing. SR is usually associated with some quantitative measure of the system's response being maximal for a certain noise intensity, depending on the other parameters A and ε. Most existing studies assume either the amplitude A, or the frequency ε, or both, to be small.

4 Stochastic Resonance

The method of choice for a mathematical analysis depends on the regime one is interested in. In order to compare the type of results obtained by the sample-path approach and other methods, we have divided this chapter into two parts as follows.

- In Section 4.1, we give a brief overview of the phenomenon of SR, discuss various measures introduced to quantify it, and present some of the results that have been obtained on their behaviour as the different parameters vary.
- In Section 4.2, we apply methods from Chapter 3 to the study of sample-path behaviour in the so-called synchronisation regime. We treat the case of relatively large values of the amplitude A, slightly below the threshold that would allow the particle to overcome the potential barrier in the deterministic case. A sufficient noise intensity then causes nearly regular transitions of the particle between both potential wells.

4.1 The Phenomenon of Stochastic Resonance

4.1.1 Origin and Qualitative Description

The concept of SR was originally introduced by Benzi, Sutera and Vulpiani [BSV81] and by Nicolis and Nicolis [NN81], in order to offer an explanation for the close-to-periodic appearance of the major Ice Ages. Various proxies indicate that during the last 700 000 years, the Earth's climate has repeatedly experienced dramatic transitions between "warm" phases, with average temperatures comparable to today's values, and Ice Ages, with temperatures about ten degrees lower. The transitions occur with a striking, though not perfect, regularity, with an average period of about 92 000 years.

The idea that this regularity might be related to (quasi-)periodic variations of the Earth's orbital parameters was put forward by Croll [Cro75] in 1864, and worked out during the first half of the twentieth century by Milankovitch [Mil41]. The slow variations of insolation, however, can only explain the rather drastic changes between climate regimes if some powerful feedbacks are involved, for example a mutual enhancement of ice cover and the Earth's albedo.

The simplest possible model for the variations of the average climate is an energy-balance model, whose sole dynamic variable is the mean temperature T of the Earth's atmosphere. Its evolution is described by

$$c \frac{\mathrm{d}T}{\mathrm{d}s} = R_{\mathrm{in}}(s) - R_{\mathrm{out}}(T,s) , \qquad (4.1.1)$$

where c is the heat capacity, and s denotes time. The incoming solar radiation $R_{\mathrm{in}}(s)$ is modelled by the periodic function

$$R_{\mathrm{in}}(s) = Q\bigl(1 + K \cos \omega s\bigr) , \qquad (4.1.2)$$

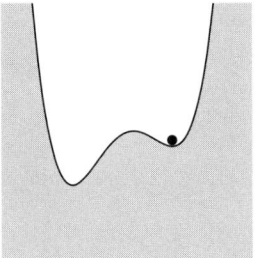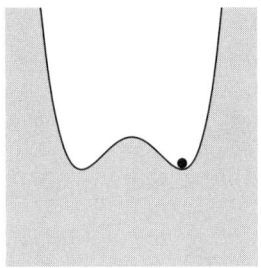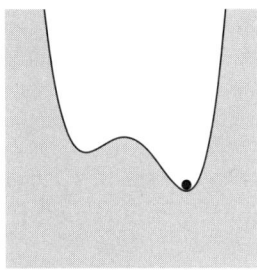

Fig. 4.1. Overdamped particle in a periodically forced double-well potential. In a climate model, the two potential wells represent, for instance, Ice Age and temperate climate. For subthreshold forcing amplitudes, resulting from variations in the Earth's orbital parameters, no transitions between wells occur, unless noise, modelling the weather, acts on the system.

where the constant Q is called *solar constant*, the amplitude K of the modulation is small, of the order 5×10^{-4}, and the period $2\pi/\omega$ equals 92 000 years. The outgoing radiation $R_{\text{out}}(T, s)$ decomposes into directly reflected radiation, and thermal emission. It thus has the form

$$R_{\text{out}}(T, s) = \alpha(T) R_{\text{in}}(s) + E(T) , \qquad (4.1.3)$$

where $\alpha(T)$ is called the Earth's albedo, and $E(T)$ its emissivity.

The emissivity, approximated by the Stefan–Boltzmann law of black-body radiation $E(T) \sim T^4$, varies little in the range of interest, and can be replaced by a constant E_0. All the richness of the model thus lies in modelling the albedo's temperature-dependence (which is influenced by factors such as size of ice sheets and vegetation coverage). The evolution equation (4.1.1) can be rewritten as

$$\frac{dT}{ds} = \frac{E_0}{c} \left[\gamma(T)(1 + K \cos \omega s) + K \cos \omega s \right] , \qquad (4.1.4)$$

where $\gamma(T) = Q(1 - \alpha(T))/E_0 - 1$. In order that two stable climate regimes can coexist, $\gamma(T)$ should have three roots, the middle root corresponding to an unstable state. The authors of [BPSV83] take a pragmatic view and choose to model $\gamma(T)$ by the cubic polynomial

$$\gamma(T) = \beta \left(1 - \frac{T}{T_1}\right)\left(1 - \frac{T}{T_2}\right)\left(1 - \frac{T}{T_3}\right) , \qquad (4.1.5)$$

where $T_1 = 278.6$ K and $T_3 = 288.6$ K are the representative temperatures of the two stable climate regimes, and $T_2 = 283.3$ K represents an intermediate, unstable regime. The parameter β determines the relaxation time τ of the system in the "temperate climate" state, taken to be 8 years, by

$$\frac{1}{\tau} = -\frac{E_0}{c} \gamma'(T_3) . \qquad (4.1.6)$$

Introducing the slow time $t = \omega s$ and the "dimensionless temperature" $x = (T - T_2)/\Delta T$, with $\Delta T = (T_3 - T_1)/2 = 5$ K, yields the rescaled equation of motion

$$\varepsilon \frac{\mathrm{d}x}{\mathrm{d}t} = -x(x - X_1)(x - X_3)(1 + K \cos t) + A \cos t . \qquad (4.1.7)$$

Here $X_1 = (T_1 - T_2)/\Delta T \simeq -0.94$ and $X_3 = (T_3 - T_2)/\Delta T \simeq 1.06$, while the adiabatic parameter ε is given by

$$\varepsilon = \omega \tau \frac{2(T_3 - T_2)}{\Delta T} \simeq 1.16 \times 10^{-3} . \qquad (4.1.8)$$

The effective driving amplitude

$$A = \frac{K}{\beta} \frac{T_1 T_2 T_3}{(\Delta T)^3} \qquad (4.1.9)$$

is approximately equal to 0.12, according to the value $E_0/c = 8.77 \times 10^{-3}/4000$ Ks^{-1} given in [BPSV83]. For simplicity, let us replace X_1 by -1, X_3 by 1, and neglect the term $K \cos 2\pi t$ in (4.1.7). This yields the equation

$$\varepsilon \frac{\mathrm{d}x}{\mathrm{d}t} = x - x^3 + A \cos t . \qquad (4.1.10)$$

The right-hand side derives from a double-well potential, and thus has two stable equilibria and one unstable equilibrium, for all $A < A_c = 2/3\sqrt{3} \simeq 0.38$. The periodic forcing is thus not sufficient to allow for transitions between the stable equilibria (Fig. 4.1).

The new idea in [BSV81, NN81] is to incorporate into the equation (4.1.7) the effect of short-timescale atmospheric fluctuations, by adding a noise term, as suggested by Hasselmann [Has76]. This yields the SDE

$$\mathrm{d}x_t = \frac{1}{\varepsilon} \left[x_t - x_t^3 + A \cos t \right] \mathrm{d}t + \frac{\sigma}{\sqrt{\varepsilon}} \mathrm{d}W_t , \qquad (4.1.11)$$

here considered on the slow timescale. For adequate parameter values, typical solutions of (4.1.11) are likely to cross the potential barrier twice per period, producing the observed sharp transitions between climate regimes. This is a manifestation of SR. Whether SR is indeed the right explanation for the appearance of Ice Ages is controversial, and hard to decide. However, models similar to (4.1.11) have been analysed in different contexts, and SR was discovered to act as an amplification mechanism in many physical systems subject to noise, for instance ring lasers. It also plays a rôle in biology where SR has been observed in neural systems, for instance.

Depending on forcing amplitude A and noise intensity σ, the following regimes are observed (Fig. 4.2):

- If $A = 0$, the study of (4.1.11) reduces to the well-known problem of barrier crossing in a static potential. The expected time between transitions is

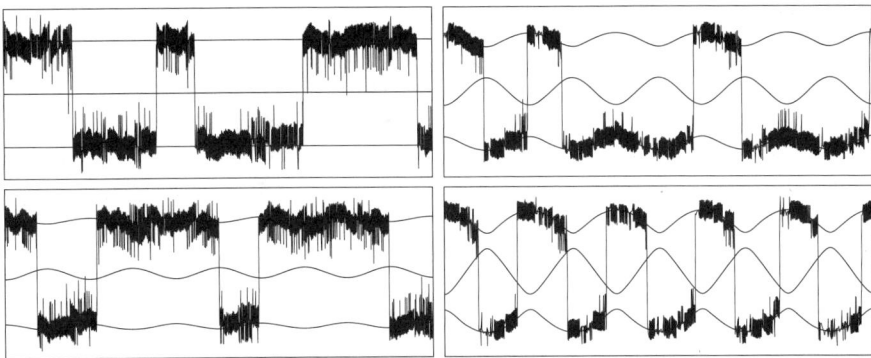

Fig. 4.2. Sample paths of Equation (4.1.11) for $\varepsilon = 0.003$ and different forcing amplitudes A and noise intensities σ. Solid lines indicate the location of the potential minima and of the saddle. From top to bottom and left to right: $A = 0$ and $\sigma = 0.3$, $A = 0.1$ and $\sigma = 0.27$, $A = 0.24$ and $\sigma = 0.2$, $A = 0.35$ and $\sigma = 0.2$.

given by Kramers' time (see Appendix A.6), which is exponentially large in $2H/\sigma^2$, where H is the potential barrier height. For small noise intensities, the distribution of waiting times between crossings is close to an exponential distribution, with parameter equal to the inverse of Kramers' time. Thus transitions asymptotically follow a Poisson point process (after a proper rescaling of time).
- If $0 < A \ll A_c$, the barrier height varies periodically, so that transitions become more likely at some moments in the cycle than at others, yielding a periodically modulated distribution of waiting times between transitions. This effect is expected to be most pronounced when the Kramers time equals half the forcing period. For smaller noise intensities, transitions are rare, while for larger intensities, they become frequent within each period.
- If $A < A_c$, but $A_c - A \ll A_c$, the potential barrier becomes very low twice per period. For sufficiently strong noise, transitions become likely twice per period, and large-amplitude oscillations appear.

Since the introduction of the model, several measures have been considered in order to quantify the effect:

- **Spectral characteristics:** The time-autocorrelation function of the signal has a periodic component, producing peaks in its Fourier transform, the power spectrum. Various quantities have been used to measure the importance of this periodic component, such as signal-to-noise ratio, or spectral power amplification.
- **Residence-time distributions:** In the absence of forcing, the rescaled waiting times between interwell transitions obey an exponential law in the limit of small noise intensity. In the presence of forcing, the distribution of residence times becomes a periodically modulated exponential, reflecting the fact that transitions are more likely to occur at certain times in the

cycle. Then resonance can be quantified by a suitable measure of concentration of residence-times within each period.
- **Sample-path properties:** In some parameter regimes, typical sample paths switch back and forth between the potential wells in a close-to-periodic way, and it can be proved that sample paths converge (in a suitable sense) towards a deterministic periodic function as the noise intensity σ goes to zero. More generally, one can obtain concentration results in a neighbourhood of such a deterministic function.

Various methods have been used to study these quantitative properties, including spectral-theoretic methods (such as expansion of the probability density in a basis of eigenfunctions of the unperturbed generator), linear response theory, and large-deviation techniques.

In the sequel, we shall give an overview of some of these results, with an emphasis on mathematically precise ones. We will concentrate on equations of the type (4.1.11). Note, however, that SR appears in other types of systems as well. Two simplified systems that have been studied in detail are

- the motion in a potential switching back and forth between two different static configurations, first considered in [BSV81];
- a two-state Markovian jump process with time-periodic transition probabilities, first considered in [ET82].

4.1.2 Spectral-Theoretic Results

For simplicity, we consider here again the special case of the system (4.1.11), written in the form

$$\mathrm{d}x_s = \bigl[-U'(x_s) + A\cos\Omega s\bigr]\,\mathrm{d}s + \sigma\,\mathrm{d}W_s\,, \qquad (4.1.12)$$

where $U(x) = \tfrac{1}{4}x^4 - \tfrac{1}{2}x^2$ is the standard static symmetric double-well potential. A first variety of ways to quantify SR are based on the process' *power spectrum*

$$S_A(\omega) = \lim_{t\to\infty}\frac{1}{2t}\left|\int_{-t}^{t}\lim_{t_0\to-\infty}\mathbb{E}^{t_0,x_0}\{x_u\}\,\mathrm{e}^{\mathrm{i}\omega u}\,\mathrm{d}u\right|^2, \qquad (4.1.13)$$

where A refers to the power spectrum's dependence on the forcing amplitude A. The limit $t_0 \to -\infty$ is taken in order that the system be in a stationary state (and thus S_A does not depend on the initial state x_0). In practice, it is often more convenient to compute the power spectrum by invoking the Wiener–Khintchin theorem, which relates the power spectrum to the process' *autocorrelation function*,

$$C_A(t,s) = \lim_{t_0\to-\infty}\mathbb{E}^{t_0,x_0}\{x_s x_t\} = C_A(s,t)\,. \qquad (4.1.14)$$

Note that, for $t > s$, $C_A(t,s)$ is given in terms of transition probabilities by

$$C_A(t,s) = \lim_{t_0 \to -\infty} \iint x\, p(x,t|y,s)\, y\, p(y,s|x_0,t_0)\, dy\, dx\,. \tag{4.1.15}$$

Then the Wiener–Khintchin theorem states that $S_A(\omega)$ is the Fourier transform of $t \mapsto C_A(t+s,s)$, averaged over s.

Let us first consider the case without periodic forcing, i.e., $A = 0$. The transition probabilities $(x,t) \mapsto p(x,t|y,s)$ are solutions of the Kolmogorov's forward, or Fokker–Planck, equation $\partial_t p = L_0^* p$ with initial condition $p(x,s|y,s) = \delta_y(x)$ (cf. Appendix A.4), where

$$(L_0^* p)(x) = \frac{\partial}{\partial x}(U'(x)p(x)) + \frac{\sigma^2}{2}\frac{\partial^2}{\partial x^2}p(x) \tag{4.1.16}$$

is the adjoint of the diffusion's generator. Note that L_0^* is essentially self-adjoint in $L^2(\mathbb{R}, W(x)^{-1}\,dx)$ with weight $W(x)^{-1}$ given by

$$W(x) = \frac{1}{N} e^{-2U(x)/\sigma^2}\,. \tag{4.1.17}$$

We choose the normalisation N in such a way that $W(x)$ is a probability density, and denote the associated scalar product by $\langle \cdot, \cdot \rangle_{W^{-1}}$. It is also well known (see, for instance, [Kol00]) that the spectrum of L_0^* consists of isolated, simple, non-positive eigenvalues

$$0 = -\lambda_0 > -\lambda_1 > -\lambda_2 > \ldots, \tag{4.1.18}$$

where

- the eigenfunction $p_0(x) = W(x)$ associated with $\lambda_0 = 0$ is the invariant density;
- the first non-vanishing eigenvalue $-\lambda_1$ satisfies

$$\lambda_1 = \frac{1}{\pi}\sqrt{U''(1)|U''(0)|}\, e^{-2H/\sigma^2}\left[1 + \mathcal{O}(\sigma^2)\right], \tag{4.1.19}$$

where $H = U(0) - U(-1) = 1/4$ is the potential barrier height; the associated eigenfunction is of the form $p_1(x) = \phi_1(x)W(x)$, where $\phi_1(x)$ is well approximated by $\operatorname{sign} x$;
- all eigenvalues λ_k for $k \geq 2$ are bounded away from 0 (by a constant independent of σ), i.e., there is a spectral gap.

The completeness of the basis of eigenfunctions immediately yields the following representation of the transition probabilities

$$p(x, s+t|y, s) = \sum_{k \geq 0} e^{-\lambda_k t} W(y)^{-1} p_k(y) p_k(x)\,. \tag{4.1.20}$$

Inserting this into the definition (4.1.15) of the autocorrelation function gives

$$C_0(s+t,s) = C_0(t,0) = \sum_{k \geqslant 0} e^{-\lambda_k |t|} \alpha_k^2 , \qquad \text{where } \alpha_k = \int x \, p_k(x) \, \mathrm{d}x .$$

(4.1.21)

The symmetry of the potential implies that eigenfunctions are even for even k and odd for odd k, so that $\alpha_k = 0$ if k is even. The power spectrum is thus the sum of Lorentzians

$$S_0(\omega) = \int e^{-i\omega t} C_0(t,0) \, \mathrm{d}t = 2 \sum_{k \in 2\mathbb{Z}+1} \alpha_k^2 \frac{\lambda_k}{\lambda_k^2 + \omega^2} .$$

(4.1.22)

Because of the spectral gap, for small noise intensities (and small ω) this sum is dominated by its first term.

Consider now the periodically perturbed system (4.1.12). Its adjoint generator is the time-periodic differential operator

$$L^*(t) = L_0^* + A L_1^*(t) , \qquad \text{where } L_1^*(t) = \cos \Omega t \frac{\partial}{\partial x} .$$

(4.1.23)

Floquet theory implies the existence of "eigenmodes" $q_k(x,t)$, depending periodically on t and satisfying

$$\left(L^*(t) - \frac{\partial}{\partial t} \right) q_k(x,t) = -\mu_k q_k(x,t) .$$

(4.1.24)

Similarly, the generator $L(t) + \partial_t$ admits dual Floquet eigenmodes that we denote by $\tilde{q}_k(x,t)$. The eigenmodes $q_k(x,t)$ and characteristic exponents μ_k depend on A and reduce to their static counterparts $p_k(x)$ and λ_k in the limit $A \to 0$. In particular, μ_0 is identically zero, and thus $q_0(x,t)$ is a time-periodic solution of the Fokker–Planck equation, describing the asymptotic dynamics of the process.

The transition probabilities can be represented in terms of Floquet modes as

$$p(x, s+t | y, s) = \sum_{k \geqslant 0} e^{-\mu_k t} \tilde{q}_k(y,s) q_k(x, s+t) .$$

(4.1.25)

A similar computation as above yields the expression

$$C_A(s+t,s) = \sum_{k \geqslant 0} e^{-\mu_k |t|} \alpha_k(s+t) \beta_k(s)$$

(4.1.26)

for the autocorrelation function, where the $\alpha_k(u)$ and $\beta_k(u)$ are periodic functions, of period $T = 2\pi/\Omega$, given by

$$\alpha_k(u) = \int x \, q_k(x,u) \, \mathrm{d}x ,$$

(4.1.27)

$$\beta_k(u) = \int x \, \tilde{q}_k(x,u) q_0(x,u) \, \mathrm{d}x .$$

(4.1.28)

4.1 The Phenomenon of Stochastic Resonance

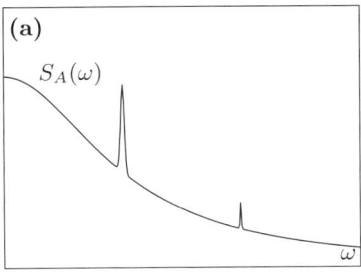
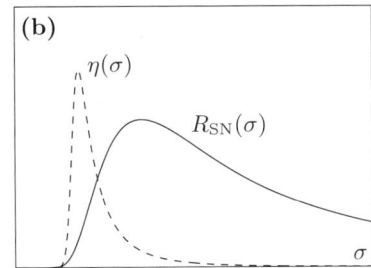

Fig. 4.3. (a) Qualitative behaviour of the power spectrum $S_A(\omega)$: the periodic forcing produces peaks in multiples of the forcing frequency Ω; (b) typical plots of the signal-to-noise ratio $R_{SN}(\sigma)$ (*solid line*) and spectral power amplification $\eta(\sigma)$ (*broken line*), as a function of the noise intensity σ.

A direct computation shows that the first dual Floquet mode is simply $\tilde{q}_0(x,t) = 1$, so that $\beta_0(u) = \alpha_0(u)$. The term $k = 0$ of the sum (4.1.26) yields a singular component of the power spectrum

$$S_A^{\text{sing}}(\omega) = \int \frac{1}{T} \int_0^T \alpha_0(s+t)\alpha_0(s)\, e^{-i\omega t}\, ds\, dt = 2\pi \sum_{\ell \in \mathbb{Z}} |\widehat{\alpha}_{0,\ell}|^2 \delta_{\ell\Omega}(\omega)\,, \tag{4.1.29}$$

where $\widehat{\alpha}_{0,\ell}$, $\ell \in \mathbb{Z}$, are the Fourier coefficients of $\alpha_0(u)$. The amplitudes of the delta-peaks in this singular component are a measure of the signal's periodicity. It is customary to consider the amplitude of the first peak,

$$S_{A,1} = \lim_{\Delta \to 0} \int_{\Omega-\Delta}^{\Omega+\Delta} S_A^{\text{sing}}(\omega)\, d\omega = 2\pi|\widehat{\alpha}_{0,1}|^2\,. \tag{4.1.30}$$

The two most commonly used measures of resonance are the *signal-to-noise ratio*

$$R_{SN} = \frac{S_{A,1}}{S_0(\Omega)} \simeq \pi \frac{|\widehat{\alpha}_{0,1}|^2}{\alpha_1^2} \frac{\lambda_1^2 + \Omega^2}{\lambda_1}\,, \tag{4.1.31}$$

and the *spectral power amplification*

$$\eta = \frac{S_{A,1}}{\pi A^2} = 2 \frac{|\widehat{\alpha}_{0,1}|^2}{A^2}\,. \tag{4.1.32}$$

Note that for fixed parameters σ, A and Ω, the quantities R_{SN} and η depend only on the expected position $\alpha_0(t)$ in the "stationary" Floquet mode $q_0(x,t)$, satisfying $\partial_t q_0(x,t) = L^*(t) q_0(x,t)$.

Several approximate calculations exist for the small-amplitude case $A \ll 1$ [Fox89, GMSS+89, JH89]. A straightforward expansion to linear order in A (which can be viewed in the framework of linear response theory) yields

$$q_0(x,t) = p_0(x) + \frac{A}{2} \sum_{k \geq 1} \left[\frac{e^{i\Omega t}}{\lambda_k + i\Omega} + \frac{e^{-i\Omega t}}{\lambda_k - i\Omega} \right] \left\langle p_k, \frac{\partial}{\partial x} p_0 \right\rangle_{W^{-1}} p_k(x) + \mathcal{O}(A^2)\,. \tag{4.1.33}$$

As a consequence,

$$\widehat{a}_{0,1} = \frac{A}{2} \sum_{k \geqslant 1} \frac{\alpha_k}{\lambda_k + i\Omega} \left\langle p_k, \frac{\partial}{\partial x} p_0 \right\rangle_{W^{-1}} + \mathcal{O}(A^2) \, . \tag{4.1.34}$$

Using the properties of the static eigenfunctions $p_k(x)$, one obtains that this sum is dominated by the term with $k = 1$, and that the matrix element $\langle p_1, \partial_x p_0 \rangle_{W^{-1}}$ is proportional to λ_1/σ^2. This yields the expressions

$$R_{\mathrm{SN}} = R_{\mathrm{SN}}(\sigma) \simeq \mathrm{const}\, \frac{A^2}{\sigma^4} \lambda_1 = \mathrm{const}\, \frac{A^2}{\sigma^4} e^{-2H/\sigma^2}\, , \tag{4.1.35}$$

$$\eta = \eta(\sigma) \simeq \mathrm{const}\, \frac{1}{\sigma^4} \frac{\lambda_1^2}{\lambda_1^2 + \Omega^2}\, , \tag{4.1.36}$$

for the signal-to-noise ratio and spectral power amplification. Each function has a unique maximum, tends to zero exponentially fast in the limit of vanishing noise, and decreases like σ^{-4} for large noise (Fig. 4.3b). The signal-to-noise ratio is maximal for $\sigma^2 = H$, while the maximal power amplification is reached when the Kramers time λ_1^{-1} is of the order of the driving period, i.e., $\sigma^2 \simeq 2H/\log \Omega^{-1}$.

Though we are not aware of completely rigorous proofs of the above results (that is, the error terms $\mathcal{O}(A^2)$ in (4.1.33) are not precisely controlled), there seems to be little doubt as to their validity, which has also been confirmed experimentally.

Another limit that can be studied perturbatively is the adiabatic limit $\Omega := \varepsilon \ll 1$. Rather than working in the basis of eigenfunctions of the symmetric double-well system, it is then advantageous to expand the periodic Floquet solution $q_0(x,t)$ in the basis of instantaneous normalised eigenfunctions $p_k(x,t)$ of $L^*(t)$. These satisfy, for each fixed time t,

$$L^*(t) p_k(x,t) = -\lambda_k(t) p_k(x,t) \, . \tag{4.1.37}$$

Let $U(x,t) = U(x) - Ax \cos \varepsilon t$ be the time-dependent potential including the effect of the periodic driving. We denote by $x_-^\star(t) < x_0^\star(t) < x_+^\star(t)$ its stationary points at time t, by $h_\pm(t) = U(x_0^\star(t), t) - U(x_\pm^\star(t), t)$ the depths of its two wells, and by $\omega_i(t)^2 = |\partial_{xx} U(x_i^\star(t), t)|$, $i \in \{-, 0, +\}$, the curvatures of U at its stationary points. Then it is known that

- $\lambda_0(t) = 0$, and $p_0(x,t) = N(t)^{-1} e^{-2U(x,t)/\sigma^2}$ is the associated stationary distribution;
- the first non-vanishing eigenvalue satisfies

$$\lambda_1(t) = \frac{\omega_0(t)}{2\pi} \left[\omega_+(t) e^{-2h_+(t)/\sigma^2} + \omega_-(t) e^{-2h_-(t)/\sigma^2} \right] [1 + \mathcal{O}(\sigma^2)] \, ; \tag{4.1.38}$$

the corresponding eigenfunction has the form $p_1(x,t) = \phi_1(x,t) p_0(x,t)$, where $\phi_1(x,t)$ is close to $-\phi_-$ for $x < x_0^\star(t)$, and close to $\phi_+ = 1/\phi_-$, for $x > x_0^\star(t)$, with ϕ_- given by

$$\phi_-^2 = \frac{\omega_-(t)}{\omega_+(t)} e^{2\Delta(t)/\sigma^2}, \qquad \Delta(t) = h_+(t) - h_-(t) \ ; \qquad (4.1.39)$$

- there is a spectral gap of order 1 between $\lambda_1(t)$ and all subsequent eigenvalues.

In the sequel, we shall work on the slow timescale εt. Substituting the decomposition $q_0(x,t) = \sum_{k \geqslant 0} c_k(t) p_k(x,t)$ into the accordingly rescaled Floquet equation yields the system

$$\varepsilon \dot{c}_k = -\lambda_k(t) c_k - \varepsilon \sum_{\ell \geqslant 0} \langle p_k, \dot{p}_\ell \rangle_{W^{-1}} c_\ell \ . \qquad (4.1.40)$$

Orthonormality of the p_k (with respect to the scalar product weighted by $W^{-1}(x,t) = p_0(x,t)^{-1}$) yields $\langle p_0, \dot{p}_\ell \rangle_{W^{-1}} = 0$ for all ℓ, so that $c_0(t) = 1$. Now the existence of a spectral gap implies that for $\varepsilon \ll 1$, all c_k with $k \geqslant 2$ are fast variables. By an infinite-dimensional variant of Tihonov's theorem, they remain thus ε-close to a slow manifold given by $c_k(t) = \mathcal{O}(\varepsilon) \ \forall k \geqslant 2$. The reduced dynamics on the slow manifold is described by the one-dimensional effective equation

$$\varepsilon \dot{c}_1 = -\lambda_1(t) c_1 - \varepsilon \langle p_1, \dot{p}_1 \rangle_{W^{-1}} c_1 - \varepsilon \langle p_1, \dot{p}_0 \rangle_{W^{-1}} \ . \qquad (4.1.41)$$

Let us now show that this equation is equivalent to the two-state Markovian jump process introduced in [ET82]. The scalar products in (4.1.41) can be evaluated by first noting that

$$\dot{p}_0(x,t) = \frac{2A}{\sigma^2} \sin t \left(\int y p_0(y,t) \, dy - x \right) p_0(x,t) \ , \qquad (4.1.42)$$

and then using the orthonormality conditions and their derivatives to obtain

$$\langle p_1, \dot{p}_0 \rangle_{W^{-1}} = -\frac{2A}{\sigma^2} \sin t \, \langle p_1, x p_0 \rangle_{W^{-1}} \ , \qquad (4.1.43)$$

$$\langle p_1, \dot{p}_1 \rangle_{W^{-1}} = \frac{A}{\sigma^2} \sin t \left[\langle p_0, x p_0 \rangle_{W^{-1}} - \langle p_1, x p_1 \rangle_{W^{-1}} \right] \ . \qquad (4.1.44)$$

The matrix elements can be estimated by the Laplace method. Below, the notation $a \cong b$ is used to indicate that $a/b = 1 + \mathcal{O}(\sigma^2)$. It is also convenient to introduce $\bar{\Delta}(t) = \Delta(t) - \frac{1}{2} \sigma^2 \log(\omega_+(t)/\omega_-(t))$. One obtains

$$\langle p_1, x p_0 \rangle_{W^{-1}} \cong \frac{x_+^\star(t) - x_-^\star(t)}{e^{\bar{\Delta}(t)/\sigma^2} + e^{-\bar{\Delta}(t)/\sigma^2}} \ , \qquad (4.1.45)$$

$$\langle p_0, x p_0 \rangle_{W^{-1}} \cong \frac{x_+^\star(t) e^{\bar{\Delta}(t)/\sigma^2} + x_-^\star(t) e^{-\bar{\Delta}(t)/\sigma^2}}{e^{\bar{\Delta}(t)/\sigma^2} + e^{-\bar{\Delta}(t)/\sigma^2}} \ , \qquad (4.1.46)$$

$$\langle p_1, x p_1 \rangle_{W^{-1}} \cong \frac{x_-^\star(t) e^{\bar{\Delta}(t)/\sigma^2} + x_+^\star(t) e^{-\bar{\Delta}(t)/\sigma^2}}{e^{\bar{\Delta}(t)/\sigma^2} + e^{-\bar{\Delta}(t)/\sigma^2}} \ . \qquad (4.1.47)$$

The reduced equation (4.1.41) thus takes the form

$$\varepsilon \dot c_1 = -\lambda_1(t)c_1 - \varepsilon \frac{A}{\sigma^2}\sin t\bigl(x_+^\star(t) - x_-^\star(t)\bigr)\frac{\sinh(\bar A(t)/\sigma^2)c_1 - 1}{\cosh(\bar A(t)/\sigma^2)}. \qquad (4.1.48)$$

To show equivalence with the two-state process, we note that the population of the right-hand potential well is given by

$$c_+(t) = \mathbb{P}^{q_0}\{x_t > x_0^\star(t)\} \cong \frac{e^{\bar A(t)/\sigma^2} + c_1(t)}{e^{\bar A(t)/\sigma^2} + e^{-\bar A(t)/\sigma^2}}. \qquad (4.1.49)$$

Conversely, $c_1(t)$ can be expressed in terms of $c_+(t)$ and $c_-(t) = 1 - c_+(t)$ by

$$c_1(t) = c_+(t)e^{-\bar A(t)/\sigma^2} - c_-(t)e^{\bar A(t)/\sigma^2}. \qquad (4.1.50)$$

A straightforward calculation, using $\dot{\bar A}(t) = -A\sin t\,(x_+^\star(t) - x_-^\star(t))$ and

$$\lambda_1(t) \cong r_+(t) + r_-(t), \qquad (4.1.51)$$

then shows that (4.1.48) is equivalent to the set of rate equations

$$\begin{aligned}\varepsilon \dot c_+ &= -r_+(t)c_+ + r_-(t)c_-, \\ \varepsilon \dot c_- &= r_+(t)c_+ - r_-(t)c_-,\end{aligned} \qquad (4.1.52)$$

with transition rates

$$r_\pm(t) \cong \frac{\omega_\pm(t)\omega_0(t)}{2\pi}e^{-2h_\pm(t)/\sigma^2}. \qquad (4.1.53)$$

These rates are interpreted as instantaneous Kramers transition rates between the two potential wells. The expected position $\alpha_0(t)$ is given by

$$\alpha_0(t) = \mathbb{E}^{q_0}\{x_t\} \cong c_+(t)x_+^\star(t) + c_-(t)x_-^\star(t), \qquad (4.1.54)$$

and satisfies the differential equation

$$\varepsilon \dot\alpha_0(t) = -\lambda_1(t)\alpha_0(t) + \bigl[r_-(t)x_+^\star(t) + r_+(t)x_-^\star(t)\bigr]\bigl[1 + \mathcal{O}(\varepsilon A)\bigr]. \qquad (4.1.55)$$

It can thus be written in the form

$$\alpha_0(t) = \frac{1 + \mathcal{O}(\varepsilon)}{1 - e^{-\Lambda/\varepsilon}}\frac{1}{\varepsilon}\int_{t-2\pi}^{t}e^{-\Lambda_1(t,s)/\varepsilon}\bigl[r_-(s)x_+^\star(s) + r_+(s)x_-^\star(s)\bigr]\,ds, \qquad (4.1.56)$$

where we have set $\Lambda_1(t,s) = \int_s^t \lambda_1(u)\,du$, and $\Lambda = \Lambda_1(2\pi,0)$. Note that Λ is of the order $e^{-2h_{\min}/\sigma^2}$, where $h_{\min}$ is the minimal depth the potential wells attain.

We can now distinguish between two situations (Fig. 4.4):

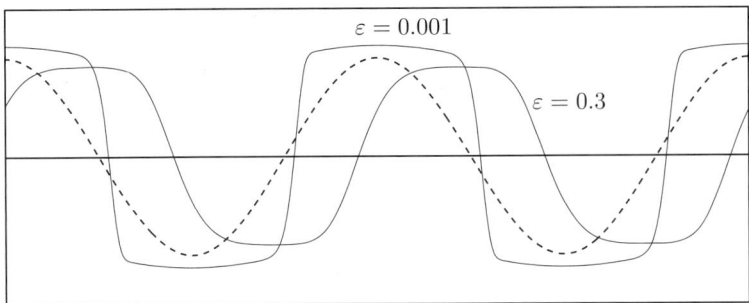

Fig. 4.4. Expected position $\alpha_0(t) = \mathbb{E}^{q_0}\{x_t\}$ (*solid lines*), for $A = 0.3$, $\sigma = 0.2$, and two different driving frequencies. The broken line is the forcing $A \cos t$. For $\varepsilon = 0.001$ (*superadiabatic regime*), $\alpha_0(t)$ tracks the deepest potential minimum. For $\varepsilon = 0.3$ (*semiadiabatic regime*), $\alpha_0(t)$ changes well near the instant of minimal barrier height.

- In the *superadiabatic regime* $\varepsilon \ll \Lambda$, the system has enough time to cross the potential barrier, and thus settles close to its static equilibrium distribution $p_0(x, t)$, while the potential slowly modifies its shape. As a consequence, $\alpha_0(t)$ adiabatically tracks its equilibrium value

$$\alpha_0^*(t) = \frac{r_-(t)x_+^*(t) + r_+(t)x_-^*(t)}{r_-(t) + r_+(t)}. \qquad (4.1.57)$$

Disregarding the motion of the potential minima $x_\pm^*(t)$, we see that $\alpha_0^*(t)$ behaves like $\tanh(\Delta(t)/\sigma^2)$, and has an amplitude of order $\tanh(A/\sigma^2)$. It follows that the signal-to-noise ratio and spectral power amplification behave like

$$R_{\mathrm{SN}}(\sigma) \simeq \mathit{const} \, \tanh^2\left(\frac{A}{\sigma^2}\right) e^{-2h_{\min}/\sigma^2}, \qquad (4.1.58)$$

$$\eta(\sigma) \simeq \mathit{const} \, \frac{1}{A^2} \tanh^2\left(\frac{A}{\sigma^2}\right). \qquad (4.1.59)$$

We recover the same small-amplitude behaviour as before, but see a saturation effect at large amplitudes, due to the fact that the oscillations are bounded by the interwell distance (cf. (4.1.54)).
- In the *semiadiabatic regime* $\mathcal{O}(\Lambda) \leqslant \varepsilon \ll 1$, the system does not react as quickly to changes of the potential. Increasing the driving frequency ε has two effects on the behaviour of $\alpha_0(t)$. Firstly, transitions between potential wells are delayed towards the instants of minimal barrier height; and secondly, the amplitude of $\alpha_0(t)$ decreases proportionally to $1/(1 + \varepsilon/\Lambda)$. In particular, the spectral power amplification now behaves like

$$\eta(\sigma) \simeq \mathit{const} \, \frac{1}{A^2} \tanh^2\left(\frac{A}{\sigma^2}\right) \frac{\Lambda^2}{\Lambda^2 + \varepsilon^2}. \qquad (4.1.60)$$

Note that $\eta(\sigma)$ still has a unique maximum, but when the amplitude A is not a small parameter, the decay of the power amplification is rather slow, and makes itself felt only for noise intensities σ^2 of order larger than A.

In the case of the switching potential, as well as for the corresponding two-state process, the spectral power amplification can easily be computed exactly. A detailed comparison between both cases is given in [IP02] for $\sigma^2 \ll A$. As expected, in some parameter regimes, the power amplification of the two-state model behaves qualitatively differently from its continuous-space counterpart, as a consequence of intrawell fluctuations.

4.1.3 Large-Deviation Results

The spectral-theoretic results discussed so far all relate to the behaviour of the expected position. It would be possible to study the properties of higher moments by similar means. However, much more information is contained in the sample-path behaviour of the process.

The theory of large deviations can be used to derive sample-path properties in the quasistatic regime, that is, for driving periods comparable to Kramers' time or larger. More precisely, consider the family of stochastic processes $\{x_t\}_t$, satisfying

$$\mathrm{d}x_t = \left[x_t - x_t^3 + A\cos(2\pi t/T(\sigma))\right] \mathrm{d}t + \sigma \, \mathrm{d}W_t \qquad (4.1.61)$$

with an initial condition x_0. In this context, the forcing period $T(\sigma)$ is assumed to scale with the noise intensity in such a way that

$$\lim_{\sigma \to 0} \sigma^2 \log T(\sigma) = \lambda , \qquad (4.1.62)$$

for some parameter $\lambda > 0$ controlling the relative value of period and "instantaneous" Kramers' time. The theory of large deviations allows one to prove convergence, in an appropriate sense, of the rescaled process $y_s = x_{sT(\sigma)}$ to a deterministic function $s \mapsto \phi(s, \lambda, x_0)$ in the limit of vanishing noise intensity σ.

Let $U(x,t) = \frac{1}{4}x^4 - \frac{1}{2}x^2 - A\cos(2\pi t/T(\sigma))x$ denote the time-dependent potential associated with the drift term in (4.1.61). Let $x^\star_\pm(t)$ again be the positions of its two minima, $x^\star_0(t)$ the position of its saddle, and let $h_\pm(t) = U(x^\star_0(t), t) - U(x^\star_\pm(t), t)$ denote the depths of the two potential wells. Consider for a moment t as a fixed parameter, so that $x^\star_\pm(t) \equiv x^\star_\pm$ and $h_\pm(t) \equiv h_\pm$. The classical Wentzell–Freidlin theory (see Appendix A.5) shows in particular that a sample path starting, say, at time $t = 0$ in $x^\star_-$, first hits $x^\star_+$ at a random time τ^+_- such that

$$\lim_{\sigma \to 0} \sigma^2 \log \mathbb{E}^{0,x^\star_-}\{\tau^+_-\} = 2h_- . \qquad (4.1.63)$$

Furthermore, for any constant $\delta > 0$,

$$\lim_{\sigma \to 0} \mathbb{P}^{0,x^\star_-}\left\{e^{(2h_- - \delta)/\sigma^2} < \tau^+_- < e^{(2h_- + \delta)/\sigma^2}\right\} = 1 . \qquad (4.1.64)$$

A sample path starting in the left-hand potential well is thus likely to remain in that same well on all timescales shorter than e^{2h_-/σ^2}. On larger timescales, it is likely to visit the right-hand well, where it typically stays for a time of order e^{2h_+/σ^2} before returning into the left-hand well. If $h_- \neq h_+$, the relative time spent in the shallower well tends to zero in the limit $\sigma \to 0$.

We return now to the time-dependent system (4.1.61). The asymptotic behaviour, in the simultaneously taken limit of weak noise and exponentially long driving period (4.1.62), is described by a deterministic function $t \mapsto \phi(t, \lambda, x)$, giving the position of the bottom of the well the particle is most likely "inhabiting" at time t, when the forcing period is of order e^{λ/σ^2}: For an initial condition $x < x_0^\star(0)$ in the left-hand well, $\phi(t, \lambda, x) = x_-^\star(t)$ until the first time τ_1, when $h_-(t)$ becomes smaller than $\lambda/2$ and the right-hand well is deeper. Then ϕ switches from $x_-^\star(t)$ to $x_+^\star(t)$. It switches back to $x_-^\star(t)$ at the first time $\tau_2 > \tau_1$ at which $h_+(t)$ falls below $\lambda/2$ and the left-hand well has again become the deeper one. Now, the pattern is repeated with switches from $x_-^\star(t)$ to $x_+^\star(t)$ at times τ_{2i-1}, and switches back from $x_+^\star(t)$ to $x_-^\star(t)$ at subsequent times τ_{2i}, $i \geqslant 1$. Thus, using $\tau_0 = 0$, we can write

$$\phi(t,\lambda,x) = \begin{cases} x_-^\star(t), & \text{if } t \in [\tau_{2i}, \tau_{2i+1}) \text{ for some } i \geqslant 0, \\ x_+^\star(t), & \text{if } t \in [\tau_{2i+1}, \tau_{2(i+1)}) \text{ for some } i \geqslant 0. \end{cases} \quad (4.1.65)$$

The definition is analogous for initial conditions $x > x_0^\star(0)$ from the right-hand well.

A typical result obtained by large-deviation theory is the following. For any fixed $S, p, \delta > 0$,

$$\lim_{\sigma \to 0} \mathbb{P}^{0,x}\left\{ \int_0^S |x_{sT(\sigma)} - \phi(sT(\sigma), \lambda, x)|^p \, ds > \delta \right\} = 0. \quad (4.1.66)$$

That is, the L^p-distance between sample paths and the limiting function ϕ converges to zero in probability as the noise intensity goes to zero (while the forcing period simultaneously diverges like e^{λ/σ^2}). If $h_{\min}$ and $h_{\max}$ denote the extremal depths of the potential wells, the following regimes can be distinguished:

- In the *superadiabatic regime* $\lambda > 2h_{\max}$, $\phi(t, \lambda, x)$ is always the deepest potential well, i.e., sample paths are always likely to be close to the deepest minimum of U. Note that this does not forbid excursions to the shallower well, which are in fact frequent, but short compared to the forcing period.
- In the intermediate regime $2h_{\min} < \lambda < 2h_{\max}$, the limiting function ϕ switches from the shallower to the deeper well some time before the barrier height reaches its minimum. This switching displays a *hysteresis* behaviour: Transitions from left to right and vice versa occur for different values of the forcing.
- For $\lambda < 2h_{\min}$, the forcing period is smaller than the minimal Kramers time, and thus no transitions between wells are observed in the weak-noise limit, during any finite number of periods.

An advantage of these large-deviation results is that they are valid in much greater generality than discussed here. They can be extended to multidimensional systems with an arbitrary (finite) number of attractors, which can be more complicated than equilibrium points. These attractors are then organised in a hierarchy of cycles, which is used to define the deterministic limiting function ϕ. A limitation of these quasistatic large-deviation estimates is the rather weak notion of convergence, which does not exclude excursions away from the limiting function. A more refined description of transitions is given by first-passage-time and residence-time distributions.

4.1.4 Residence-Time Distributions

The term *residence time* refers to the time elapsing between two successful transitions from one potential well to the other. In the absence of forcing, a transition time can be defined as the first time τ_1 at which a sample path crosses the saddle, say from the left-hand to the right-hand well. The time of the next *successful* crossing is then the smallest time $\tau_2 > \tau_1$ at which x_t reaches the saddle again, after having left a suitably chosen neighbourhood of the saddle and having visited the right-hand well between times τ_1 and τ_2 (this is done in order to mask the effect of rapid fluctuations of sample paths around the saddle). Continuing in this way, one obtains a sequence $\{\tau_n\}_{n \geq 1}$ of successive crossing times; in the properly rescaled vanishing-noise limit, their distribution is given by a Poisson point process. The *residence-time distribution* is the large-n asymptotics of the distribution of differences $\tau_{n+1} - \tau_n$, and tends to an exponential distribution in the limit $\sigma \to 0$.

Consider now the periodically forced system

$$\mathrm{d}x_t = \left[x_t - x_t^3 + A\cos(2\pi t/T)\right]\mathrm{d}t + \sigma\,\mathrm{d}W_t\;. \tag{4.1.67}$$

If $\sigma = 0$ and $A < A_c = 2/3\sqrt{3}$, the system has exactly three periodic orbits, two of them being stable and tracking the potential wells, the third one being unstable and tracking the saddle. We denote this last periodic orbit, which separates the basins of attraction of the two stable ones, by $x^{\mathrm{per}}(t)$. In the adiabatic case $T \gg 1$, $x^{\mathrm{per}}(t)$ tracks the saddle at $x_0^\star(t)$ at a distance of order $1/T$, while for decreasing T, the orbit $x^{\mathrm{per}}(t)$ oscillates with an amplitude decreasing like $(1 + 1/T)^{-1}$.

In the presence of forcing, the noise-induced transitions between wells are characterised by the first-passage times $\{\tau_n\}_{n \geq 1}$ through the unstable periodic orbit $x^{\mathrm{per}}(t)$. Again, the residence-time distribution is defined as the asymptotic distribution of differences $\tau_{n+1} - \tau_n$. Stochastic resonance can then be quantified by the deviation of this distribution from an exponential one.

The problem of determining the residence-time distribution can be split into two tasks:

- Find the (conditional) first-passage density

4.1 The Phenomenon of Stochastic Resonance

$$p(t|s) = \frac{\partial}{\partial t} \mathbb{P}^{s, x^{\text{per}}(s)}\{\tau < t\}, \qquad (4.1.68)$$

where τ is the time of the first successful crossing from the left-hand to the right-hand well, following a preceding transition in the opposite direction at time s.

- Find the density $\psi(s)$ of asymptotic arrival *phases*, which is a T-periodic function of average $1/T$, satisfying

$$\psi(t) = \int_{-\infty}^{t} p(t|s)\psi(s - T/2)\,\mathrm{d}s. \qquad (4.1.69)$$

The shift by $T/2$ reflects the fact that the two potential wells move half a period out of phase, i.e., if $\psi(s)$ is the density of arrival phases for transitions from left to right, then $\psi(s - T/2)$ is the density of arrival phases for transitions from right to left.

The density $q(t)$ of the residence-time distribution is then obtained by determining the density of the first-passage time when starting with an initial condition chosen according to the distribution of arrival phases, i.e., by averaging $p(s+t|s)$ over the arrival phase s:

$$q(t) = \int_0^T p(s+t|s)\psi(s - T/2)\,\mathrm{d}s. \qquad (4.1.70)$$

In the adiabatic regime $T \gg 1$, the residence-time distribution should be well-approximated by the residence-time distribution of the two-state system (4.1.52), which can be computed explicitly. In particular, for small driving amplitude $A \ll \sigma^2$, one obtains [ZMJ90, CFJ98]

$$q(t) \simeq \frac{1}{N} e^{-\kappa t/T_K^0}\bigl[1 - c_1 \sin(2\pi t/T) - c_2 \cos(2\pi t/T)\bigr], \qquad (4.1.71)$$

where $T_K^0 \simeq e^{2H/\sigma^2}$ is the (static) Kramers time, and κ, c_1, c_2 are constants related to A/σ^2 and T/T_K^0, with c_1 and c_2 being of order $(A/\sigma^2)^2$.

In the weak-noise regime $\sigma^2 \ll A$, on the other hand, quite a precise characterisation of the residence-time distribution can be obtained from results on the exit problem (see Appendix A.6). Equation (4.1.67) can be rewritten as a two-dimensional system on the cylinder $\mathbb{R} \times (\mathbb{R}/T\mathbb{Z})$ as

$$\begin{aligned}\mathrm{d}x_t &= \bigl[x_t - x_t^3 + A\cos(2\pi y_t/T)\bigr]\mathrm{d}t + \sigma\,\mathrm{d}W_t, \\ \mathrm{d}y_t &= \mathrm{d}t.\end{aligned} \qquad (4.1.72)$$

Consider the set $\mathcal{D} = \{(x,y) : x < x^{\text{per}}(y)\}$, and fix an initial condition $(x_0, y_0) \in \mathcal{D}$. The exit problem consists in determining the distribution of the first-exit time τ of (x_t, y_t) from $\mathcal{D}$, and of the first-exit location $(x_\tau, y_\tau) \in \partial\mathcal{D}$. Note that in our particular case, $y_\tau = y_0 + \tau \pmod{T}$ and $x_\tau = x^{\text{per}}(y_\tau)$.

A first approach to the exit problem relies on the theory of large deviations, which relates properties of the exit law to the behaviour of the large-deviation rate function and the quasipotential induced, see in particular Theorem A.6.2. Two difficulties are inherent to the present situation:

- The system (4.1.72) is degenerate, in the sense that the noise only acts on the x-variable.
- The quasipotential is constant on the boundary $\partial \mathcal{D}$, owing to the fact that translations along the periodic orbit $x^{\text{per}}(t)$ do not contribute to the "cost" of a path in terms of the action functional, because $\partial \mathcal{D}$ is invariant under the deterministic flow. This is a special case of a *characteristic boundary*.

The first difficulty turns out not to be too serious, because the distribution of exit locations depends little on whether or not there is noise acting on the y-variable. The second difficulty is more serious, as it implies that on the level of exponential asymptotics, all points on the boundary are reached with the same probability — an apparent contradiction with the quasistatic approach of the previous section. The explanation lies, of course, in the subexponential behaviour of the exit law, which becomes highly nonuniform in the adiabatic limit.

Exit problems with characteristic boundary have been studied in detail. In particular, Day [Day90b] discovered a striking phenomenon, that he called *cycling*: As the noise intensity converges to zero, the distribution of exit locations does not converge, but rotates around the boundary, with an angular velocity proportional to $|\log \sigma|$.

Though the theory of large deviations is not sufficiently precise to compute the subexponential behaviour of the exit distribution, it is helpful inasmuch as it shows that paths leaving the set $\mathcal{D}$ through a given subset of $\partial \mathcal{D}$ most likely track minimisers of the action functional. Generically, for paths reaching $\partial \mathcal{D}$ after n periods, the action functional has n minimisers, distinguished by the number of periods they spend near the unstable orbit before crossing it.

The most precise result useful for our purpose is the following. Assume the amplitude A is of order 1, and $\sigma^2 \ll A$. Let $a(t) = -\partial_{xx} U(x^{\text{per}}(t), t)$ be the curvature of the potential at $x^{\text{per}}(t)$, and

$$\lambda = \frac{1}{T} \int_0^T a(t) \, dt \qquad (4.1.73)$$

the characteristic (or Lyapunov) exponent of the unstable orbit. Then, for any $\Delta \geqslant \sqrt{\sigma}$, and $t \geqslant t_0$, one can show [BG04, BG05a, BG05b] that

$$\mathbb{P}^{x_0, t_0}\{\tau \in [t, t+\Delta]\} = \int_t^{t+\Delta} \bar{p}(s|t_0) \, ds \, [1 + r(\sigma)], \qquad (4.1.74)$$

where $r(\sigma) = \mathcal{O}(\sqrt{\sigma})$ and

$$\bar{p}(t|t_0) = \frac{1}{N(t_0)} P_{\lambda T}\bigl(\theta(t) - |\log \sigma|\bigr) \frac{\theta'(t)}{\lambda T_{\text{K}}(\sigma)} e^{-[\theta(t) - \theta(t_0)]/\lambda T_{\text{K}}(\sigma)} f_{\text{trans}}(t|t_0). \qquad (4.1.75)$$

Because of the restriction on Δ, this result does not quite prove that the first-exit time density is given by (4.1.75), but only allows for a coarse-grained description of $p(t|t_0)$. For simplicity, in the sequel we shall identify $p(t|t_0)$ and $\bar p(t|t_0)$. The following notations are used in (4.1.75):

- $T_K(\sigma)$ is the analogue of Kramers' time in the autonomous case; it has the form
$$T_K(\sigma) = \frac{C}{\sigma} e^{\overline{V}/\sigma^2}, \qquad (4.1.76)$$
where $\overline{V}$ is the constant value of the quasipotential on the boundary. The prefactor has order σ^{-1} rather than 1, due to the fact that most paths reach $x^{\mathrm{per}}(t)$ through a bottleneck of width σ around a minimiser of the action functional. (Note that the width of these bottlenecks would be larger if $|A|$ were not of order 1.)
- $P_{\lambda T}(y)$ is a universal λT-periodic function, given by
$$P_{\lambda T}(y) = \frac{\lambda T}{2} \sum_{k=-\infty}^{\infty} A(y - k\lambda T) \quad \text{with} \quad A(z) = 2e^{-2z} \exp\left\{-\frac{1}{2} e^{-2z}\right\}. \qquad (4.1.77)$$
It thus consists of a superposition of identical asymmetric peaks, shifted by a distance λT. The kth term of the sum is the contribution of a minimiser of the action functional tracking the unstable orbit during k periods before crossing it.[1] The shape of the peaks $A(z)$ is given by the density of a Gumbel distribution with scale parameter $1/2$, describing the distribution of the largest extreme of a large number of independent trials. The average of $P_{\lambda T}(y)$ over one period is equal to 1, and its Fourier series is given by
$$P_{\lambda T}(y) = \sum_{q \in \mathbb{Z}} 2^{\pi i q/\lambda T} \Gamma\left(1 + \frac{\pi i q}{\lambda T}\right) e^{2\pi i q y/\lambda T}, \qquad (4.1.78)$$
where Γ denotes the Euler Gamma function.
- $\theta(t)$ contains the model-dependent part of the distribution; it is an increasing function of t, satisfying $\theta(t + T) = \theta(t) + \lambda T$, and is given by
$$\theta(t) = \mathrm{const} + \int_0^t a(s)\,\mathrm{d}s - \frac{1}{2} \log \frac{v(t)}{v(0)}, \qquad (4.1.79)$$
where $v(t)$ is the unique periodic solution of the differential equation $\dot v(t) = 2a(t)v(t) + 1$, that is,
$$v(t) = \frac{1}{e^{2\lambda T} - 1} \int_t^{t+T} \exp\left\{\int_s^{t+T} 2a(u)\,\mathrm{d}u\right\} \mathrm{d}s. \qquad (4.1.80)$$

---
[1] If we neglect boundary effects, we may assume that there are $n = \lfloor t/T \rfloor$ minimisers of the action functional, and k should thus actually range from 1 to n. However, extending the sum to all integers only yields a small correction which can be hidden in the error term $r(\sigma)$.

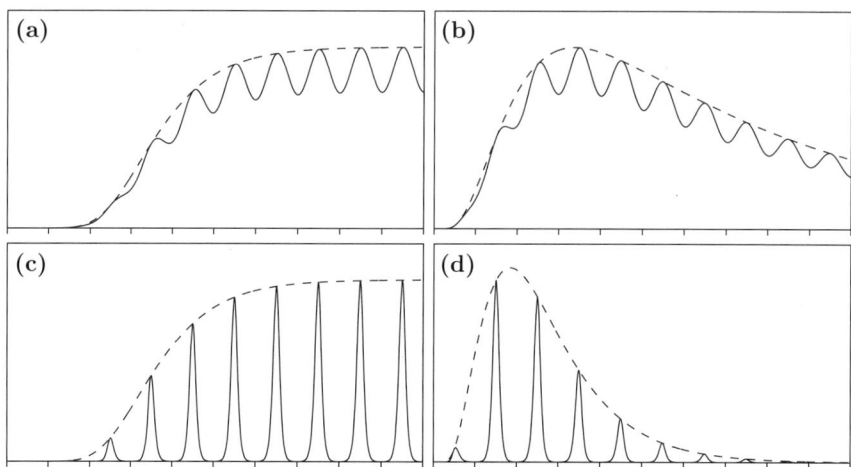

Fig. 4.5. Plots of residence-time distributions $q(t)$ (*full curves*) as a function of t/T, shown for ten periods. The vertical scale is different between plots. The *broken lines* show the average behaviour of $q(t)$, that is, without the periodic modulation $Q_{\lambda T}(t/T)$, scaled vertically in order to match the peak height. Parameter values are $\bar{V} = 0.5$, $\lambda = 1$ and **(a)** $T = 2$, $\sigma = 0.2$, **(b)** $T = 2$, $\sigma = 0.45$, **(c)** $T = 10$, $\sigma = 0.2$ and **(d)** $T = 10$, $\sigma = 0.45$.

- $f_{\text{trans}}(t|t_0)$ accounts for the initial transient behaviour of the system; it is an increasing function satisfying

$$f_{\text{trans}}(t|t_0) = \begin{cases} \mathcal{O}\left(\exp\left\{-\dfrac{L}{\sigma^2}\dfrac{e^{-\lambda(t-t_0)}}{1-e^{-2\lambda(t-t_0)}}\right\}\right) & \text{for } \lambda(t-t_0) < 2|\log\sigma|, \\ 1 - \mathcal{O}\left(\dfrac{e^{-\lambda(t-t_0)}}{\sigma^2}\right) & \text{for } \lambda(t-t_0) \geqslant 2|\log\sigma|, \end{cases}$$
(4.1.81)

where L is a constant, describing the rate at which the distribution in the initial well approaches metastable equilibrium.

The function $\theta(t)$ yields a natural parametrisation of time. Redefining time in such a way that $\theta(t) = \lambda t$, one obtains the simpler expression of the (approximate) first-exit time density

$$\bar{p}(t|t_0) = \dfrac{1}{N(t_0)} P_{\lambda T}\left(\lambda t - |\log\sigma|\right) \dfrac{1}{T_{\text{K}}(\sigma)} e^{-(t-t_0)/T_{\text{K}}(\sigma)} f_{\text{trans}}(t|t_0) \,. \quad (4.1.82)$$

After a transient phase of order $|\log\sigma|$, $\bar{p}(t|t_0)$ becomes a periodically modulated exponential. The cycling phenomenon manifests itself as the argument $\lambda t - |\log\sigma|$ of the periodic modulation.

The expression (4.1.82) gives an approximation of the first-passage-time distribution for an initial condition inside the left-hand well. Note, however, that it depends on the starting point x_0 only via $f_{\text{trans}}(t|t_0)$ which can be neglected after a short time, and that the periodic part does not depend on the initial time t_0. In order to apply (4.1.70) to compute the residence-time distribution, we should consider initial conditions on the unstable periodic orbit. But as only sample paths visiting the left-hand well contribute, we may approximate $p(t|t_0)$ by $\bar{p}(t|t_0)$, at the cost of making a small error on t_0.

For $T \ll T_K(\sigma)$, one easily obtains that the normalisation is given by $N(t_0) = 1 + \mathcal{O}(T/T_K(\sigma)) + \mathcal{O}(r(\sigma))$, and that the solution of (4.1.69) is of the form

$$\psi(t) = \frac{1}{T} P_{\lambda T}\big(\lambda t - |\log \sigma|\big)\big[1 + \mathcal{O}(T/T_K(\sigma)) + \mathcal{O}(r(\sigma))\big]. \quad (4.1.83)$$

It follows that the residence-time distribution has the expression

$$q(t) = Q_{\lambda T}(t/T)\frac{1}{T_K(\sigma)} e^{-t/T_K(\sigma)} \tilde{f}_{\text{trans}}(t)\big[1 + \mathcal{O}(T/T_K(\sigma)) + \mathcal{O}(r(\sigma))\big],$$
$$(4.1.84)$$

where the transient term $\tilde{f}_{\text{trans}}(t)$ behaves like $f_{\text{trans}}(t|0)$ and $Q_{\lambda T}$ is a periodic modulation described by the 1-periodic function

$$Q_{\lambda T}(t) = \frac{1}{T}\int_0^T P_{\lambda T}\big(\lambda(tT+s) - |\log \sigma|\big) P_{\lambda T}\big(\lambda(s-T/2) - |\log \sigma|\big) \,ds$$
$$= \left(\frac{\lambda T}{2}\right)^2 \sum_{k=-\infty}^{\infty} \int_{-\infty}^{\infty} A\big(\lambda T(t+u-k)\big) A\big(\lambda T(u-1/2)\big) \,du$$
$$= \frac{\lambda T}{2} \sum_{k=-\infty}^{\infty} \frac{1}{\cosh^2(\lambda T(t+1/2-k))}. \quad (4.1.85)$$

After a transient phase of duration $|\log \sigma|/\lambda$, the residence-time distribution approaches a periodically modulated exponential distribution. The periodic modulation $t \mapsto Q_{\lambda T}(t/T)$ has maxima in odd multiples of $T/2$, and the larger the period, the more sharply peaked the function $s \mapsto Q_{\lambda T}(s)$ becomes, so that one recovers the strongly localised residence times of the quasistatic regime.

An important aspect of the expression (4.1.84) of the residence-time distribution is that is controlled by two independent dimensionless parameters, λT and $T/T_K(\sigma)$. While λT governs the concentration of residence *phases*, $T/T_K(\sigma)$ determines the speed of asymptotic decay of $q(t)$. For $T \ll T_K(\sigma)$, this decay is very slow, so that sample paths have an appreciable probability to stay in the same potential well for several periods. As T approaches the Kramers time $T_K(\sigma)$, the decay becomes faster (relatively to the period), until the first peak of $q(t)$ at $T/2$ dominates.[2] In this *synchronisation regime*, the system is likely to switch wells twice per period.

---

[2] Though $\psi(t)$ departs from its expression (4.1.83) when $T/T_K(\sigma)$ is not small, one can show that this only results in a deformation of order $1/\lambda T_K(\sigma)$ of $Q_{\lambda T}$, which is very small in the regime $\sigma^2 \ll A$ considered here.

4.2 Stochastic Synchronisation: Sample-Paths Approach

Consider again the standard example (4.0.1) of an overdamped Brownian particle in a periodically forced double-well potential. When the minimal Kramers time, associated with the lowest value of the potential barrier height, is close to half the forcing period, the process may display almost regular transitions between the two potential wells, once per period in each direction. Thus, for fixed forcing amplitude $A \in (0, A_c)$, the forcing period allowing for such a synchronisation behaviour must be exponentially large in $1/\sigma^2$.

The situation is different when the driving amplitude A approaches its critical value A_c: If $A_c - A$ is allowed to decrease with the noise intensity σ, synchronisation can be observed for moderately large forcing periods. In terms of slow–fast vector fields, the situation of close-to-threshold driving amplitude corresponds to the existence of *avoided bifurcations* between the stable equilibrium branches (the bottoms of the potential wells) and the unstable equilibrium branch (the saddle). In this section, we show how the methods of Chapter 3 can be extended to such situations of avoided bifurcations, in order to yield results on the sample-path behaviour in the synchronisation regime.

4.2.1 Avoided Transcritical Bifurcation

Consider the stochastic differential equation

$$\mathrm{d}x_s = f(x_s, \varepsilon s)\,\mathrm{d}s + \sigma\,\mathrm{d}W_s \qquad (4.2.1)$$

first for the particular case of the drift term

$$f(x,t) = f(x,t,\varepsilon) = x - x^3 - A(\varepsilon)\cos t . \qquad (4.2.2)$$

We assume that the driving frequency ε is small, and that its amplitude $A(\varepsilon)$ satisfies $A(\varepsilon) = A_c - \delta(\varepsilon)$ for some $\delta = \delta(\varepsilon)$ tending to zero with ε. (Recall that $A_c = 2/3\sqrt{3}$ is the critical driving amplitude above which the potential ceases always to have two wells.) As in the previous chapters, we will study the SDE (4.2.1) on the slow timescale given by $t = \varepsilon s$.

Observe that $\partial_x f(x,t)$ vanishes in $\pm x_c$, where $x_c = 1/\sqrt{3}$, and that

$$\begin{aligned} f(x_c + y, t) &= A_c - A(\varepsilon)\cos t - \sqrt{3}y^2 - y^3 \\ &= \delta(\varepsilon) + \frac{1}{2}(A_c - \delta(\varepsilon))t^2 - \sqrt{3}y^2 + \mathcal{O}(t^4) + \mathcal{O}(y^3) . \end{aligned} \qquad (4.2.3)$$

Near the point $(x_c, 0)$, the drift term vanishes for $\sqrt{3}y^2 \simeq \delta(\varepsilon) + const\ t^2$.

- If $\delta = 0$, we are in the situation of the transcritical bifurcation discussed in Section 3.5.1. We know (see in particular Fig. 3.13) that in the weak-noise regime $\sigma \ll \varepsilon^{3/4}$, sample paths are unlikely to cross the saddle, while for $\sigma \gg \varepsilon^{3/4}$, they are likely to cross it, at a time of order $-\sigma^{2/3}$.

- If $\delta > 0$, the equilibrium branches of f no longer meet, but are separated by a minimal distance of order $\delta^{1/2}$. As a consequence, one expects that a larger threshold noise intensity $\sigma_c = \sigma_c(\varepsilon)$ is necessary to make transitions likely. It will turn out that σ_c scales like $(\delta \vee \varepsilon)^{3/4}$.

We will now consider the SDE (4.2.1) for more general drift terms, satisfying the following conditions.[3]

Assumption 4.2.1 (Avoided transcritical bifurcation).

- Domain and differentiability: $f \in \mathcal{C}^3(\mathcal{D}, \mathbb{R})$, where $\mathcal{D} = (-d, d) \times \mathbb{R}$, with $d > 0$. We further assume that f and all its partial derivatives up to order 3 are uniformly bounded in $\mathcal{D}$ by a constant M.
- Periodicity: There is a $T > 0$ such that $f(x, t+T) = f(x,t)$ for all $(x,t) \in \mathcal{D}$.
- Equilibrium branches: There are continuous functions $x_-^\star(t) < x_0^\star(t) < x_+^\star(t)$ from $\mathbb{R}$ to $(-d, d)$ such that $f(x, t) = 0$ in $\mathcal{D}$ if and only if $x = x_i^\star(t)$, with $i \in \{-, 0, +\}$.[4]
- Stability: The equilibrium branches $x_\pm^\star(t)$ are stable and the equilibrium branch $x_0^\star(t)$ is unstable, that is, for all $t \in \mathbb{R}$,
$$a_\pm^\star(t) := \partial_x f(x_\pm^\star(t), t) < 0 , \qquad (4.2.4)$$
$$a_0^\star(t) := \partial_x f(x_0^\star(t), t) > 0 .$$
- Avoided bifurcation at $t = 0$: There exists $x_c \in (-d, d)$ such that
$$\partial_{xx} f(x_c, 0) < 0 ,$$
$$\partial_x f(x_c, t) \leqslant \mathcal{O}(t^2) , \qquad (4.2.5)$$
$$f(x_c, t) = \delta + a_1 t^2 + \mathcal{O}(t^3) ,$$
where $a_1 > 0$ and $\partial_{xx} f(x_c, 0)$ are fixed (of order 1), while $\delta = \delta(\varepsilon) \geqslant 0$ tends to zero as $\varepsilon \to 0$.
- Avoided bifurcation at $t = t_c$: Similar relations as (4.2.5), but with opposite signs, imply the existence of an avoided bifurcation near a point (x_c', t_c), where $t_c \neq 0$.
- Absence of other avoided bifurcations: There is a constant $c_0 > 0$ such that $x_+^\star(t) - x_0^\star(t)$ and $x_0^\star(t) - x_-^\star(t)$, as well as the linearisations (4.2.4), are bounded away from 0 if $t \pmod{T}$ is not in a c_0-neighbourhood of 0 or t_c. Furthermore, f is bounded away from 0 outside a c_0-neighbourhood of the three equilibrium branches.

---

[3]Part of these assumptions concern the local behaviour near the avoided bifurcation points, and are essential for the following results. Some other assumptions, on the global behaviour, such as periodicity and number of equilibrium branches, are made for convenience of the discussion, but are not needed in the proof of the theorems.

[4]If the drift term f depends not only on x and t but also on ε, we need to assume in addition that the equilibrium branches remain bounded away from $\pm d$ as ε varies.

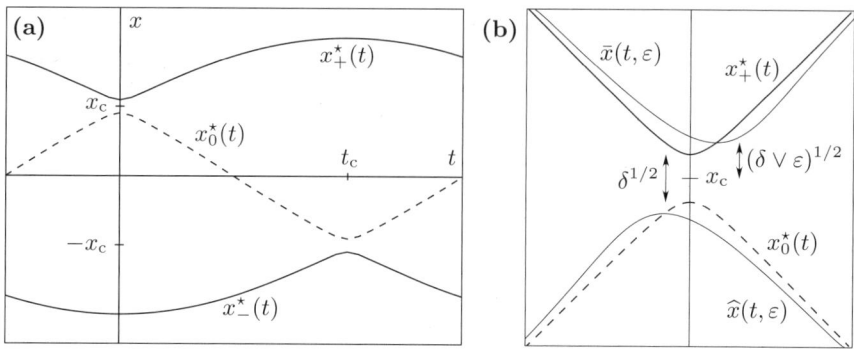

Fig. 4.6. (a) Location of equilibrium branches in Assumption 4.2.1. There are two avoided transcritical bifurcations, at $(0, x_c)$ and at $(t_c, -x_c)$. (b) Equilibrium branches and associated adiabatic solutions near the avoided transcritical bifurcation point $(0, x_c)$.

These assumptions imply in particular that

$$x_+^\star(t) - x_c \asymp x_c - x_0^\star(t) \asymp \delta^{1/2} \vee |t| \tag{4.2.6}$$

near $t = 0$ (Fig. 4.6), while

$$-a_+^\star(t) \asymp a_0^\star(t) \asymp \delta^{1/2} \vee |t| . \tag{4.2.7}$$

Similar relations hold for the branches $x_-^\star(t)$ and $x_0^\star(t)$ near $t = t_c$. For simplicity, we will further assume that each of the three equilibrium curves has exactly one minimum and one maximum per period.

We first need to establish a few properties of the adiabatic solutions $\bar{x}(t, \varepsilon)$ and $\widehat{x}(t, \varepsilon)$ of the deterministic equation $\varepsilon \dot{x} = f(x, t)$, tracking, respectively, the bottom $x_+^\star(t)$ of the right-hand potential well, and the saddle at $x_0^\star(t)$. These particular solutions can be chosen to be T-periodic (this is easily proved by studying qualitative properties of the system's Poincaré map).

Proposition 4.2.2. *The solution $\bar{x}(t, \varepsilon)$ crosses $x_+^\star(t)$ exactly twice during each period. There is a constant $c_1 > 0$ such that*

$$\bar{x}(t, \varepsilon) - x_+^\star(t) \asymp \begin{cases} \dfrac{\varepsilon}{|t|} & \text{for } -c_0 \leqslant t \leqslant -c_1(\delta \vee \varepsilon)^{1/2}, \\ -\dfrac{\varepsilon}{|t|} & \text{for } c_1(\delta \vee \varepsilon)^{1/2} \leqslant t \leqslant c_0, \end{cases} \tag{4.2.8}$$

and thus $\bar{x}(t, \varepsilon) - x_+^\star(t) \asymp |t|$ on these intervals. Furthermore,

$$\bar{x}(t, \varepsilon) - x_c \asymp (\delta \vee \varepsilon)^{1/2} \quad \text{for } |t| \leqslant c_1(\delta \vee \varepsilon)^{1/2} . \tag{4.2.9}$$

Otherwise, that is, if $|t \pmod{T}| > c_1$, $\bar{x}(t, \varepsilon) = x_+^\star(t) + \mathcal{O}(\varepsilon)$.

The linearisation of f at $\bar{x}(t,\varepsilon)$ satisfies

$$\bar{a}(t) := \partial_x f(\bar{x}(t,\varepsilon),t) \asymp -\bigl(t^2 \vee \delta \vee \varepsilon\bigr)^{1/2} \qquad (4.2.10)$$

for $|t| \leqslant c_0$, and is ε-close to $a_+^\star(t)$ otherwise.

The solution $\widehat{x}(t,\varepsilon)$ crosses $x_0^\star(t)$ exactly twice in each period. It satisfies analogous relations as above, with opposite signs, when compared to $x_0^\star(t)$.

Proof.

- The assertion on the number of crossings follows from the fact that deterministic solutions are increasing when lying between $x_0^\star(t)$ and $x_+^\star(t)$, and decreasing when lying above $x_+^\star(t)$. Thus in a given period, any solution can cross $x_+^\star(t)$ at most once for increasing $x_+^\star(t)$, and once for decreasing $x_+^\star(t)$.
- The first relation in (4.2.8) is proved exactly as in Theorem 2.2.9, by analysing the behaviour of $y = x - x_+^\star(t)$. The proof of the second relation is similar, once the order of $\bar{x}(c_1(\delta \vee \varepsilon)^{1/2}, \varepsilon)$ is known.
- For $\delta \leqslant \varepsilon$, (4.2.9) can be established by carrying out the space–time scaling $x - x_c = (\gamma^3 \varepsilon / a_1)^{1/2} z$ and $t = (\gamma \varepsilon / a_1)^{1/2} s$, with $\gamma = (2a_1/|\partial_{xx} f(x_c, 0)|)^{1/2}$, which yields a perturbation of order $\varepsilon^{1/2}$ of the Riccati equation

$$\frac{dz}{ds} = \frac{\delta}{\gamma \varepsilon} + s^2 - z^2 \,. \qquad (4.2.11)$$

One shows that solutions of this equation stay of order 1 for s of order 1, implying that $\bar{x}(t,\varepsilon)$ stays of order $\varepsilon^{1/2}$ for s of order $\varepsilon^{1/2}$. Note that this also implies that $\bar{x}(t,\varepsilon)$ has to cross $x_+^\star(t)$ at a time of order $\varepsilon^{1/2}$.
- For $\delta > \varepsilon$, one can proceed as in the proof of Tihonov's theorem to show that $\bar{x}(t,\varepsilon)$ tracks $x_+^\star(t)$ adiabatically at a distance of order $\varepsilon/\delta^{1/2}$ at most, which is smaller than $\delta^{1/2}$.
- The assertion on $\bar{a}(t)$ follows from the fact that $\partial_x f(x_c + y, t) \asymp -y$ near $(y, t) = (0, 0)$ because of the assumption (4.2.5).
- Finally, the properties of $\widehat{x}(t,\varepsilon)$ are proved in the same way by considering the time-reversed equation, in which stable and unstable branches are interchanged. □

4.2.2 Weak-Noise Regime

We return now to the dynamics in the presence of noise. Equation (4.2.1) can be rewritten in slow time $t = \varepsilon s$ as

$$dx_t = \frac{1}{\varepsilon} f(x_t, t)\,dt + \frac{\sigma}{\sqrt{\varepsilon}}\,dW_t \,. \qquad (4.2.12)$$

For sufficiently weak noise, we expect sample paths starting at some time $t_0 \in [-T+c_0, -c_0]$ in the right-hand potential well to remain close to $\bar{x}(t,\varepsilon)$. In perfect analogy with all stable situations discussed in Chapter 3, we introduce the function

$$\zeta(t) = \zeta(t_0)e^{2\bar{a}(t,t_0)/\varepsilon} + \frac{1}{\varepsilon}\int_{t_0}^{t} e^{2\bar{a}(t,s)/\varepsilon}\,ds\,, \qquad (4.2.13)$$

where $\bar{a}(t,s) = \int_s^t \bar{a}(u)\,du$ denotes the accumulated curvature over the time interval $[s,t]$, measured at the position of the deterministic particle at $\bar{x}(t,\varepsilon)$. We will use $\zeta(t)$ to characterise the deviation of typical sample paths from $\bar{x}(t,\varepsilon)$. Note that one can choose the initial condition $\zeta(t_0)$ in such a way that $\zeta(t)$ is periodic, which yields a convenient extension of the definition of $\zeta(t)$ to all times $t \in \mathbb{R}$. Using Lemma 2.2.8 and (4.2.10), one obtains

$$\zeta(t) \asymp \frac{1}{|\bar{a}(t)|} \asymp \frac{1}{(t^2 \vee \delta \vee \varepsilon)^{1/2}} \qquad (4.2.14)$$

for $|t| \leqslant c_0$. Also note that, otherwise, $\zeta(t)$ is of order 1. Hence the set

$$\mathcal{B}(h) = \{(x,t)\colon t \in \mathbb{R},\ |x - \bar{x}(t,\varepsilon)| < h\sqrt{\zeta(t)}\} \qquad (4.2.15)$$

has a maximal width of order $h/(\delta \vee \varepsilon)^{1/4}$, reached near multiples of the forcing period T. The function $\hat{\zeta}(t) = \sup_{t_0 \leqslant s \leqslant t} \zeta(s)$ first grows like $1/|t|$ for $t_0 \leqslant t \leqslant -c_1(\delta \vee \varepsilon)^{1/2}$, and then stays of constant order $1/(\delta \vee \varepsilon)^{1/2}$ for all subsequent times.

The following analogue of Theorems 3.3.3 and 3.5.1 is proved in exactly the same manner as these two results.

Theorem 4.2.3 (Stochastic avoided transcritical bifurcation – stable phase). *Let $x_0 = \bar{x}(t_0,\varepsilon)$. There exist constants $h_0, r_0, c, c_2 > 0$ such that, for all $h \leqslant h_0 \hat{\zeta}(t)^{-3/2}$ and $t \in [t_0, T-c_0]$ satisfying*

$$r(h/\sigma,\varepsilon) := \frac{\sigma}{h} + \frac{1}{\varepsilon}e^{-ch^2/\sigma^2} \leqslant r_0\,, \qquad (4.2.16)$$

one has

$$\mathcal{C}_{h/\sigma}(t,\varepsilon)e^{-\kappa_- h^2/2\sigma^2} \leqslant \mathbb{P}^{t_0,x_0}\{\tau_{\mathcal{B}(h)} < t\} \leqslant \mathcal{C}_{h/\sigma}(t,\varepsilon)e^{-\kappa_+ h^2/2\sigma^2}\,, \qquad (4.2.17)$$

where the exponents $\kappa_\pm$ are given by

$$\begin{aligned}\kappa_+ &= 1 - c_2 h\hat{\zeta}(t)^{3/2}\,,\\ \kappa_- &= 1 + c_2 h\hat{\zeta}(t)^{3/2}\,,\end{aligned} \qquad (4.2.18)$$

and the prefactor satisfies

$$\mathcal{C}_{h/\sigma}(t,\varepsilon) = \sqrt{\frac{2}{\pi}\frac{|\bar{a}(t,t_0)|}{\varepsilon}}\frac{h}{\sigma} \qquad (4.2.19)$$
$$\times \left[1 + \mathcal{O}\!\left(\frac{\varepsilon[|\log\varepsilon| + \log(1+h/\sigma)]}{|\bar{a}(t,t_0)|} + r(h/\sigma,\varepsilon) + h\hat{\zeta}(t)^{3/2}\right)\right].$$

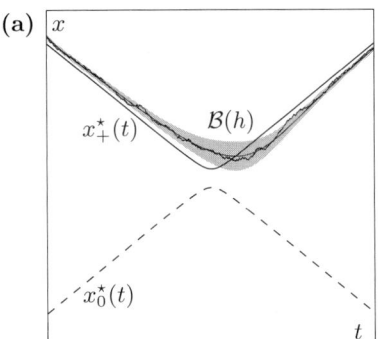 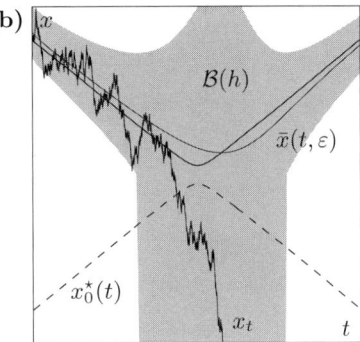

Fig. 4.7. Behaviour of sample paths near a dynamic avoided transcritical bifurcation. (a) In the weak-noise regime $\sigma < (\delta \vee \varepsilon)^{3/4}$, paths remain concentrated in the shaded set $\mathcal{B}(h)$ centred in the deterministic solution; (b) in the strong-noise regime $\sigma > (\delta \vee \varepsilon)^{3/4}$, paths are likely to overcome the unstable solution some time before the bifurcation.

As in the particular case $\delta = 0$, one can distinguish between two regime:

- *Regime I:* $\sigma < \sigma_{\rm c} = (\delta \vee \varepsilon)^{3/4}$. In this case, the theorem can be applied for all times $t \leqslant T - c_0$, and shows that paths are likely to remain in $\mathcal{B}(h)$ whenever $\sigma \ll h \leqslant h_0 \sigma_{\rm c}$. The maximal spreading of typical paths has order $\sigma/(\delta \vee \varepsilon)^{1/4}$ (Fig. 4.7a). In particular, the probability that paths cross the saddle during one period is bounded by

$$C_{h_0\sigma_c/\sigma}(t,\varepsilon)e^{-\mathrm{const}\,\sigma_c^2/\sigma^2}. \tag{4.2.20}$$

- *Regime II:* $\sigma \geqslant \sigma_{\rm c} = (\delta \vee \varepsilon)^{3/4}$. In this case, the theorem can only be applied up to times of order $-\sigma^{2/3}$, when the spreading of paths reaches order $\sigma^{2/3}$, so that paths may reach the saddle with appreciable probability (Fig. 4.7b).

For $\delta > \varepsilon$, the numerator $\sigma_{\rm c}^2 = \delta^{3/2}$ in the exponent of the crossing probability indeed corresponds to the minimal potential barrier height, since $x_+^\star(0) - x_0^\star(0) \asymp \delta^{1/2}$, and the potential $U(x,t)$, associated with the drift term $f(x,t)$, satisfies

$$U(x_{\rm c} + y, t) - U(x_{\rm c}, t) \asymp y^3 - \delta y \tag{4.2.21}$$

near $(y,t) = (0,0)$, as a consequence of (4.2.5). Thus the driving is sufficiently slow for the crossing probability to be determined by the (static) Kramers rate associated to the minimal barrier height.

On the other hand, for all $\delta < \varepsilon$, the behaviour is the same as for $\delta = 0$. This is a dynamical effect: The system behaves as if there were an effective potential barrier of height $\varepsilon^{3/2}$, because the driving is not slow enough for the particle to take full advantage of the barrier becoming low. This effect is already present in the behaviour of the deterministic particle which approaches the saddle not closer than order $\varepsilon^{1/2}$, cf. (4.2.9).

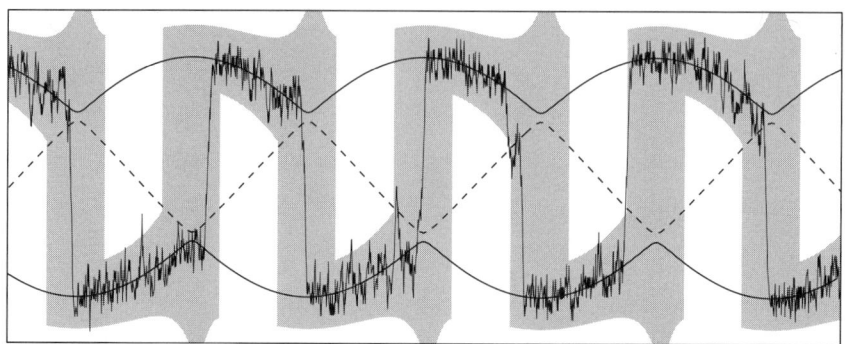

Fig. 4.8. A sample path of Equation (4.2.1) for the drift term (4.2.2) in the strong-noise regime $\sigma > (\delta \vee \varepsilon)^{3/4}$. The path switches from one potential well to the other during time windows of width $\sigma^{2/3}$ around each avoided transcritical bifurcation point.

4.2.3 Synchronisation Regime

We now complete the analysis of the strong-noise regime $\sigma \geqslant \sigma_c = (\delta \vee \varepsilon)^{3/4}$. Theorem 4.2.3 shows that paths starting in the right-hand well are unlikely to cross the saddle before times of order $-\sigma^{2/3}$. The study of the subsequent behaviour is similar to the analysis of the strong-noise regime for the saddle–node bifurcation, see Section 3.3.2. We can introduce constants d_0, d_1, d_2 such that $d_0 < d_1 < x_c < d_2 < d$, and

$$\begin{aligned} f(x,t) &\asymp -1 & &\text{for } d_0 \leqslant x \leqslant d_1 \text{ and } |t| \leqslant c_0, \\ \partial_{xx} f(x,t) &\leqslant 0 & &\text{for } d_1 \leqslant x \leqslant d_2 \text{ and } |t| \leqslant c_0. \end{aligned} \qquad (4.2.22)$$

Then the following result is proved in exactly the same way as Theorem 3.3.4.

Theorem 4.2.4 (Stochastic avoided transcritical bifurcation – the strong-noise regime). *Let $t_0 \in [-T + c_0, -c_0]$ be fixed, and $x_0 = \bar{x}(t_0, \varepsilon)$. There exist constants $r_0, c, c_3 > 0$ such that, whenever $t \in [t_1, 0] = [-c_3 \sigma^{2/3}, 0]$ and h satisfy the conditions*

$$\bar{x}(s,\varepsilon) + h\sqrt{\zeta(s)} \leqslant d_2 \qquad \forall s \in [t_0, t], \qquad (4.2.23)$$

and

$$r(h/\sigma, \varepsilon) := \frac{\sigma}{h} + \frac{1}{\varepsilon} e^{-ch^2/\sigma^2} \leqslant r_0, \qquad (4.2.24)$$

there is a constant $\kappa > 0$ such that

$$\mathbb{P}^{t_0, x_0}\{x_s > d_0 \; \forall s \in [t_0, t]\} \leqslant \frac{3}{2} \exp\left\{-\kappa \frac{|\bar{\alpha}(t, t_1)|}{\varepsilon(|\log \sigma| \vee \log h/\sigma)}\right\} + C_{h/\sigma}(t,\varepsilon) e^{-h^2/2\sigma^2}, \qquad (4.2.25)$$

where

$$C_{h/\sigma}(t,\varepsilon) = \sqrt{\frac{2}{\pi}\frac{|\bar{a}(t,t_0)|}{\varepsilon}\frac{h}{\sigma}} \qquad (4.2.26)$$

$$\times \left[1 + \mathcal{O}\left(\frac{\varepsilon[|\log \varepsilon| + \log(1+h/\sigma)]}{|\bar{a}(t,t_0)|} + r(h/\sigma,\varepsilon)\right)\right].$$

The numerator in the exponent in (4.2.25) behaves like $\sigma^{2/3}(t-t_1)$, and is thus of order $\sigma^{4/3}$ as soon as $t - t_1$ reaches order $\sigma^{2/3}$. Therefore, the probability not to cross the saddle is of order

$$\exp\left\{-\text{const}\,\frac{\sigma^{4/3}}{\varepsilon|\log \sigma|}\right\} + \frac{\text{const}}{\varepsilon}\mathrm{e}^{-\text{const}/\sigma^{4/3}}. \qquad (4.2.27)$$

The second term on the right-hand side is due to the possibility that paths escape through the upper boundary of $\mathcal{D}$; generically, Condition (4.2.23) only allows h to take values up to order $\sigma^{1/3}$. In special cases, like for the drift term (4.2.2), one can take arbitrarily large d_2, and the second term can be neglected.

Once x_t has reached d_0, one can apply Theorem 3.1.11 to show that x_t will reach an h-neighbourhood of the bottom of the left-hand potential well in a time of order $\varepsilon|\log h|$. From there on, the Markov property allows one to restart the process in the left-hand well, and one can use a similar description for the dynamics near $x_-^\star(t)$, until the next avoided bifurcation occurs at time t_c. In particular, x_t is likely to jump back into the right-hand well at a time of order $\sigma^{2/3}$ before t_c. As a result, sample paths jump back and forth between the potential wells during time windows of width $\sigma^{2/3}$ around times kT and $t_c + kT$ (Fig. 4.8).

4.2.4 Symmetric Case

The techniques applied to the study of bistable systems with avoided saddle–node bifurcations can also be applied to periodically forced systems with other types of avoided bifurcations. Consider for instance the SDE

$$\mathrm{d}x_t = \frac{1}{\varepsilon}\left[a(t)x_t - x_t^3\right]\mathrm{d}t + \frac{\sigma}{\sqrt{\varepsilon}}\mathrm{d}W_t, \qquad (4.2.28)$$

where

$$a(t) = \delta(\varepsilon) + 1 - \cos t. \qquad (4.2.29)$$

The drift term derives from the symmetric double-well potential $U(x,t) = \frac{1}{4}x^4 - \frac{1}{2}a(t)x^2$, which has the well bottoms in $\pm a(t)^{1/2}$. The wells are separated by a barrier of height $\frac{1}{4}a(t)^2$, which becomes low periodically. One can thus view the instants of lowest barrier height as the occurrence of an *avoided pitchfork bifurcation* at the origin.

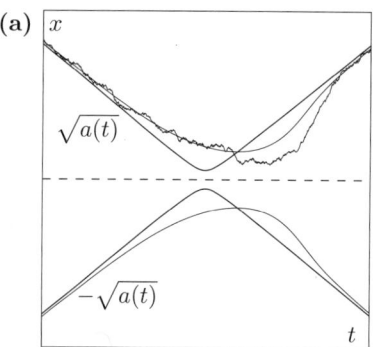 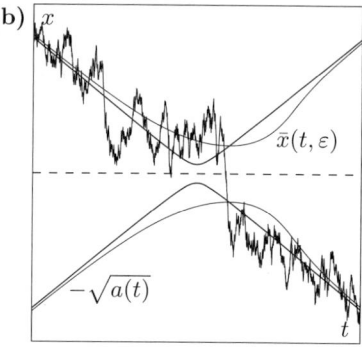

Fig. 4.9. Behaviour of sample paths near a dynamic avoided pitchfork bifurcation. (a) In the weak-noise regime $\sigma < \delta \vee \varepsilon^{2/3}$, paths remain concentrated near the deterministic solution; (b) in the strong-noise regime $\sigma > \delta \vee \varepsilon^{2/3}$, paths are likely to reach the saddle some time before the bifurcation.

In this case, one shows that the deterministic adiabatic solution $\bar{x}(t,\varepsilon)$ tracking the right-hand potential well behaves like $|t| \vee \delta^{1/2} \vee \varepsilon^{1/3}$ near $t = 0$. The set $\mathcal{B}(h)$, defined as in (4.2.15), has width $h\zeta(t)^{1/2}$, where $\zeta(t)$ again relates to the linearisation of the drift term, and behaves as

$$\zeta(t) \asymp \frac{1}{t^2 \vee \delta \vee \varepsilon^{2/3}} \qquad (4.2.30)$$

near $t = 0$. Thus its maximum process $\hat\zeta(t) = \sup_{t_0 \leqslant s \leqslant t} \zeta(s)$ first grows like $1/t^2$ for $t_0 \leqslant t \leqslant -(\delta^{1/2} \vee \varepsilon^{1/3})$, and then stays of constant order $1/(\delta \vee \varepsilon^{2/3})$ for all subsequent times.

The analogue of Theorem 4.2.3 then states that, for $h \leqslant h_0 \hat\zeta(t)^{-1}$,

$$C_{h/\sigma}(t,\varepsilon) e^{-\kappa_- h^2/2\sigma^2} \leqslant \mathbb{P}^{t_0,x_0}\{\tau_{\mathcal{B}(h)} < t\} \leqslant C_{h/\sigma}(t,\varepsilon) e^{-\kappa_+ h^2/2\sigma^2} \qquad (4.2.31)$$

holds with $\kappa_\pm = 1 \mp c_2 h \hat\zeta(t)$, and a similar prefactor $C_{h/\sigma}(t,\varepsilon)$.[5] There are thus two regimes to be considered:

- *Regime I:* $\sigma < \sigma_c = \delta \vee \varepsilon^{2/3}$. In this case, the theorem can be applied for all times t from a neighbourhood of the avoided bifurcation at $t = 0$, and shows that paths are likely to remain in $\mathcal{B}(h)$ whenever $\sigma \ll h \leqslant \sigma_c$. The maximal spreading of typical paths has order $\sigma/(\delta^{1/2} \vee \varepsilon^{1/3})$ (Fig. 4.9a). In particular, the probability that paths cross the saddle is bounded by

$$C_{h_0 \sigma_c/\sigma}(t,\varepsilon) e^{-\mathrm{const}\, \sigma_c^2/\sigma^2} \ . \qquad (4.2.32)$$

- *Regime II:* $\sigma \geqslant \sigma_c = \delta \vee \varepsilon^{2/3}$. In this case, the theorem can only be applied up to time $-\sigma^{1/2}$, when the spreading of paths reaches order $\sigma^{1/2}$, so that paths may reach the saddle with appreciable probability (Fig. 4.9b).

---

[5] The different power of $\hat\zeta(t)$ is due to the cubic form of the nonlinearity, compare Theorem 3.4.3.

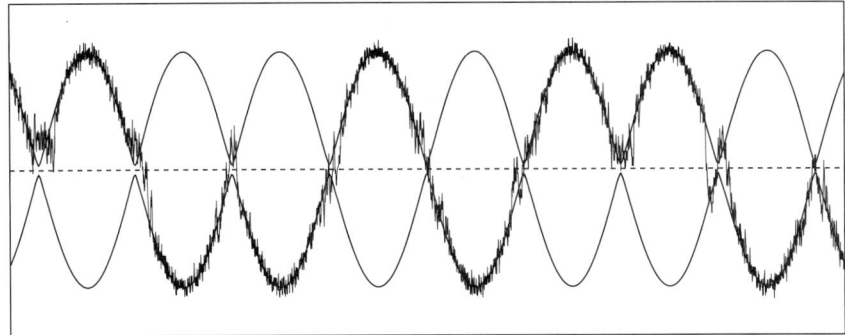

Fig. 4.10. A sample path of Equation (4.2.28) in the strong-noise regime $\sigma > \delta \vee \varepsilon^{2/3}$. The path has a probability close to $1/2$ to switch from one potential well to the other during time windows of width $\sigma^{1/2}$ around each avoided pitchfork bifurcation point.

In the strong-noise regime $\sigma \geqslant \sigma_c$, the analogue of Theorem 4.2.4 states that if $t \in [t_1, t_2] = [-c_2 \sigma^{1/2}, c_2 \sigma^{1/2}]$, then the probability for the process x_t not to reach the saddle up to time t satisfies

$$\mathbb{P}^{t_0,x_0}\{x_s > 0 \,\forall s \in [t_0,t]\} \leqslant 2\exp\left\{-\kappa\frac{|\bar{a}(t,t_1)|}{\varepsilon|\log\sigma|}\right\} + C_{h/\sigma}(t,\varepsilon)e^{-\kappa_+ h^2/2\sigma^2}. \tag{4.2.33}$$

The denominator in the first exponent is simpler as in (4.2.25), because we only consider the probability not to reach the saddle. The exponent behaves like $\sigma(t - t_1)$, and is of order $\sigma^{3/2}$ as soon as $t - t_1$ reaches order $\sigma^{1/2}$. The probability not to reach the saddle is thus of order

$$\exp\left\{-\text{const}\,\frac{\sigma^{3/2}}{\varepsilon|\log\sigma|}\right\}. \tag{4.2.34}$$

The new aspect of this symmetric situation is that once paths have reached the saddle, they have probability $1/2$ to end up in the left-hand or the right-hand potential well, once the avoided bifurcation has been passed. Paths may in fact cross the saddle several times during the time window $[t_1, t_2]$, but an analogue of Theorem 3.4.5 shows that they are likely to leave a neighbourhood of the saddle after time t_2.

As a result, in the strong-noise regime, paths choose to switch or not to switch wells, each time the potential barrier is low, according to a Bernoulli process (Fig. 4.10).

Bibliographic Comments

The mechanism of SR has been introduced by Benzi, Sutera and Vulpiani in [BSV81], and by Nicolis and Nicolis in [NN81]. There are many review articles on SR, treating theoretical approaches, applications, as well as historical aspects, e.g. [MPO94, MW95, WM95, GHJM98, WJ98, Hän02, WSB04].

The earliest theoretical approaches were given in [ET82] and [CCRSJ81]. Many spectral-theoretic results on stochastic resonance go back to the end of the eighties. The signal-to-noise ratio and power amplification have first been computed for small driving amplitude in [GMSS$^+$89, Fox89, JH89], and in [MW89] for the two-state system. The superadiabatic regime has been considered in [JH91], and the semiadiabatic regime in [Tal99, TŁ04]. The two-state Markov process has later been reconsidered in more detail in [IP02] for transition probabilities switching their value twice per period, and in [HI02] for time-continuously changing transition probabilities.

The theory of large deviations has been applied by Freidlin to the quasistatic regime in [Fre00, Fre01]. The method exposed in these papers applies to a very general class of multidimensional multistable systems, where it allows to organise the attractors in a hierarchy of cycles, which is used to determine the limiting quasideterministic dynamics. See also [HI05].

Residence-time distributions were first considered in [ET82] for the two-state process, and, e.g., in [ZMJ90, CFJ98] for continuous systems. The results presented in Section 4.1.4 have been developed in [BG04, BG05a, BG05b]. They use insights by Day on the exit problem with characteristic boundary [Day90a, Day92, Day94]. See also [GT84, GT85] for related studies of the behaviour near unstable orbits, and [MS96, MS01, LRH00a, LRH00b, DGM$^+$01] for results on the escape through an oscillating barrier, obtained by extensions of the Wentzell–Freidlin theory.

The notion of noise-induced synchronisation has been first introduced in [SNA95, NSASG98]. The sample-path results for the synchronisation regime presented in Section 4.2 have been first derived in [BG02e], although with less precise prefactors.

5
Multi-Dimensional Slow–Fast Systems

In this chapter, we turn to the analysis of the effect of noise on general systems of fully coupled, multidimensional slow–fast differential equations, of the form

$$\varepsilon \dot{x} = f(x, y) \,, \qquad (5.0.1)$$
$$\dot{y} = g(x, y) \,.$$

To allow for the possibility that several independent sources of noise act on each variable, we will perturb this system by adding position-dependent noise terms, resulting from multidimensional white noise. The intensity of the noise acting on the fast and slow components, respectively, is measured by small parameters σ and σ'. We will choose a scaling of the form

$$\mathrm{d}x_t = \frac{1}{\varepsilon} f(x_t, y_t) \, \mathrm{d}t + \frac{\sigma}{\sqrt{\varepsilon}} F(x_t, y_t) \, \mathrm{d}W_t \,, \qquad (5.0.2)$$
$$\mathrm{d}y_t = g(x_t, y_t) \, \mathrm{d}t + \sigma' G(x_t, y_t) \, \mathrm{d}W_t \,.$$

In this way, σ^2 and $(\sigma')^2$ each measure the relative importance of diffusion and drift, respectively, for the fast and slow variables. We think of $\sigma = \sigma(\varepsilon)$ and $\sigma' = \sigma'(\varepsilon)$ as being functions of ε, and we will further assume that $\rho(\varepsilon) = \sigma'(\varepsilon)/\sigma(\varepsilon)$ is bounded above as $\varepsilon \to 0$. This implies that the noise acting on the slow variable does not dominate the noise acting on the fast variable.

Compared to the one-dimensional situation examined in Chapter 3, several new difficulties have to be dealt with in general systems of the form (5.0.2):

- The multidimensional nature of x allows for more complicated asymptotic deterministic dynamics than stationary.
- The explicit dependence of $\mathrm{d}y_t$ on the fast variable x_t yields a richer slow dynamics.
- The fact that noise also acts on the slow variable effectively introduces an additional stochastic term acting on the dynamics of x_t.

144   5 Multi-Dimensional Slow–Fast Systems

As a result of these difficulties, we will not obtain an equally precise description as in the one-dimensional case. Still, our approach is based on similar considerations as in Chapter 3.

- In Section 5.1, we consider the dynamics near uniformly asymptotically stable slow manifolds. First we prove an extension of Theorem 3.1.10 on the concentration of sample paths in an appropriately constructed neighbourhood of the slow manifold. Then we discuss the possibility of reducing the dynamics to an effective equation, involving only slow variables. Finally, we give some more precise estimates on the dynamics near the slow manifold.
- Section 5.2 is dedicated to the situation arising when the associated (or frozen) system admits an asymptotically stable periodic orbit for each value of the slow variable.
- In Section 5.3, we give some results on multidimensional systems with bifurcations. We first establish concentration and reduction results near invariant manifolds, allowing, to some extent, to reduce the dynamics to an effective equation, which involves only bifurcating modes. Then we consider the effect of noise on dynamic Hopf bifurcations.

5.1 Slow Manifolds

Let $\{W_t\}_{t \geqslant 0}$ be a k-dimensional Brownian motion. In this section, we consider the slow–fast system of SDEs

$$
\begin{aligned}
\mathrm{d}x_t &= \frac{1}{\varepsilon} f(x_t, y_t) \, \mathrm{d}t + \frac{\sigma}{\sqrt{\varepsilon}} F(x_t, y_t) \, \mathrm{d}W_t , \\
\mathrm{d}y_t &= g(x_t, y_t) \, \mathrm{d}t + \sigma' G(x_t, y_t) \, \mathrm{d}W_t ,
\end{aligned}
\tag{5.1.1}
$$

with $\sigma' = \rho\sigma$, in the case where the deterministic system admits a uniformly asymptotically stable slow manifold $\mathcal{M}$. More precisely, we will make the following assumptions.

Assumption 5.1.1 (Asymptotically stable slow manifold).

- Domain and differentiability: *There are integers $n, m, k \geqslant 1$ such that $f \in \mathcal{C}^2(\mathcal{D}, \mathbb{R}^n)$, $g \in \mathcal{C}^2(\mathcal{D}, \mathbb{R}^m)$ and $F \in \mathcal{C}^1(\mathcal{D}, \mathbb{R}^{n \times k})$, $G \in \mathcal{C}^1(\mathcal{D}, \mathbb{R}^{m \times k})$, where $\mathcal{D}$ is an open subset of $\mathbb{R}^n \times \mathbb{R}^m$. We further assume that f, g, F, G and all their partial derivatives up to order 2, respectively 1, are uniformly bounded in norm in $\mathcal{D}$ by a constant M.*
- Slow manifold: *There is a connected open subset $\mathcal{D}_0 \subset \mathbb{R}^m$ and a continuous function $x^\star : \mathcal{D}_0 \to \mathbb{R}^n$ such that*

$$
\mathcal{M} = \{(x, y) \in \mathcal{D} : x = x^\star(y), y \in \mathcal{D}_0\}
\tag{5.1.2}
$$

is a slow manifold of the deterministic system, that is, $(x^\star(y), y) \in \mathcal{D}$ and $f(x^\star(y), y) = 0$ for all $y \in \mathcal{D}_0$.

- Stability: *The slow manifold is uniformly asymptotically stable, that is, all eigenvalues of the Jacobian matrix*

$$A^\star(y) = \partial_x f(x^\star(y), y) \tag{5.1.3}$$

have negative real parts, uniformly bounded away from 0 for $y \in \mathcal{D}_0$.

We will also need a non-degeneracy assumption on F, which we will specify in Assumption 5.1.4 below. (It is sufficient, but not at all necessary, to assume that the diffusion matrix FF^{T} be positive definite.) Recall that under the above assumptions, Fenichel's theorem (Theorem 2.1.8) ensures the existence of an invariant manifold

$$\mathcal{M}_\varepsilon = \{(x, y) \in \mathcal{D} \colon x = \bar{x}(y, \varepsilon), y \in \mathcal{D}_0\}, \tag{5.1.4}$$

where $\bar{x}(y, \varepsilon) = x^\star(y) + \mathcal{O}(\varepsilon)$.

We will start, in Section 5.1.1, by constructing a neighbourhood $\mathcal{B}(h)$ of $\mathcal{M}_\varepsilon$, extending the definition of Section 3.1.1, and proving a generalisation of Theorem 3.1.10 on concentration of sample paths in this set. Section 5.1.2 contains the relatively long proof of this result. Section 5.1.3 considers the problem of reducing the dynamics near $\mathcal{M}_\varepsilon$ to an effective m-dimensional equation, and Section 5.1.4 gives more precise results on the dynamics in $\mathcal{B}(h)$.

5.1.1 Concentration of Sample Paths

In order to define the domain of concentration $\mathcal{B}(h)$, we first consider a linear approximation of the system (5.1.1) near the adiabatic manifold $\mathcal{M}_\varepsilon$. To this end, we use Itô's formula (cf. Theorem A.2.7) to obtain that the deviation $\xi_t = x_t - \bar{x}(y_t, \varepsilon)$ of sample paths from the adiabatic manifold satisfies the SDE

$$\begin{aligned}
\mathrm{d}\xi_t &= \mathrm{d}x_t - \partial_y \bar{x}(y_t, \varepsilon) \, \mathrm{d}y_t + \mathcal{O}((\sigma')^2) \, \mathrm{d}t \\
&= \frac{1}{\varepsilon}\big[f(\bar{x}(y_t,\varepsilon) + \xi_t, y_t) - \varepsilon \partial_y \bar{x}(y_t,\varepsilon) g(\bar{x}(y_t,\varepsilon) + \xi_t, y_t) + \mathcal{O}(\varepsilon(\sigma')^2)\big] \, \mathrm{d}t \\
&\quad + \frac{\sigma}{\sqrt{\varepsilon}}\big[F(\bar{x}(y_t,\varepsilon) + \xi_t, y_t) - \rho\sqrt{\varepsilon} \partial_y \bar{x}(y_t,\varepsilon) G(\bar{x}(y_t,\varepsilon) + \xi_t, y_t)\big] \, \mathrm{d}W_t \, .
\end{aligned} \tag{5.1.5}$$

Note that the new drift term vanishes when $\xi_t = 0$ and $\sigma' = 0$, because of the equation (2.1.26) satisfied by $\bar{x}(y, \varepsilon)$.

The linear approximation is obtained by considering (5.1.5) to lowest order in ξ_t, neglecting the Itô-term $\mathcal{O}(\varepsilon(\sigma')^2)$, and replacing y_t by its deterministic counterpart y_t^{det}. This yields the system

$$\begin{aligned}
\mathrm{d}\xi_t^0 &= \frac{1}{\varepsilon} A(y_t^{\mathrm{det}}, \varepsilon)\xi_t^0 \, \mathrm{d}t + \frac{\sigma}{\sqrt{\varepsilon}} F_0(y_t^{\mathrm{det}}, \varepsilon) \, \mathrm{d}W_t \, , \\
\mathrm{d}y_t^{\mathrm{det}} &= g(\bar{x}(y_t^{\mathrm{det}}, \varepsilon), y_t^{\mathrm{det}}) \, \mathrm{d}t \, ,
\end{aligned} \tag{5.1.6}$$

where we have introduced the matrices

$$A(y,\varepsilon) = \partial_x f(\bar{x}(y,\varepsilon), y) - \varepsilon \partial_y \bar{x}(y,\varepsilon) \partial_x g(\bar{x}(y,\varepsilon), y) ,$$
$$F_0(y,\varepsilon) = F(\bar{x}(y,\varepsilon), y) - \rho\sqrt{\varepsilon} \partial_y \bar{x}(y,\varepsilon) G(\bar{x}(y,\varepsilon), y) . \tag{5.1.7}$$

Note, in particular, that $A(y,0) = A^\star(y)$ and $F_0(y,0) = F(x^\star(y), y)$.

We now analyse the solution of the system (5.1.6) for a given initial condition $(\xi_0^0, y_0^{\mathrm{det}}) = (0, y_0^{\mathrm{det}})$. It can be represented as the Itô integral

$$\xi_t^0 = \frac{\sigma}{\sqrt{\varepsilon}} \int_0^t U(t,s) F_0(y_s^{\mathrm{det}}, \varepsilon) \, \mathrm{d}W_s , \tag{5.1.8}$$

where $U(t,s)$ denotes the principal solution of the homogeneous linear system $\varepsilon \dot{\xi} = A(y_t^{\mathrm{det}}, \varepsilon) \xi$. For fixed time t, ξ_t^0 is a Gaussian random variable of zero mean and covariance matrix

$$\mathrm{Cov}(\xi_t^0) = \frac{\sigma^2}{\varepsilon} \int_0^t U(t,s) F_0(y_s^{\mathrm{det}}, \varepsilon) F_0(y_s^{\mathrm{det}}, \varepsilon)^{\mathrm{T}} U(t,s)^{\mathrm{T}} \, \mathrm{d}s . \tag{5.1.9}$$

The key observation is that $X_t = \sigma^{-2} \mathrm{Cov}(\xi_t^0)$ obeys the slow–fast ODE

$$\varepsilon \dot{X} = A(y,\varepsilon) X + X A(y,\varepsilon)^{\mathrm{T}} + F_0(y,\varepsilon) F_0(y,\varepsilon)^{\mathrm{T}} ,$$
$$\dot{y} = g(\bar{x}(y,\varepsilon), y) . \tag{5.1.10}$$

This system admits a slow manifold of equation $X = X^\star(y,\varepsilon)$, where $X^\star(y,\varepsilon)$ is a solution of the Lyapunov equation[1]

$$A(y,\varepsilon) X + X A(y,\varepsilon)^{\mathrm{T}} + F_0(y,\varepsilon) F_0(y,\varepsilon)^{\mathrm{T}} = 0 . \tag{5.1.11}$$

Let us recall a fundamental result on such equations.

Lemma 5.1.2. *Let A and B be square matrices of respective dimension n and m. Denote their eigenvalues by $a_1, \ldots, a_n$ and $b_1, \ldots, b_m$. Define a linear operator $L : \mathbb{R}^{n \times m} \to \mathbb{R}^{n \times m}$ by*

$$L(X) = AX + XB . \tag{5.1.12}$$

Then the nm eigenvalues of L are given by $\{a_i + b_j\}_{i=1,\ldots,n, j=1,\ldots,m}$, and thus L is invertible if and only if A and $-B$ have no common eigenvalue.

If, moreover, all a_i and b_j have negative real parts, then for any matrix $C \in \mathbb{R}^{n \times m}$, the unique solution of the equation $AX + XB + C = 0$ is given by

$$X = \int_0^\infty \mathrm{e}^{As} C \mathrm{e}^{Bs} \, \mathrm{d}s . \tag{5.1.13}$$

---

[1] Equivalently, one may consider the slow manifold to be given by the solution $X^\star(y)$ of Equation (5.1.11) with $\varepsilon = 0$. This will yield the same invariant manifold $\bar{X}(t,\varepsilon) = X^\star(y) + \mathcal{O}(\varepsilon)$.

Proof. First note that the integral (5.1.13) exists whenever all a_i and b_j have negative real parts. The fact that X solves the equation $AX + XB + C = 0$ is checked by a direct computation, see [Bel60, p. 175]. The assertion on the eigenvalues of L follows from the theory of Kronecker products, see [Bel60, Chapter 12, in particular Theorem 4]. □

Assumption 5.1.1 ensures that the eigenvalues of $A(y, \varepsilon)$ have strictly negative real parts for sufficiently small ε. Lemma 5.1.2 thus shows that (5.1.11) admits a unique solution, given by the symmetric matrix

$$X^\star(y,\varepsilon) = \int_0^\infty e^{sA(y,\varepsilon)} F_0(y,\varepsilon) F_0(y,\varepsilon)^\mathrm{T} e^{sA(y,\varepsilon)^\mathrm{T}} \, ds \, . \tag{5.1.14}$$

This expression is called a *controllability Grammian* in control theory. Furthermore, the lemma shows that $\{X^\star(y,\varepsilon)\colon y \in \mathcal{D}_0\}$ is a uniformly asymptotically stable slow manifold. Thus Fenichel's theorem implies the existence of an invariant manifold $\{\bar{X}(y,\varepsilon)\colon y \in \mathcal{D}_0\}$, with

$$\bar{X}(y,\varepsilon) = X^\star(y,\varepsilon) + \mathcal{O}(\varepsilon) \, . \tag{5.1.15}$$

In fact, the remainder has the form $\varepsilon X_1(y,\varepsilon) + \mathcal{O}(\varepsilon^2)$, where $X_1(y,\varepsilon)$ obeys the Lyapunov equation

$$A(y,\varepsilon)X_1 + X_1 A(y,\varepsilon)^\mathrm{T} = \dot{X}^\star(y,\varepsilon) = \partial_y X^\star(y,\varepsilon) g(\bar{x}(y,\varepsilon), y) \, . \tag{5.1.16}$$

For sufficiently differentiable coefficients, the adiabatic manifold can be further expanded into a series in ε, each term of which satisfies a similar Lyapunov equation.

Next we invoke a result from control theory to give a condition under which $\bar{X}(y,\varepsilon)$ is invertible.

Lemma 5.1.3. *The Grammian $X^\star(y,\varepsilon)$ is invertible if and only if the matrix*

$$\begin{bmatrix} F_0(y,\varepsilon) & A(y,\varepsilon)F_0(y,\varepsilon) & \ldots & A(y,\varepsilon)^{n-1} F_0(y,\varepsilon) \end{bmatrix} \in \mathbb{R}^{n \times nk} \tag{5.1.17}$$

has full rank, that is, the pair $(A(y,\varepsilon), F_0(y,\varepsilon))$ is controllable.

Proof. The integrand in (5.1.14) being positive semi-definite, $X^\star(y,\varepsilon)$ is singular if and only if there exists a row vector $w \neq 0$ such that

$$w e^{sA(y,\varepsilon)} F_0(y,\varepsilon) = 0 \qquad \forall s \geqslant 0 \, . \tag{5.1.18}$$

Evaluating successive derivatives of this relation in $s = 0$ shows, by analyticity of $s \mapsto e^{sA(y,\varepsilon)}$, that (5.1.18) is equivalent to

$$w A(y,\varepsilon)^\ell F_0(y,\varepsilon) = 0 \qquad \forall \ell = 0, 1, 2, \ldots \, . \tag{5.1.19}$$

By the Cayley–Hamilton theorem, this relation holds for all $\ell \geqslant 0$ if and only if it holds for $\ell = 0, \ldots, n-1$. This in turn is equivalent to the matrix in (5.1.17) not having full rank. □

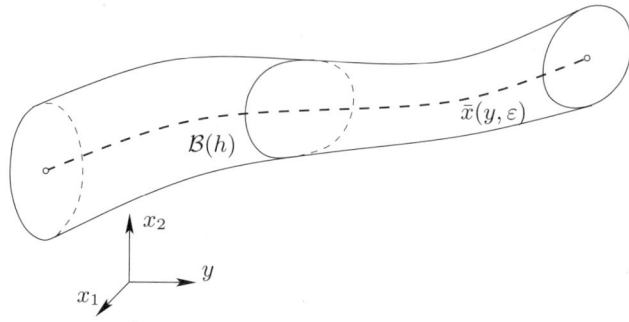

Fig. 5.1. Example of a domain of concentration $\mathcal{B}(h)$ of paths around an adiabatic manifold $\mathcal{M}_\varepsilon$, given by an adiabatic solution $\bar{x}(t,\varepsilon)$, for one slow dimension and two fast dimensions.

If $X^\star(y,\varepsilon)$ is invertible, the adiabatic manifold $\overline{X}(y,\varepsilon)$ is also invertible for sufficiently small ε. In the sequel, we shall make the following, slightly stronger non-degeneracy assumption:

Assumption 5.1.4 (Non-degeneracy of noise term). *The operator norms $\|\overline{X}(y,\varepsilon)\|$ and $\|\overline{X}(y,\varepsilon)^{-1}\|$ are uniformly bounded for $y \in \mathcal{D}_0$.*

Under this assumption, we may safely define the set

$$\mathcal{B}(h) = \left\{ (x,y) \in \mathcal{D} \colon y \in \mathcal{D}_0, \langle x - \bar{x}(y,\varepsilon), \overline{X}(y,\varepsilon)^{-1}(x - \bar{x}(y,\varepsilon)) \rangle < h^2 \right\}. \tag{5.1.20}$$

This set is a union of ellipsoids, centred in the adiabatic manifold $\mathcal{M}_\varepsilon$, whose shape is determined by the symmetric matrix $\overline{X}(y,\varepsilon)$. We offer two graphical representations of the situation. Fig. 5.1 shows a case with $n = 2$, $m = 1$, in which $\mathcal{B}(h)$ is a tube-like object surrounding an adiabatic solution of the deterministic system. Fig. 5.2 shows a case with $n = 1$, $m = 2$, where $\mathcal{B}(h)$ has the shape of a mattress of variable thickness surrounding $\mathcal{M}_\varepsilon$.

Example 5.1.5 (Gradient system). Assume that $f(x,y) = -\nabla_x U(x,y)$ derives from a potential U, and $F(x,y) = \mathbb{1}$ (with $k = n$). Then $A^\star(y)$ and $A(y,\varepsilon)$, being Hessian matrices of U, are automatically symmetric, and $A(y,\varepsilon)$ commutes with $X^\star(y,\varepsilon)$. Hence the Lyapunov equation admits the solution

$$X^\star(y,\varepsilon) = -\frac{1}{2} A(y,\varepsilon)^{-1}, \tag{5.1.21}$$

so that $X^\star(y,\varepsilon)^{-1} = -2A(y,\varepsilon)$. The (mutually orthogonal) eigenvectors of $A(y,\varepsilon)$ and its eigenvalues define respectively the principal curvature directions of the potential at y, and its principal curvatures. As a consequence, the set $\mathcal{B}(h)$ is wider in those directions in which the potential is flatter, capturing the effect that we expect larger spreading of sample paths in these directions.

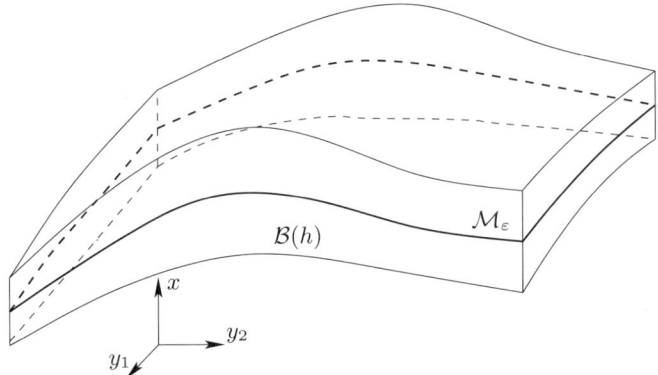

Fig. 5.2. Example of a domain of concentration $\mathcal{B}(h)$ of paths around an adiabatic manifold $\mathcal{M}_\varepsilon$, for two slow dimensions and one fast dimension.

We return now to the original system (5.1.1). For a given initial condition $(x_0, y_0) \in \mathcal{B}(h)$, we define the first-exit times

$$\tau_{\mathcal{B}(h)} = \inf\{t > 0 \colon (x_t, y_t) \notin \mathcal{B}(h)\},$$
$$\tau_{\mathcal{D}_0} = \inf\{t > 0 \colon y_t \notin \mathcal{D}_0\}. \qquad (5.1.22)$$

We can now state the main result of this section.

Theorem 5.1.6 (Multidimensional stochastic stable case). *Under Assumptions 5.1.1 and 5.1.4, there exist constants $\varepsilon_0, \Delta_0, h_0, c, c_1, L > 0$ such that the following relations hold for all $\varepsilon \leqslant \varepsilon_0$, $\Delta \leqslant \Delta_0$, $h \leqslant h_0$, all $\gamma \in (0,1)$ and all $t \geqslant 0$.*

- *Upper bound: Let $x_0 = \bar{x}(y_0, \varepsilon)$. Then*

$$\mathbb{P}^{x_0, y_0}\{\tau_{\mathcal{B}(h)} < t \wedge \tau_{\mathcal{D}_0}\} \leqslant C^+_{h/\sigma, n, m, \gamma, \Delta}(t, \varepsilon) e^{-\kappa_+ h^2/2\sigma^2}, \qquad (5.1.23)$$

where the exponent κ_+ is uniform in time and satisfies

$$\kappa_+ = \gamma\bigl[1 - c_1\bigl(h + \Delta + m\varepsilon\rho^2\sigma^2/h^2 + e^{-c/\varepsilon}/(1-\gamma)\bigr)\bigr], \qquad (5.1.24)$$

and the prefactor is given by

$$C^+_{h/\sigma, n, m, \gamma, \Delta}(t, \varepsilon) = L\frac{(1+t)^2}{\Delta\varepsilon}\bigl[(1-\gamma)^{-n} + e^{n/4} + e^{m/4}\bigr]\left(1 + \frac{h^2}{\sigma^2}\right). \qquad (5.1.25)$$

- *Lower bound: There exists a time $t_0 > 0$, independent of ε, such that for all $t > 0$,*

$$\mathbb{P}^{x_0, y_0}\{\tau_{\mathcal{B}(h)} < t\} \geqslant C^-_{h/\sigma, n, m, \Delta}(t, \varepsilon) e^{-\kappa_- h^2/2\sigma^2}, \qquad (5.1.26)$$

where the exponent κ_- is uniform in time for $t \geqslant t_0$ and satisfies

$$\kappa_- = 1 + c_1\bigl(h + e^{-c(t \wedge t_0)/\varepsilon}\bigr)\,, \qquad (5.1.27)$$

and the prefactor is given by

$$C^-_{h/\sigma,n,m,\Delta}(t,\varepsilon) = \frac{1}{L}\left[1 - \left(e^{n/4} + \frac{e^{m/4}}{\Delta\varepsilon}\right)e^{-h^2/4\sigma^2}\right]. \qquad (5.1.28)$$

- General initial condition: *There exist $\delta_0 > 0$ and $t_1 \asymp \varepsilon|\log h|$ such that, whenever $\delta \leqslant \delta_0$ and we choose an initial condition (x_0, y_0) such that $y_0 \in \mathcal{D}_0$ and $\xi_0 = x_0 - \bar{x}(y_0, \varepsilon)$ satisfies $\langle \xi_0, \bar{X}(y_0,\varepsilon)^{-1}\xi_0\rangle < \delta^2$, then, for all $t \geqslant t_2 \geqslant t_1$,*

$$\mathbb{P}^{x_0,y_0}\bigl\{\exists s \in [t_2, t] : (x_s, y_s) \notin \mathcal{B}(h)\bigr\} \leqslant C^+_{h/\sigma,n,m,\gamma,\Delta}(t,\varepsilon)e^{-\bar{\kappa}_+ + h^2/2\sigma^2}\,, \qquad (5.1.29)$$

with the same prefactor given by (5.1.25), and an exponent

$$\bar{\kappa}_+ = \gamma\bigl[1 - c_1\bigl(h + \Delta + m\varepsilon\rho^2\sigma^2/h^2 + \delta e^{-c(t_2 \wedge 1)/\varepsilon}/(1-\gamma)\bigr)\bigr]\,. \qquad (5.1.30)$$

The upper bound (5.1.23) shows that sample paths are likely to stay in $\mathcal{B}(h)$ during exponentially long time spans (roughly of the order $\varepsilon e^{h^2/2\sigma^2}$), unless the slow dynamics causes y_t to leave the set $\mathcal{D}_0$, in which the slow manifold is defined. For instance, it might drive the system towards a bifurcation point, where the slow manifold ceases to be attracting, and possibly even ceases to exist altogether. In Section 5.1.4, we shall provide more precise estimates on the deviation of sample paths from deterministic solutions, which give informations on the distribution of $\tau_{\mathcal{D}_0}$. If, on the other hand, the domain $\mathcal{D}_0$ is positively invariant under the deterministic flow, then the first-exit time $\tau_{\mathcal{D}_0}$ is typically exponentially large, and so is $\tau_{\mathcal{B}(h)}$.

The small parameters Δ and γ can be chosen arbitrarily in their domains, but optimal bounds are attained for Δ small and γ close to 1: This gives the best possible exponent, at the affordable cost of blowing up the prefactor. Since the exponents contain error terms of order h anyway, a convenient choice is $\Delta = h$, $\gamma = 1 - h$, which yields a prefactor of order $h^{-(n+1)}$ and no additional error in the exponents.

The bounds C^+ and C^- on the prefactor are certainly not optimal, in particular the time-dependence of the actual prefactor can reasonably be expected to be linear, as in the one-dimensional case. Note that an upper bound with the same exponent, but a smaller prefactor, holds for the probability that the endpoint (x_t, y_t) does not lie in $\mathcal{B}(h)$ at any fixed time t, cf. Lemma 5.1.14. The fact that C^+ grows exponentially with the spatial dimension is to be expected, as it reflects the fact that the tails of n-dimensional Gaussians only show their typical decay outside a ball of radius scaling with the square root of n.

Finally, Estimate (5.1.29) shows that if the initial condition lies in a sufficiently small neighbourhood of the adiabatic manifold, after a time of order $\varepsilon|\log h|$, typical sample paths will have reached $\mathcal{B}(h)$, and they start behaving as if they had started on the adiabatic manifold.

5.1.2 Proof of Theorem 5.1.6

The first step of the proof is to perform the change of variables

$$x_t = \bar{x}(y_t^{\text{det}} + \eta_t, \varepsilon) + \xi_t ,$$
$$y_t = y_t^{\text{det}} + \eta_t \qquad (5.1.31)$$

in the original system (5.1.1). Applying the Itô formula, and making a Taylor expansion of the resulting system, one obtains the SDE

$$d\xi_t = \frac{1}{\varepsilon}\left[A(y_t^{\text{det}})\xi_t + b(\xi_t, \eta_t, t)\right] dt + \frac{\sigma}{\sqrt{\varepsilon}}\left[F_0(y_t^{\text{det}}) + F_1(\xi_t, \eta_t, t)\right] dW_t ,$$
$$d\eta_t = \left[C(y_t^{\text{det}})\xi_t + B(y_t^{\text{det}})\eta_t + c(\xi_t, \eta_t, t)\right] dt$$
$$\qquad + \sigma'\left[G_0(y_t^{\text{det}}) + G_1(\xi_t, \eta_t, t)\right] dW_t , \qquad (5.1.32)$$

where we have omitted to indicate the ε-dependence of the functions for simplicity. The matrices $A(y) = A(y, \varepsilon)$ and $F_0(y) = F_0(y, \varepsilon)$ are those defined in (5.1.7), and

$$C(y) = C(y, \varepsilon) = \partial_x g(\bar{x}(y, \varepsilon), y) ,$$
$$B(y) = B(y, \varepsilon) = C(y)\partial_y \bar{x}(y, \varepsilon) + \partial_y g(\bar{x}(y, \varepsilon), y) , \qquad (5.1.33)$$
$$G_0(y) = G_0(y, \varepsilon) = G(\bar{x}(y, \varepsilon), y) .$$

The other functions appearing in (5.1.32) are remainders, satisfying

$$\|F_1(\xi, \eta, t)\| \leqslant M_1(\|\xi\| + \|\eta\|) ,$$
$$\|G_1(\xi, \eta, t)\| \leqslant M_1(\|\xi\| + \|\eta\|) ,$$
$$\|b(\xi, \eta, t)\| \leqslant M_2(\|\xi\|^2 + \|\xi\|\|\eta\| + m\varepsilon\rho^2\sigma^2) , \qquad (5.1.34)$$
$$\|c(\xi, \eta, t)\| \leqslant M_2(\|\xi\|^2 + \|\eta\|^2) ,$$

for some constants M_1, M_2 depending on M and the dimensions n, m. The term $m\varepsilon\rho^2\sigma^2$ is due to the contribution of the diffusion term to the new drift coefficient when applying Itô's formula.

Remark 5.1.7. The term $F_1(\xi, \eta, t)$ could be eliminated from (5.1.32) by a change of variables of the form $\tilde{\xi} = h(\xi, \eta, t)$ if one could find a function h satisfying the equation $\partial_\xi h = 1\!\!1 + F_1(\xi, \eta, t)F_0(y_t^{\text{det}})^{-1}$. However, such a function does not exist in general.

The solution of the first equation in (5.1.32) can be represented as

$$\xi_t = U(t)\xi_0 + \frac{\sigma}{\sqrt{\varepsilon}}\int_0^t U(t, s)F_0(y_s^{\text{det}}) dW_s \qquad (5.1.35)$$
$$+ \frac{\sigma}{\sqrt{\varepsilon}}\int_0^t U(t, s)F_1(\xi_s, \eta_s, s) dW_s + \frac{1}{\varepsilon}\int_0^t U(t, s)b(\xi_s, \eta_s, s) ds ,$$

where $U(t,s)$ is again the fundamental solution of $\varepsilon\dot\xi = A(y_t^{\text{det}})\xi$ and $U(t) = U(t,0)$. Our strategy will be to control the behaviour of the first two terms on the right-hand side as precisely as possible, and to treat the other two terms as small perturbations. We will proceed in three steps:

1. First we consider the dynamics on a small time interval of length $\Delta\varepsilon$, where the process ξ_t can be approximated by a Gaussian martingale, and η_t remains small.
2. Then we extend the description to timescales of order 1, patching together intervals of length $\Delta\varepsilon$.
3. Finally, we consider arbitrarily long time intervals (up to $\tau_{\mathcal{D}_0}$) by restarting the process, with the help of the Markov property, at multiples of a fixed time T.

Timescales of length $\Delta\varepsilon$

We start by considering the dynamics on a fixed interval $[s,t]$ of length $\Delta\varepsilon$, assuming $y_u^{\text{det}} \in \mathcal{D}_0$ for all $u \leqslant t$. We denote by K_+^2 and K_-^2 the upper bounds on $\|\bar X(y)\|$ and $\|\bar X(y)^{-1}\|$, respectively, valid for $y \in \mathcal{D}_0$. Further let $K_0 > 0$ be such that $\|U(t,u)\|$ is bounded by a constant times $\exp\{-K_0(t-u)/\varepsilon\}$ for all $t \geqslant u \geqslant 0$. The existence of such a K_0 follows from the asymptotic stability of the slow manifold.

Instead of directly estimating $\langle \xi_u, \bar X(y_u)^{-1}\xi_u\rangle$, we start by considering the behaviour of $\langle \xi_u, \bar X(y_u^{\text{det}})^{-1}\xi_u\rangle$, which can be written as

$$\langle \xi_u, \bar X(y_u^{\text{det}})^{-1}\xi_u\rangle = \|Q_s(u)\Upsilon_u\|^2 , \qquad (5.1.36)$$

where $\Upsilon_u = U(s,u)\xi_u$ and $Q_s(u)$ is the symmetric matrix satisfying

$$Q_s(u)^2 = U(u,s)^{\text{T}} \bar X(y_u^{\text{det}})^{-1} U(u,s) . \qquad (5.1.37)$$

We now split Υ_u into the sum $\Upsilon_u = \Upsilon_u^0 + \Upsilon_u^1 + \Upsilon_u^2$, with

$$\Upsilon_u^0 = U(s)\xi_0 + \frac{\sigma}{\sqrt\varepsilon} \int_0^u U(s,v)F_0(y_v^{\text{det}})\,\mathrm{d}W_v , \qquad (5.1.38)$$

$$\Upsilon_u^1 = \frac{\sigma}{\sqrt\varepsilon} \int_0^u U(s,v)F_1(\xi_v,\eta_v,v)\,\mathrm{d}W_v , \qquad (5.1.39)$$

$$\Upsilon_u^2 = \frac{1}{\varepsilon} \int_0^u U(s,v)b(\xi_v,\eta_v,v)\,\mathrm{d}v . \qquad (5.1.40)$$

Let us first focus on the Gaussian martingale Υ_u^0. Since (5.1.10) implies

$$\begin{aligned}\varepsilon\frac{\mathrm{d}}{\mathrm{d}u}&Q_s(u)^{-2}\\ &= U(s,u)\Big[\varepsilon\frac{\mathrm{d}}{\mathrm{d}u}\bar X(y_u^{\text{det}}) - A(y_u^{\text{det}})\bar X(y_u^{\text{det}}) - \bar X(y_u^{\text{det}})A(y_u^{\text{det}})^{\text{T}}\Big]U(s,u)^{\text{T}}\\ &= U(s,u)F_0(y_u^{\text{det}})F_0(y_u^{\text{det}})^{\text{T}}U(s,u)^{\text{T}} ,\end{aligned} \qquad (5.1.41)$$

the covariance of $Q_s(u)\Upsilon_u^0$ has the form

$$\operatorname{Cov}(Q_s(u)\Upsilon_u^0) = Q_s(u)\frac{\sigma^2}{\varepsilon}\int_0^u U(s,v)F_0(y_v^{\det})F_0(y_v^{\det})^{\mathrm{T}}U(s,v)^{\mathrm{T}}\,dv\,Q_s(u)$$
$$= \sigma^2\left[\mathbb{1} - Q_s(u)Q_s(0)^{-2}Q_s(u)\right]. \qquad (5.1.42)$$

For $u \in [s,t]$, $\|Q_s(u)\|$ is of order 1, while $Q_s(0)^{-2} = U(s)\bar{X}(y_0^{\det})U(s)^{\mathrm{T}}$ is exponentially small in $K_0 s/\varepsilon$. Thus $Q_s(u)\Upsilon_u^0$ behaves like the scaled Brownian motion σW_u for large u. However, $Q_s(u)\Upsilon_u^0$ is not a martingale, so that we first need to establish a bound on the supremum of $Q_t(u)\Upsilon_u^0$, and later use the fact that $Q_s(u)$ and $Q_t(u)$ are close to each other, in order to estimate the supremum of $Q_s(u)\Upsilon_u^0$.

Lemma 5.1.8. *For every $\gamma \in (0,1)$ and $H_0 > \alpha h > 0$, the bound*

$$\mathbb{P}^{\xi_0,0}\left\{\sup_{s \leq u \leq t}\|Q_s(t)\Upsilon_u^0\| \geq H_0\right\} \leq \frac{1}{(1-\gamma)^{n/2}}\exp\left\{-\gamma\frac{H_0^2 - \alpha^2 h^2}{2\sigma^2}\right\} \qquad (5.1.43)$$

holds uniformly for all ξ_0 such that $\langle \xi_0, \bar{X}(y_0)^{-1}\xi_0\rangle \leq \alpha^2 h^2$.

Proof. Let $\hat{\gamma} = \gamma/\sigma^2$. The process $\{\exp\{\hat{\gamma}\|Q_s(t)\Upsilon_u^0\|^2\}\}_u$ being a positive submartingale, Doob's submartingale inequality (cf. Lemma B.1.2) allows the left-hand side of (5.1.43) to be bounded by

$$\mathbb{P}^{\xi_0,0}\left\{\sup_{s \leq u \leq t} e^{\hat{\gamma}\|Q_s(t)\Upsilon_u^0\|^2} \geq e^{\hat{\gamma}H_0^2}\right\} \leq e^{-\hat{\gamma}H_0^2}\mathbb{E}^{\xi_0,0}\left\{e^{\hat{\gamma}\|Q_s(t)\Upsilon_t^0\|^2}\right\}. \qquad (5.1.44)$$

The Gaussian random variable $Q_s(t)\Upsilon_t^0$ has expectation $E = Q_s(t)U(s)\xi_0$ and covariance $\Sigma = \sigma^2[\mathbb{1} - RR^{\mathrm{T}}]$, where $R = Q_s(t)U(s)\bar{X}(y_0^{\det})^{1/2}$. Using completion of squares to compute the Gaussian integral, we find

$$\mathbb{E}^{\xi_0,0}\left\{e^{\hat{\gamma}\|Q_s(t)\Upsilon_t^0\|^2}\right\} = \frac{e^{\hat{\gamma}\langle E,(\mathbb{1}-\hat{\gamma}\Sigma)^{-1}E\rangle}}{(\det[\mathbb{1} - \hat{\gamma}\Sigma])^{1/2}}. \qquad (5.1.45)$$

Now $\det[\mathbb{1} - \hat{\gamma}\Sigma] \geq (1 - \hat{\gamma}\sigma^2)^n = (1-\gamma)^n$, and $\|RR^{\mathrm{T}}\| = \|R^{\mathrm{T}}R\| \in (0,1)$. This follows from the fact that $\bar{X}(y_t^{\det}) - U(t)\bar{X}(y_0^{\det})U(t)^{\mathrm{T}}$, being the covariance (5.1.9) of ξ_t^0, is positive definite. It follows that

$$\langle E, (\mathbb{1} - \hat{\gamma}\Sigma)^{-1}E\rangle = \langle \bar{X}(y_0^{\det})^{-1/2}\xi_0, R^{\mathrm{T}}(\mathbb{1} - \hat{\gamma}\Sigma)^{-1}R\bar{X}(y_0^{\det})^{-1/2}\xi_0\rangle$$
$$\leq \alpha^2 h^2 \|R^{\mathrm{T}}(\mathbb{1} - \hat{\gamma}\sigma^2[\mathbb{1} - RR^{\mathrm{T}}])^{-1}R\| \qquad (5.1.46)$$
$$\leq \alpha^2 h^2 ([1 - \hat{\gamma}\sigma^2]\|R^{\mathrm{T}}R\|^{-1} + \hat{\gamma}\sigma^2)^{-1} \leq \alpha^2 h^2$$

for all ξ_0 satisfying $\langle \xi_0, \bar{X}(y_0)^{-1}\xi_0\rangle \leq \alpha^2 h^2$. □

Remark 5.1.9. Properties of the Bessel process $\|W_t\|$ may yield an alternative way to bound the probability in (5.1.43).

Next we examine the effect of the position-dependent diffusion coefficient $F_1(\xi, \eta, t)$. To this end, we introduce the functions

$$\Psi(t) = \frac{1}{\varepsilon} \int_0^t \|U(t,u)^{\mathrm{T}} \overline{X}(y_t^{\mathrm{det}})^{-1} U(t,u)\| \, \mathrm{d}u \,,$$

$$\Phi(t) = \frac{1}{\varepsilon} \int_0^t \mathrm{Tr}\big[U(t,u)^{\mathrm{T}} \overline{X}(y_t^{\mathrm{det}})^{-1} U(t,u) \big] \, \mathrm{d}u \,, \qquad (5.1.47)$$

$$\Theta(t) = \frac{1}{\varepsilon} \int_0^t \|U(t,u)\| \, \mathrm{d}u \,.$$

Note that $\Psi(t)$ and $\Theta(t)$ are of order 1, while $\Phi(t) \leqslant n\Psi(t)$ is of order n. We also introduce stopping times

$$\tau_\eta = \inf\{u > 0 \colon \|\eta_u\| \geqslant h_1\} \,, \qquad (5.1.48)$$

$$\tau_\xi = \inf\{u > 0 \colon \langle \xi_u, \overline{X}(y_u^{\mathrm{det}})^{-1} \xi_u \rangle \geqslant h^2\} \,, \qquad (5.1.49)$$

where $h_1 < h$ is a constant.

Lemma 5.1.10. *The bound*

$$P_1 := \mathbb{P}^{\xi_0,0}\left\{ \sup_{s \leqslant u \leqslant t \wedge \tau_\xi \wedge \tau_\eta} \|Q_s(t) \Upsilon_u^1\| \geqslant H_1 \right\}$$

$$\leqslant \exp\left\{ -\frac{(H_1^2 - \sigma^2 M_1^2 (K_+ h + h_1)^2 \Phi(t))^2}{8\sigma^2 M_1^2 (K_+ h + h_1)^2 H_1^2 \Psi(t)} \right\} \qquad (5.1.50)$$

holds uniformly for all ξ_0 such that $\langle \xi_0, \overline{X}(y_0)^{-1} \xi_0 \rangle \leqslant h^2$.

Proof. Let τ denote the stopping time

$$\tau = \tau_\xi \wedge \tau_\eta \wedge \inf\{u \geqslant 0 \colon \|Q_s(t) \Upsilon_u^1\| \geqslant H_1\} \,, \qquad (5.1.51)$$

and define, for a given γ_1, the stochastic process $\Xi_u = \mathrm{e}^{\gamma_1 \|Q_s(t) \Upsilon_u^1\|^2}$. $(\Xi_u)_u$ being a positive submartingale, another application of Doob's submartingale inequality yields

$$P_1 \leqslant \mathrm{e}^{-\gamma_1 H_1^2} \mathbb{E}^{\xi_0,0}\{\Xi_{t \wedge \tau}\} \,. \qquad (5.1.52)$$

Itô's formula (together with the fact that $(\mathrm{d}W_u)^{\mathrm{T}} R^{\mathrm{T}} R \, \mathrm{d}W_u = \mathrm{Tr}(R^{\mathrm{T}} R) \, \mathrm{d}u$ for any matrix $R \in \mathbb{R}^{n \times k}$) shows that Ξ_u obeys the SDE

$$\mathrm{d}\Xi_u = 2\gamma_1 \frac{\sigma}{\sqrt{\varepsilon}} \Xi_u (\Upsilon_u^1)^{\mathrm{T}} Q_s(t)^2 U(s,u) F_1(\xi_u, \eta_u, u) \, \mathrm{d}W_u$$

$$+ \gamma_1 \frac{\sigma^2}{\varepsilon} \Xi_u \, \mathrm{Tr}\big[R_1^{\mathrm{T}} R_1 + 2\gamma_1 R_2^{\mathrm{T}} R_2 \big] \, \mathrm{d}u \,, \qquad (5.1.53)$$

where

$$R_1 = Q_s(t)U(s,u)F_1(\xi_u, \eta_u, u) ,$$
$$R_2 = (\Upsilon_u^1)^{\mathrm{T}} Q_s(t)^2 U(s,u) F_1(\xi_u, \eta_u, u) .$$
(5.1.54)

The first term in the trace can be estimated as

$$\begin{aligned}\operatorname{Tr}[R_1^{\mathrm{T}} R_1] &= \operatorname{Tr}[R_1 R_1^{\mathrm{T}}] \\ &\leqslant M_1^2 (\|\xi_u\| + \|\eta_u\|)^2 \operatorname{Tr}[Q_s(t)^{\mathrm{T}} U(s,u) U(s,u)^{\mathrm{T}} Q_s(t)] \\ &\leqslant M_1^2 (\|\xi_u\| + \|\eta_u\|)^2 \operatorname{Tr}[U(t,u)^{\mathrm{T}} \bar{X}(y_t^{\mathrm{det}})^{-1} U(t,u)] ,\end{aligned}$$
(5.1.55)

while the second term satisfies the bound

$$\begin{aligned}\operatorname{Tr}[R_2^{\mathrm{T}} R_2] &= \|F_1(\xi_u,\eta_u,u)^{\mathrm{T}} U(s,u)^{\mathrm{T}} Q_s(t)^2 \Upsilon_u^1\|^2 \\ &\leqslant M_1^2 (\|\xi_u\|+\|\eta_u\|)^2 \|U(s,u)^{\mathrm{T}} Q_s(t)\|^2 \|Q_s(t)\Upsilon_u^1\|^2 \\ &= M_1^2 (\|\xi_u\|+\|\eta_u\|)^2 \|U(t,u)^{\mathrm{T}} \bar{X}(y_t^{\mathrm{det}})^{-1} U(t,u)\| \|Q_s(t)\Upsilon_u^1\|^2 .\end{aligned}$$
(5.1.56)

Using the fact that $\|\xi_u\| \leqslant K_+ h$, $\|\eta_u\| \leqslant h_1$ and $\|Q_s(t)\Upsilon_u^1\| \leqslant H_1$ hold for all $0 \leqslant u \leqslant t \wedge \tau$, we obtain

$$\mathbb{E}^{\xi_0,0}\{\Xi_{u\wedge\tau}\}$$
(5.1.57)
$$\leqslant 1 + \gamma_1 \frac{\sigma^2}{\varepsilon} M_1^2 (K_+ h + h_1)^2 \int_0^u \mathbb{E}^{\xi_0,0}\{\Xi_{v\wedge\tau}\}$$
$$\times \left[\operatorname{Tr}[U(t,v)^{\mathrm{T}} \bar{X}(y_t^{\mathrm{det}})^{-1} U(t,v)] + 2\gamma_1 H_1^2 \|U(t,v)^{\mathrm{T}} \bar{X}(y_t^{\mathrm{det}})^{-1} U(t,v)\|\right] dv ,$$

and Gronwall's inequality (Lemma B.3.1) yields

$$\mathbb{E}^{\xi_0,0}\{\Xi_{t\wedge\tau}\} \leqslant \exp\left\{\gamma_1 \sigma^2 M_1^2 (K_+ h + h_1)^2 [\Phi(t) + 2\gamma_1 H_1^2 \Psi(t)]\right\} . \quad (5.1.58)$$

Now, (5.1.52) implies that P_1 is bounded above by

$$\exp\left\{-\gamma_1 (H_1^2 - \sigma^2 M_1^2 (K_+ h + h_1)^2 \Phi(t)) + 2\gamma_1^2 \sigma^2 M_1^2 (K_+ h + h_1)^2 H_1^2 \Psi(t)\right\} ,$$
(5.1.59)
and (5.1.50) follows by optimising over γ_1. □

Combining Lemmas 5.1.8 and 5.1.10 with the decomposition (5.1.38)–(5.1.40), we obtain the following estimate on the probability of the quantity $\langle \xi_u, \bar{X}(y_u^{\mathrm{det}})^{-1} \xi_u \rangle$ making a large excursion.

Proposition 5.1.11. *For all $\alpha \in [0,1)$, all $\gamma \in (0,1)$ and all $\mu > 0$,*

$$\sup_{\xi_0 \,:\, \langle \xi_0, \bar{X}(y_0)^{-1}\xi_0\rangle \leqslant \alpha^2 h^2} \mathbb{P}^{\xi_0,0}\left\{\sup_{s\leqslant u \leqslant t\wedge\tau_\eta} \langle \xi_u, \bar{X}(y_u^{\mathrm{det}})^{-1}\xi_u\rangle \geqslant h^2\right\}$$
$$\leqslant \frac{e^{m\varepsilon\rho^2}}{(1-\gamma)^{n/2}} \exp\left\{-\gamma \frac{h^2}{2\sigma^2}\left[1 - \alpha^2 - M_0(\Delta + (1+\mu)h + (h+h_1)\Theta(t))\right]\right\}$$
$$+ e^{\Phi(t)/4\Psi(t)} \exp\left\{-\frac{h^2}{\sigma^2} \frac{\mu^2(1-M_0\Delta)}{8M_1^2(K_+ + h_1/h)^2 \Psi(t)}\right\} \quad (5.1.60)$$

holds for all $h < 1/\mu$, with a constant M_0 depending only on the linearisation A of f, K_+, K_-, M, $\|F_0\|_\infty$, and on the dimensions n and m via M.

Proof. Recall that $\langle \xi_u, \overline{X}(y_u^{\text{det}})^{-1}\xi_u\rangle = \|Q_s(u)\Upsilon_u\|^2$. Since we can only estimate $\|Q_s(t)\Upsilon_u\|$, we need to compare $Q_s(u)$ with $Q_s(t)$. Relation (5.1.41) implies

$$Q_s(u)^2 Q_s(t)^{-2} = \mathbb{1} + Q_s(u)^2 \frac{1}{\varepsilon}\int_u^t U(s,v) F_0(y_v^{\text{det}},\varepsilon) F_0(y_v^{\text{det}},\varepsilon)^{\mathrm{T}} U(s,v)^{\mathrm{T}}\,\mathrm{d}v$$
$$= \mathbb{1} + \mathcal{O}(\Delta)\,, \qquad (5.1.61)$$

and hence $\|Q_s(u)\Upsilon_u\|$ can only exceed h if $\|Q_s(t)\Upsilon_u\|$ exceeds H, where

$$H := h\left(\sup_{s\leqslant u\leqslant t}\|Q_s(u)Q_s(t)^{-1}\|\right)^{-1} = h\bigl(1 - \mathcal{O}(\Delta)\bigr)\,. \qquad (5.1.62)$$

Thus for any decomposition $H = H_0 + H_1 + H_2$, the probability on the left-hand side of (5.1.60) is bounded by

$$\mathbb{P}^{\xi_0,0}\left\{\sup_{s\leqslant u\leqslant t\wedge\tau_\xi\wedge\tau_\eta}\|Q_s(t)\Upsilon_u\| \geqslant H\right\} \leqslant P_0 + P_1 + P_2\,, \qquad (5.1.63)$$

where

$$P_i = \mathbb{P}^{\xi_0,0}\left\{\sup_{s\leqslant u\leqslant t\wedge\tau_\xi\wedge\tau_\eta}\|Q_s(t)\Upsilon_u^i\| \geqslant H_i\right\} \qquad (5.1.64)$$

for $i = 0, 1, 2$. The terms P_0 and P_1 have been estimated in Lemma 5.1.8 and Lemma 5.1.10, respectively. As for P_2, since

$$\sup_{s\leqslant u\leqslant t\wedge\tau_\xi\wedge\tau_\eta}\|Q_s(t)\Upsilon_u^2\| \qquad (5.1.65)$$
$$\leqslant K_- M_2(K_+^2 h^2 + K_+ h h_1 + m\varepsilon\rho^2\sigma^2)(1+\mathcal{O}(\Delta))\Theta(t) := \overline{H}_2\,,$$

P_2 vanishes whenever $H_2 > \overline{H}_2$. Thus we take $H_2 = 2\overline{H}_2$, $H_1 = \mu h H$ for $0 < \mu < 1/h$ and $H_0 = H - H_1 - H_2$. When estimating H_0^2, we may assume $M_0 h \Theta(t) < 1$, the bound (5.1.60) being trivial otherwise. □

To complete the discussion of the dynamics on timescales of order $\Delta\varepsilon$, we need to estimate the probability that η_u becomes large. Let $V(u,v)$ be the principal solution of $\dot{\eta} = B(y_u^{\text{det}},\varepsilon)\eta$. The quantities

$$\chi^{(1)}(t) = \sup_{0\leqslant s\leqslant t}\int_0^s\Bigl(\sup_{u\leqslant v\leqslant s}\|V(s,v)\|\Bigr)\,\mathrm{d}u\,, \qquad (5.1.66)$$

$$\chi^{(2)}(t) = \sup_{0\leqslant s\leqslant t}\int_0^s\Bigl(\sup_{u\leqslant v\leqslant s}\|V(s,v)\|^2\Bigr)\,\mathrm{d}u \qquad (5.1.67)$$

are measures of the rate of expansion of orbits in the slow directions. The following bound is proved in a similar way as Proposition 5.1.11.

5.1 Slow Manifolds

Proposition 5.1.12. *There exists a constant $c_\eta > 0$ such that for all choices of $h_1 > 0$ satisfying $h_1 \leqslant c_\eta \chi^{(1)}(t)^{-1}$,*

$$\sup_{\xi_0 \,:\, \langle \xi_0, \bar{X}(y_0)^{-1}\xi_0\rangle \leqslant h^2} \mathbb{P}^{\xi_0, 0}\left\{ \sup_{s \leqslant u \leqslant t \wedge \tau_{\mathcal{B}(h)}} \|\eta_u\| \geqslant h_1 \right\}$$
$$\leqslant 2e^{m/4} \exp\left\{ \frac{m\varepsilon\rho^2}{(\rho^2+\varepsilon)\chi^{(2)}(t)} \right\} \quad (5.1.68)$$
$$\times \exp\left\{ -\kappa_0 \frac{h_1^2(1-\mathcal{O}(\Delta\varepsilon))}{\sigma^2(\rho^2+\varepsilon)\chi^{(2)}(t)}\left[1 - M_0' \chi^{(1)}(t) h_1\left(1 + K_+^2 \frac{h^2}{h_1^2}\right)\right] \right\},$$

where $\kappa_0 > 0$ is a constant depending only on $\|F\|_\infty$, $\|G\|_\infty$, $\|C\|_\infty$ and U, while the constant M_0' depends only on M, $\|C\|_\infty$ and U. Note that c_η may depend on the dimensions n and m via M.

Proof. The solution of (5.1.32) satisfied by η_u can be split into four parts, $\eta_u = \eta_u^0 + \eta_u^1 + \eta_u^2 + \eta_u^3$, where

$$\eta_u^0 = \sigma' \int_0^u V(u,v)[G_0(y_v^{\mathrm{det}}) + G_1(\xi_v, \eta_v, v)]\,\mathrm{d}W_v ,$$
$$\eta_u^1 = \frac{\sigma}{\sqrt{\varepsilon}} \int_0^u S(u,v)[F_0(y_v^{\mathrm{det}}) + F_1(\xi_v, \eta_v, v)]\,\mathrm{d}W_v , \quad (5.1.69)$$
$$\eta_u^2 = \int_0^u V(u,v)c(\xi_v, \eta_v, v)\,\mathrm{d}v ,$$
$$\eta_u^3 = \frac{1}{\varepsilon}\int_0^u S(u,v)b(\xi_v, \eta_v, v)\,\mathrm{d}v ,$$

with

$$S(u,v) = \int_v^u V(u,w)C(y_w^{\mathrm{det}})U(w,v)\,\mathrm{d}w . \quad (5.1.70)$$

Let $\tau = \tau_{\mathcal{B}(h)} \wedge \tau_\eta$. It follows immediately from the definitions of $\tau_{\mathcal{B}(h)}$, τ_η and the bounds (5.1.34) that

$$\|\eta_{u\wedge\tau}^2\| \leqslant M_2(1 + \mathcal{O}(\Delta\varepsilon))\chi^{(1)}(t)(K_+^2 h^2 + h_1^2) ,$$
$$\|\eta_{u\wedge\tau}^3\| \leqslant M'\chi^{(1)}(t)(K_+^2 h^2 + K_+ h h_1 + m\varepsilon\rho^2\sigma^2) \quad (5.1.71)$$

for all $u \in [s,t]$. Here M' depends only on M_2, U and $\|C\|_\infty$. Furthermore, using similar ideas as in the proof of Lemma 5.1.10, it is straightforward to establish for all $H_0, H_1 > 0$ that

$$\mathbb{P}^{\xi_0,0}\left\{ \sup_{s\leqslant u\leqslant t\wedge\tau} \|\eta_u^0\| \geqslant H_0 \right\} \leqslant e^{m/4}\exp\left\{ -\frac{H_0^2(1-\mathcal{O}(\Delta\varepsilon))}{8(\sigma')^2\|G_0 + G_1\|_\infty^2 \chi^{(2)}(t)} \right\},$$
$$\mathbb{P}^{\xi_0,0}\left\{ \sup_{s\leqslant u\leqslant t\wedge\tau} \|\eta_u^1\| \geqslant H_1 \right\} \leqslant e^{m/4}\exp\left\{ -\frac{H_1^2(1-\mathcal{O}(\Delta\varepsilon))}{8\sigma^2\varepsilon c_S\|F_0 + F_1\|_\infty^2 \chi^{(2)}(t)} \right\},$$
$$(5.1.72)$$

where c_S is a constant depending only on S. Then (5.1.68) is obtained by taking, e.g., $H_0 = H_1 = \frac{1}{2}h_1 - 2(M_2 + M')\chi^{(1)}(t)(K_+^2 h^2 + h_1^2 + m\varepsilon\rho^2\sigma^2)$, and using $h_1 \leqslant c_\eta \chi^{(1)}(t)^{-1}$, where we may choose $c_\eta \leqslant 1/(2M_0')$. □

Estimate (5.1.68) implies that $\|\eta_u\|$ is unlikely to become much larger than $\sigma(\rho + \sqrt{\varepsilon})\chi^{(2)}(t)^{1/2}$ for $u \leqslant t$.

Timescales of length 1

We still consider the situation where $y_0 \in \mathcal{D}_0$, but drop the requirement that $y_u^{\text{det}} \in \mathcal{D}_0$ for all $u \leqslant t$. The uniform-hyperbolicity assumption implies the existence of a $\delta > 0$ such that the slow manifold is also uniformly asymptotically stable in the δ-neighbourhood $\mathcal{D}_0^+(\delta)$ of $\mathcal{D}_0$. Then the deterministic first-exit time
$$\tau_{\mathcal{D}_0}^{\text{det}} = \tau_{\mathcal{D}_0}^{\text{det}}(y_0) = \inf\{u \geqslant 0 \colon y_u^{\text{det}} \notin \mathcal{D}_0^+(\delta)\} \tag{5.1.73}$$
is strictly positive, and $\tau_{\mathcal{B}(h)} \wedge \tau_\eta \leqslant \tau_{\mathcal{D}_0}^{\text{det}}$ whenever $h_1 \leqslant \delta$.

We can now prove a version of Theorem 5.1.6 on timescales of order 1.

Proposition 5.1.13. *Fix a time $t > 0$ and a constant $h > 0$ in such a way that $h \leqslant c_1 \chi^{(1)}(t \wedge \tau_{\mathcal{D}_0}^{\text{det}})^{-1}$ for a sufficiently small constant $c_1 > 0$ and $\chi^{(2)}(t \wedge \tau_{\mathcal{D}_0}^{\text{det}}) \leqslant (\rho^2 + \varepsilon)^{-1}$. Then, for any $\alpha \in [0, 1)$, any $\gamma \in (0, 1)$ and any sufficiently small Δ,*
$$C_{n,m}^-(t,\varepsilon) e^{-\kappa_-(0)h^2/2\sigma^2} \leqslant \mathbb{P}^{\xi_0,0}\{\tau_{\mathcal{B}(h)} < t\} \leqslant C_{n,m,\gamma}^+(t,\varepsilon) e^{-\kappa_+(\alpha)h^2/\sigma^2} \tag{5.1.74}$$
holds uniformly for all ξ_0 satisfying $\langle \xi_0, \overline{X}(y_0)^{-1}\xi_0 \rangle \leqslant \alpha^2 h^2$. Here the exponents are given by
$$\kappa_+(\alpha) = \gamma\left[1 - \alpha^2 - \mathcal{O}(\Delta) - \mathcal{O}(m\varepsilon\rho^2\sigma^2/h^2) - \mathcal{O}((1 + \chi^{(1)}(t \wedge \tau_{\mathcal{D}_0}^{\text{det}}))h)\right],$$
$$\kappa_-(0) = 1 + \mathcal{O}(h) + \mathcal{O}\bigl(e^{-K_0 t/\varepsilon}\bigr), \tag{5.1.75}$$
and the prefactors satisfy
$$C_{n,m,\gamma}^+(t,\varepsilon) = \left\lceil \frac{t}{\Delta\varepsilon} \right\rceil \left[\frac{1}{(1-\gamma)^{n/2}} + \bigl(e^{n/4} + 2e^{m/4}\bigr)e^{-\kappa_+(0)h^2/\sigma^2}\right], \tag{5.1.76}$$
$$C_{n,m}^-(t,\varepsilon) = \left(\sqrt{\frac{2}{\pi}}\frac{h}{\sigma} \wedge 1\right)e^{-\mathcal{O}(m\varepsilon\rho^2)} - \left(e^{n/4} + 4\left\lceil \frac{t}{\Delta\varepsilon} \right\rceil e^{m/4}\right)e^{-\tilde{\kappa}h^2/2\sigma^2},$$
where $\tilde{\kappa} = 1 - \mathcal{O}(e^{-K_0 t/\varepsilon}) - \mathcal{O}(m\varepsilon\rho^2\sigma^2/h^2) - \mathcal{O}((1 + \chi^{(1)}(t \wedge \tau_{\mathcal{D}_0}^{\text{det}}))h)$.

Proof. We first establish the upper bound. Fix an initial condition $(\xi_0, 0)$ satisfying $\langle \xi_0, \overline{X}(y_0)^{-1}\xi_0 \rangle \leqslant \alpha^2 h^2$, and observe that

5.1 Slow Manifolds

$$\mathbb{P}^{\xi_0,0}\{\tau_{\mathcal{B}(h)} < t\} \tag{5.1.77}$$
$$\leqslant \mathbb{P}^{\xi_0,0}\{\tau_{\mathcal{B}(h)} < t \wedge \tau_\eta\} + \mathbb{P}^{\xi_0,0}\{\tau_\eta < t \wedge \tau_{\mathcal{B}(h)}\}$$
$$= \mathbb{P}^{\xi_0,0}\{\tau_{\mathcal{B}(h)} < t \wedge \tau_{\mathcal{D}_0}^{\det} \wedge \tau_\eta\} + \mathbb{P}^{\xi_0,0}\{\tau_\eta < t \wedge \tau_{\mathcal{D}_0}^{\det} \wedge \tau_{\mathcal{B}(h)}\}\;.$$

To estimate the first term on the right-hand side, we introduce a partition $0 = u_0 < u_1 < \cdots < u_K = t$ of the time interval $[0,t]$, defined by $u_k = k\Delta\varepsilon$ for $0 \leqslant k < K = \lceil t/(\Delta\varepsilon) \rceil$. Thereby we obtain

$$\mathbb{P}^{\xi_0,0}\{\tau_{\mathcal{B}(h)} < t \wedge \tau_{\mathcal{D}_0}^{\det} \wedge \tau_\eta\} \leqslant \sum_{k=1}^{K} \mathbb{P}^{\xi_0,0}\{u_{k-1} \leqslant \tau_{\mathcal{B}(h)} < u_k \wedge \tau_{\mathcal{D}_0}^{\det} \wedge \tau_\eta\}\;.$$
(5.1.78)

Before we estimate the summands on the right-hand side of (5.1.78), note that by the boundedness assumption on $\|\overline{X}(y)\|$ and $\|\overline{X}^{-1}(y)\|$, we have $\overline{X}(y_u)^{-1} = \overline{X}(y_u^{\det})^{-1} + \mathcal{O}(h_1)$ for $u \leqslant \tau_{\mathcal{D}_0}^{\det} \wedge \tau_\eta$. Thus the bound obtained in Proposition 5.1.11 can also be applied to estimate first-exit times from $\mathcal{B}(h)$ itself:

$$\mathbb{P}^{\xi_0,0}\{u_{k-1} \leqslant \tau_{\mathcal{B}(h)} < u_k \wedge \tau_{\mathcal{D}_0}^{\det} \wedge \tau_\eta\} \tag{5.1.79}$$
$$\leqslant \mathbb{P}^{\xi_0,0}\Big\{\sup_{u_{k-1} \leqslant u < u_k \wedge \tau_{\mathcal{D}_0}^{\det} \wedge \tau_\eta} \langle \xi_u, \overline{X}(y_u^{\det})^{-1}\xi_u \rangle \geqslant h^2(1 - \mathcal{O}(h_1))\Big\}\;.$$

The second term on the right-hand side of (5.1.77) can be similarly estimated by Proposition 5.1.12. Choosing

$$\mu^2 = 8M_1^2\big[K_+ + h_1/(h(1-\mathcal{O}(h_1)))\big]^2 \Psi(t \wedge \tau_{\mathcal{D}_0}^{\det})/[1 - \mathcal{O}(h_1) - M_0\Delta] \tag{5.1.80}$$

and $h_1 = h/\sqrt{\kappa_0}$ in the resulting expression, we see that the Gaussian part of ξ_t gives the major contribution to the probability. Thus we obtain that the probability in (5.1.77) is bounded by

$$\left\lceil \frac{t}{\Delta\varepsilon} \right\rceil \Bigg[\frac{e^{m\varepsilon\rho^2}}{(1-\gamma)^{n/2}} \exp\Big\{-\gamma\frac{h^2}{2\sigma^2}[1-\alpha^2 - \mathcal{O}(\Delta) - \mathcal{O}(h)]\Big\} + e^{n/4}e^{-h^2/\sigma^2}$$
$$+ 2e^{m/4}\exp\Big\{-\frac{h^2(1-\mathcal{O}(\chi^{(1)}(t \wedge \tau_{\mathcal{D}_0}^{\det})h) - \mathcal{O}(\Delta\varepsilon) - \mathcal{O}(m\varepsilon\rho^2\sigma^2/h^2))}{\sigma^2(\rho^2+\varepsilon)\chi^{(2)}(t \wedge \tau_{\mathcal{D}_0}^{\det})}\Big\}\Bigg],$$
(5.1.81)

where we have used the fact that $\Phi(t) \leqslant n\Psi(t)$, while $\Psi(t)$ and $\Theta(t)$ are at most of order 1. The prefactor $e^{m\varepsilon\rho^2}$ can be absorbed into the error term in the exponent. This completes the proof of the upper bound in (5.1.74).

The lower bound is again a consequence of the fact that the Gaussian part of ξ_t gives the major contribution to the probability in (5.1.74). To check this, we split the probability as follows:

$$\mathbb{P}^{\xi_0,0}\{\tau_{\mathcal{B}(h)} < t\} \tag{5.1.82}$$
$$\geqslant \mathbb{P}^{\xi_0,0}\{\tilde{\tau}_\xi < t, \tau_\eta \geqslant t\} + \mathbb{P}^{\xi_0,0}\{\tau_{\mathcal{B}(h)} < t, \tau_\eta < t\}$$
$$= \mathbb{P}^{\xi_0,0}\{\tilde{\tau}_\xi < t \wedge \tau_\eta\} - \mathbb{P}^{\xi_0,0}\{\tilde{\tau}_\xi < \tau_\eta < t\} + \mathbb{P}^{\xi_0,0}\{\tau_{\mathcal{B}(h)} < t, \tau_\eta < t\}$$
$$\geqslant \mathbb{P}^{\xi_0,0}\{\tilde{\tau}_\xi < t \wedge \tau_\eta\} - \mathbb{P}^{\xi_0,0}\{\tau_\eta < t \wedge \tau_{\mathcal{B}(h)}\},$$

where

$$\tilde{\tau}_\xi = \inf\{u \geqslant 0 \colon \langle \xi_u, \overline{X}(y_u^{\text{det}})^{-1}\xi_u\rangle \geqslant h^2(1 + \mathcal{O}(h_1))\}, \tag{5.1.83}$$

and the $\mathcal{O}(h_1)$-term stems from estimating $\overline{X}(y_u)^{-1}$ by $\overline{X}(y_u^{\text{det}})^{-1}$ as in (5.1.79). The first term on the last line of (5.1.82) can be estimated as in the proof of Proposition 5.1.11: A lower bound is obtained trivially by considering the endpoint instead of the whole path, and instead of applying Lemma 5.1.8, the Gaussian contribution can be estimated below by a straightforward calculation. The non-Gaussian parts are estimated *above* as before and are of smaller order. Finally, we need an upper bound for the probability that $\tau_\eta < t \wedge \tau_{\mathcal{B}(h)}$, which can be obtained from Proposition 5.1.12. □

The lower bound in (5.1.74) implies the lower bound (5.1.26) of Theorem 5.1.6, as well as the upper bound (5.1.23) on timescales of order 1. It remains to prove the upper bound on longer timescales, and the assertion on general initial conditions.

Longer timescales

The approach used so far fails to control the dynamics on timescales on which $\chi^{(i)}(t) \gg 1$, because it uses in an essential way the fact that $\eta_t = y_t - y_t^{\text{det}}$ remains small. Our strategy in order to describe the paths on longer timescales is to compare them to different deterministic solutions on time intervals $[0, T]$, $[T, 2T]$, ..., where T is a possibly large constant such that Proposition 5.1.13 holds on time intervals of length T, provided y_t remains in $\mathcal{D}_0$.

We will also need an estimate on the probability that $\langle \xi_T, \overline{X}(y_T)^{-1}\xi_T\rangle$ exceeds h^2. Proposition 5.1.13 provides, of course, such an estimate, but since it applies to the whole path, it does not give optimal bounds for the endpoint. An improved bound is given by the following lemma.

Lemma 5.1.14 (Endpoint estimate). *If T and h satisfy the conditions $h \leqslant c_1 \chi^{(1)}(T \wedge \tau_{\mathcal{D}_0}^{\text{det}})^{-1}$ and $\chi^{(2)}(T \wedge \tau_{\mathcal{D}_0}^{\text{det}}) \leqslant (\rho^2 + \varepsilon)^{-1}$, we have, for every $\gamma \in (0, 1)$,*

$$\sup_{\xi_0 \,:\, \langle \xi_0, \overline{X}(y_0)^{-1}\xi_0\rangle \leqslant h^2} \mathbb{P}^{\xi_0,0}\Big\{\langle \xi_T, \overline{X}(y_T)^{-1}\xi_T\rangle \geqslant h^2, \tau_{\mathcal{D}_0} \geqslant T\Big\}$$
$$\leqslant \widehat{C}_{n,m,\gamma}(T, \varepsilon) e^{-\kappa' h^2/2\sigma^2}, \tag{5.1.84}$$

where

$$\kappa' = \gamma\bigl[1 - \mathcal{O}(\Delta) - \mathcal{O}(h) - \mathcal{O}\bigl(\mathrm{e}^{-2K_0 T/\varepsilon}/(1-\gamma)\bigr)\bigr]\,, \tag{5.1.85}$$

$$\widehat{C}_{n,m,\gamma}(T,\varepsilon) = \frac{\mathrm{e}^{m\varepsilon\rho^2}}{(1-\gamma)^{n/2}} + 4C^+_{n,m,\gamma}(T,\varepsilon)\mathrm{e}^{-\kappa_+(0)h^2/2\sigma^2}\,. \tag{5.1.86}$$

Proof. Let $\xi_t = \xi_t^0 + \xi_t^1 + \xi_t^2$, where ξ_t^0 denotes the Gaussian contribution, that is, the first two terms on the right-hand side of (5.1.35), and the two other terms stand for the contributions of F_1 and b, respectively. We introduce the notations $\tilde{\tau}_\xi$ and $\tilde{\tau}_\eta$ for stopping times defined like τ_ξ and τ_η in (5.1.48) and (5.1.48), but with h and h_1 replaced by $2h$ and $2h_1$, respectively. The probability in (5.1.84) is bounded by

$$\mathbb{P}^{\xi_0,0}\bigl\{\langle\xi_T, \bar{X}(y_T^{\det})^{-1}\xi_T\rangle \geqslant h^2(1-\mathcal{O}(h_1)), \tilde{\tau}_\eta > T\bigr\} + \mathbb{P}^{\xi_0,0}\bigl\{\tilde{\tau}_\eta \leqslant T\bigr\}\,. \tag{5.1.87}$$

Let $H^2 = h^2(1 - \mathcal{O}(h_1))$. As in the proof of Proposition 5.1.11, the first term can be further decomposed as

$$\begin{aligned}
\mathbb{P}^{\xi_0,0}&\bigl\{\langle\xi_T, \bar{X}(y_T^{\det})^{-1}\xi_T\rangle \geqslant H^2, \tilde{\tau}_\eta > T\bigr\} \\
&\leqslant \mathbb{P}^{\xi_0,0}\bigl\{\|\bar{X}(y_T^{\det})^{-1/2}\xi_T^0\| \geqslant H_0\bigr\} + \mathbb{P}^{\xi_0,0}\bigl\{\tilde{\tau}_\eta > T, \tilde{\tau}_\xi \leqslant T\bigr\} \\
&\quad + \mathbb{P}^{\xi_0,0}\bigl\{\|\bar{X}(y_T^{\det})^{-1/2}\xi_T^1\| \geqslant H_1, \tilde{\tau}_\eta > T, \tilde{\tau}_\xi > T\bigr\} \\
&\quad + \mathbb{P}^{\xi_0,0}\bigl\{\|\bar{X}(y_T^{\det})^{-1/2}\xi_T^2\| \geqslant H_2, \tilde{\tau}_\eta > T, \tilde{\tau}_\xi > T\bigr\}\,,
\end{aligned} \tag{5.1.88}$$

where we choose H_1, H_2 twice as large as in the proof of Proposition 5.1.11, while $H_0 = H - H_1 - H_2$.

The first term on the right-hand side can be estimated as in Lemma 5.1.8, with the difference that, the expectation of ξ_T^0 being exponentially small in T/ε, it leads only to a correction of order $\mathrm{e}^{-2K_0 T/\varepsilon}/(1-\gamma)$ in the exponent. The second and the third term can be estimated by Proposition 5.1.13 and Lemma 5.1.10, the only difference lying in a larger absolute value of the exponent, because we enlarged h and h_1. The last term vanishes by our choice of H_2. Finally, the second term in (5.1.87) can be estimated by splitting according to the value of $\tau_{\mathcal{B}(2h)}$ and applying Lemma 5.1.10 and Proposition 5.1.13. □

We are now ready to establish an improved estimate on the distribution of $\tau_{\mathcal{B}(h)}$. As we will restart the process $y_t^{\det}$ whenever t is a multiple of T, we need the assumptions made in the previous section to hold uniformly in the initial condition $y_0 \in \mathcal{D}_0$. Therefore we will introduce replacements for some of the notations introduced before. Note that $\chi^{(1)}(t) = \chi^{(1)}_{y_0}(t)$ and $\chi^{(2)}(t) = \chi^{(2)}_{y_0}(t)$ depend on y_0 via the principal solution V. We define

$$\widehat{\chi}^{(1)}(t) = \sup_{y_0 \in \mathcal{D}_0} \chi^{(1)}_{y_0}\bigl(t \wedge \tau^{\det}_{\mathcal{D}_0}(y_0)\bigr)\,, \tag{5.1.89}$$

$$\widehat{\chi}^{(2)}(t) = \sup_{y_0 \in \mathcal{D}_0} \chi^{(2)}_{y_0}\bigl(t \wedge \tau^{\det}_{\mathcal{D}_0}(y_0)\bigr)\,. \tag{5.1.90}$$

In the same spirit, the $\chi^{(i)}(T)$-dependent $\mathcal{O}(\cdot)$-terms in the definitions of $\kappa_+(\alpha)$, κ' and the prefactors like $C_{n,m,\gamma}^+(T,\varepsilon)$ are modified.

We fix a time T of order 1 satisfying $\widehat{\chi}^{(2)}(T) \leqslant (\rho^2 + \varepsilon)^{-1}$. T is chosen in such a way that whenever $h \leqslant c_1 \widehat{\chi}^{(1)}(T)^{-1}$, Proposition 5.1.13 (and Lemma 5.1.14) apply. Note that larger T would be possible unless ρ is of order 1, but for larger T the constraint on h becomes more restrictive which is not desirable. Having chosen T, we define the probabilities

$$P_k(h) = \mathbb{P}^{0,0}\{\tau_{\mathcal{B}(h)} < kT \wedge \tau_{\mathcal{D}_0}\},$$
$$Q_k(h) = \mathbb{P}^{0,0}\{\langle \xi_{kT}, \overline{X}(y_{kT})^{-1}\xi_{kT}\rangle \geqslant h^2, \tau_{\mathcal{D}_0} \geqslant kT\}. \tag{5.1.91}$$

Proposition 5.1.13 provides a bound for $P_1(h)$, and Lemma 5.1.14 provides a bound for $Q_1(h)$. Subsequent bounds are computed by induction, and the following proposition describes one induction step.

Proposition 5.1.15. *Let $\hat{\kappa} \leqslant \kappa_+(0) \wedge \kappa'$. Assume that for some $k \in \mathbb{N}$,*

$$P_k(h) \leqslant D_k e^{-\hat{\kappa} h^2/2\sigma^2}, \tag{5.1.92}$$
$$Q_k(h) \leqslant \widehat{D}_k e^{-\hat{\kappa} h^2/2\sigma^2}. \tag{5.1.93}$$

Then the same bounds hold for k replaced by $k+1$, provided

$$D_{k+1} \geqslant D_k + C_{n,m,\gamma}^+(T,\varepsilon)\widehat{D}_k \frac{\gamma}{\gamma - \hat{\kappa}} e^{(\gamma - \hat{\kappa})h^2/2\sigma^2}, \tag{5.1.94}$$
$$\widehat{D}_{k+1} \geqslant \widehat{D}_k + \widehat{C}_{n,m,\gamma}(T,\varepsilon). \tag{5.1.95}$$

Proof. Let $\rho_k = \langle \xi_{kT}, \overline{X}(y_{kT})^{-1}\xi_{kT}\rangle^{1/2}$. We start by establishing (5.1.95). The Markov property allows for the decomposition

$$Q_{k+1}(h) \leqslant \mathbb{P}^{0,0}\{\tau_{\mathcal{B}(h)} < kT, \tau_{\mathcal{D}_0} \geqslant kT\}$$
$$+ \mathbb{E}^{0,0}\{1_{\{\tau_{\mathcal{B}(h)} \geqslant kT\}}\mathbb{P}^{kT,(\xi_{kT},0)}\{\rho_{k+1} \geqslant h, \tau_{\mathcal{D}_0} \geqslant (k+1)T\}\}$$
$$\leqslant Q_k(h) + \widehat{C}_{n,m,\gamma}(T,\varepsilon)e^{-\hat{\kappa} h^2/2\sigma^2}, \tag{5.1.96}$$

where the superscript in $\mathbb{P}^{kT,(\xi_{kT},0)}\{\cdot\}$ indicates that at time kT we also restart the process of the deterministic slow variables y_t^{det} in the point $y_{kT} \in \mathcal{D}_0$. In the second line, we used Lemma 5.1.14. This shows (5.1.95).

As for (5.1.94), we again start from a decomposition, similar to (5.1.96):

$$P_{k+1}(h) = \mathbb{P}^{0,0}\{\tau_{\mathcal{B}(h)} < kT \wedge \tau_{\mathcal{D}_0}\} \tag{5.1.97}$$
$$+ \mathbb{E}^{0,0}\{1_{\{\tau_{\mathcal{B}(h)} \geqslant kT\}}\mathbb{P}^{kT,(\xi_{kT},0)}\{\tau_{\mathcal{B}(h)} < (k+1)T \wedge \tau_{\mathcal{D}_0}\}\}.$$

Proposition 5.1.13 allows us to estimate

$$P_{k+1}(h) \leqslant P_k(h) + \mathbb{E}^{0,0}\{1_{\{\rho_k \leqslant h\}}[\varphi(\rho_k) \wedge 1] \mid \tau_{\mathcal{D}_0} \geqslant kT\}\mathbb{P}^{0,0}\{\tau_{\mathcal{D}_0} \geqslant kT\}, \tag{5.1.98}$$

with
$$\varphi(\rho) = C^+_{n,m,\gamma}(T,\varepsilon)e^{(\gamma-\hat{\kappa})h^2/2\sigma^2}e^{-\gamma(h^2-\rho^2)/2\sigma^2}. \tag{5.1.99}$$

Assumption (5.1.93) yields
$$\mathbb{P}^{0,0}\{\rho_k < \rho \mid \tau_{\mathcal{D}_0} \geq kT\} \geq \varphi_k(\rho), \tag{5.1.100}$$

where
$$\varphi_k(\rho) := \left(1 - \widehat{D}_k e^{-\hat{\kappa}\rho^2/2\sigma^2}\right)/\mathbb{P}^{0,0}\{\tau_{\mathcal{D}_0} \geq kT\}. \tag{5.1.101}$$

Then (5.1.94) follows by estimating the expectation in (5.1.98) with the help of Lemma B.2.1 of Appendix B, taking $\tau = \rho_k^2$, $s = \rho^2$, $t = h^2$, $g(\rho^2) = \varphi(\rho) \wedge 1$ and $G(\rho^2) = \varphi_k(\rho)$. □

The following bounds can now be obtained directly by induction on k, using as initial values the bounds $P_1(h)$ from Proposition 5.1.13 and $Q_1(h)$ from Lemma 5.1.14.

Corollary 5.1.16. *Assume that $y_0 \in \mathcal{D}_0$, $x_0 = \bar{x}(y_0,\varepsilon)$. Then, for every time $t > 0$, we have*
$$\mathbb{P}^{x_0,y_0}\{\tau_{\mathcal{B}(h)} < t \wedge \tau_{\mathcal{D}_0}\} \tag{5.1.102}$$
$$\leq C^+_{n,m,\gamma}(T,\varepsilon)\left[1 + \widehat{C}_{n,m,\gamma}(T,\varepsilon)\left(\frac{1}{2} + \frac{t}{T}\right)^2 \frac{\gamma}{2(\gamma-\hat{\kappa})}\right]e^{-(2\hat{\kappa}-\gamma)h^2/2\sigma^2}.$$

In addition, the distribution of the endpoint ξ_t satisfies
$$\mathbb{P}^{x_0,y_0}\{\langle \xi_t, \overline{X}(y_t)^{-1}\xi_t \rangle \geq h^2, \tau_{\mathcal{D}_0} \geq t\} \leq \widehat{C}_{n,m,\gamma}(T,\varepsilon)\left[\frac{t}{T}\right]e^{-\hat{\kappa}h^2/2\sigma^2}. \tag{5.1.103}$$

We can now complete the proof of Theorem 5.1.6.

Proof (of Theorem 5.1.6). We first optimise our choice of $\hat{\kappa}$, taking into account the constraint $\hat{\kappa} \leq \kappa_+(0) \wedge \kappa'$. By doing so, we find that
$$\frac{\gamma}{2(\gamma-\hat{\kappa})}e^{-(2\hat{\kappa}-\gamma)h^2/2\sigma^2} \leq \frac{2h^2}{\sigma^2}e^{-\kappa_+ h^2/2\sigma^2}, \tag{5.1.104}$$

where we have set
$$\kappa_+ = \gamma\left[1 - \mathcal{O}(h) - \mathcal{O}(\Delta) - \mathcal{O}(m\varepsilon\rho^2) - \mathcal{O}(e^{-c/\varepsilon}/(1-\gamma))\right]. \tag{5.1.105}$$

Simplifying the prefactor in (5.1.102) finally yields the upper bound (5.1.23). As for general initial conditions, cf. (5.1.29), one starts by establishing
$$\sup_{\xi_0\,:\,\langle\xi_0,\overline{X}(y_0)^{-1}\xi_0\rangle \leq \delta^2} \mathbb{P}^{\xi_0,0}\left\{\sup_{0\leq s \leq t \wedge \tau_{\mathcal{D}_0}} \frac{\langle \xi_s, \overline{X}(y_s)^{-1}\xi_s \rangle}{(h + c_0\delta e^{-K_0 s/\varepsilon})^2} \geq 1\right\}$$
$$\leq \left[\frac{t}{\Delta\varepsilon}\right]\left[\frac{1}{(1-\gamma)^{n/2}} + \left(e^{n/4} + 2e^{m/4}\right)e^{-\bar{\kappa}h^2/2\sigma^2}\right]e^{-\bar{\kappa}h^2/2\sigma^2}, \tag{5.1.106}$$

where $\bar{\kappa} = \gamma[1 - \mathcal{O}(h) - \mathcal{O}(\Delta) - \mathcal{O}(m\varepsilon\rho^2\sigma^2/h^2) - \mathcal{O}(\delta)]$. The proof differs from the proof of Proposition 5.1.13 only in the definition of the set in which sample paths are concentrated. This shows that after a time t_1 of order $\varepsilon|\log h|$, the paths are likely to have reached $\mathcal{B}(h)$. Mimicking the proof of Corollary 5.1.16, one can show that after any time $t_2 \geqslant t_1$, the probability of leaving $\mathcal{B}(h)$ behaves as if the process had started on the adiabatic manifold, i.e.,

$$\mathbb{P}^{\xi_0,0}\left\{\sup_{t_2 \leqslant s \leqslant t \wedge \tau_{\mathcal{D}_0}} \langle \xi_s, \overline{X}(y_s)^{-1}\xi_s \rangle \geqslant h^2\right\} \leqslant C^+_{n,m,\gamma,\Delta}(t,\varepsilon) e^{-\bar{\kappa}_+ h^2/2\sigma^2},$$
(5.1.107)

uniformly for all ξ_0 such that $\langle \xi_0, \overline{X}(y_0)^{-1}\xi_0 \rangle \leqslant \delta^2$. Here the prefactor is the same as for the upper bound (5.1.23), and $\bar{\kappa}_+$ is given by (5.1.30). □

5.1.3 Reduction to Slow Variables

So far, we have shown that sample paths of the slow–fast SDE

$$\begin{aligned} dx_t &= \frac{1}{\varepsilon} f(x_t, y_t)\, dt + \frac{\sigma}{\sqrt{\varepsilon}} F(x_t, y_t)\, dW_t, \\ dy_t &= g(x_t, y_t)\, dt + \sigma' G(x_t, y_t)\, dW_t \end{aligned}$$
(5.1.108)

remain concentrated in a σ-neighbourhood of an asymptotically stable adiabatic manifold $x = \bar{x}(y,\varepsilon)$, as long as the slow dynamics permits. This naturally leads to the idea to approximate the system by the *reduced stochastic system*

$$dy^0_t = g(\bar{x}(y^0_t, \varepsilon), y^0_t)\, dt + \sigma' G(\bar{x}(y^0_t, \varepsilon), y^0_t)\, dW_t,$$
(5.1.109)

obtained by "projection" of the original system (5.1.108) on the adiabatic manifold $\mathcal{M}_\varepsilon$. The question is how well the process $\{y^0_t\}_t$ approximates solutions $\{y_t\}_t$ of (5.1.108), and, in particular, whether this approximation is more accurate than the approximation by the *reduced deterministic system*

$$dy^{\text{det}}_t = g(\bar{x}(y^{\text{det}}_t, \varepsilon), y^{\text{det}}_t)\, dt.$$
(5.1.110)

The following theorem provides estimates on the respective deviations between y_t, y^0_t and y^{det}_t.

Theorem 5.1.17 (Reduction to slow variables). *Let the system* (5.1.108) *satisfy Assumptions 5.1.1 and 5.1.4. Fix an initial condition* $(x_0, y_0) = (\bar{x}(y_0, \varepsilon), y_0)$ *on the adiabatic manifold. Let* $\chi^{(1)}(t)$ *and* $\chi^{(2)}(t)$ *be the quantities defined in* (5.1.66) *and* (5.1.67). *Then there exist constants* κ_1, κ_2, L *such that, for any choice of* h, h_1 *at most up to order* $\chi^{(1)}(t)^{-1}$,

5.1 Slow Manifolds   165

$$\mathbb{P}^{x_0,y_0}\left\{\sup_{0\leqslant s\leqslant t\wedge\tau_{\mathcal{B}(h)}} \|y_s - y_s^0\| \geqslant h\right\}$$

$$\leqslant L\left(1+\frac{t}{\varepsilon}\right)e^{m/4}\left[\exp\left\{-\frac{\kappa_1 h^2}{[(\sigma')^2 h^2 + \sigma^2\varepsilon](1+\chi^{(2)}(t))}\right\}\right.$$

$$\left.+ \exp\left\{-\frac{\kappa_2 h_1^2}{(\sigma')^2(1+\chi^{(2)}(t))}\right\}\right] \qquad (5.1.111)$$

and

$$\mathbb{P}^{y_0}\left\{\sup_{0\leqslant s\leqslant t\wedge\tau_{\mathcal{D}_0}} \|y_s^0 - y_s^{\text{det}}\| \geqslant h_1\right\}$$

$$\leqslant L\left(1+\frac{t}{\varepsilon}\right)e^{m/4}\exp\left\{-\frac{\kappa_2 h_1^2}{(\sigma')^2(1+\chi^{(2)}(t))}\right\}. \qquad (5.1.112)$$

Proof. The proof resembles the one of Proposition 5.1.12. To show (5.1.112), one first observes that $\eta_t^0 = y_t^0 - y_t^{\text{det}}$ obeys an SDE of the form

$$\mathrm{d}\eta_t^0 = \left[B(y_t^{\text{det}})\eta_t^0 + \tilde{c}(y_t^{\text{det}}, \eta_t^0)\right]\mathrm{d}t + \sigma'\widehat{G}(y_t^{\text{det}}, \eta_t^0)\,\mathrm{d}W_t, \qquad (5.1.113)$$

where $B(y)$ is given in (5.1.33), $\tilde{c}(y_t^{\text{det}}, \eta_t^0)$ is a nonlinear term of order $\|\eta_t^0\|^2$, and $G(y_t^{\text{det}}, \eta_t^0) = G(\bar{x}(y_t^{\text{det}} + \eta_t^0, \varepsilon), y_t^{\text{det}} + \eta_t^0)$. Thus

$$\eta_t^0 = \int_0^t V(t,s)\tilde{c}(y_s^{\text{det}}, \eta_s^0)\,\mathrm{d}s + \sigma'\int_0^t V(t,s)\widehat{G}(y_s^{\text{det}}, \eta_s^0)\,\mathrm{d}W_s. \qquad (5.1.114)$$

The norm of the two integrals is then bounded as in (5.1.71) and (5.1.72) on short timescales, and on longer timescales by adding contributions of short intervals. The proof of (5.1.111) is similar. The probability can be estimated above by the sum of (5.1.112), and the probability of $\|y_s - y_s^0\|$ exceeding h despite $\|y_s^0 - y_s^{\text{det}}\|$ remaining small. This last contribution is then estimated with the help of an SDE satisfied by $y_s - y_s^0$, in which the noise term has order $\sigma'h$. □

The probability (5.1.111) becomes small as soon as $h \gg \sigma\sqrt{\varepsilon}(1+\chi^{(2)}(t))^{1/2}$ (taking h_1 large enough for the first term on the right-hand side to dominate). Thus one can say that the reduced stochastic system provides an approximation of the slow dynamics to order $\sigma\sqrt{\varepsilon}(1+\chi^{(2)}(t))^{1/2}$. The behaviour of $\chi^{(2)}(t)$ depends on the deterministic reduced dynamics: In stable situations, for instance in the vicinity of an asymptotically stable fixed point or periodic orbit, it remains bounded. In unstable situations, however, it can grow exponentially fast, with an exponent given by the (local) Lyapunov exponent of the orbit, and thus the approximation is good only on timescales of order $|\log(\sigma\sqrt{\varepsilon})|$.

The probability (5.1.112) becomes small as soon as $h_1 \gg \sigma'(1+\chi^{(2)}(t))^{1/2}$, where $\sigma' = \rho\sigma$. Thus for $\sigma' > \sigma\sqrt{\varepsilon}$ (that is, $\rho > \sqrt{\varepsilon}$), the reduced stochastic system (5.1.109) provides a better approximation of the slow dynamics than the reduced deterministic system (5.1.110). This can be understood as being an effect of the stronger intensity of noise acting on the slow dynamics.

5.1.4 Refined Concentration Results

In this section, we improve the results of Theorem 5.1.6 and Theorem 5.1.17 by giving a more precise description of the deviation of sample paths from deterministic solutions $(x_t^{\mathrm{det}}, y_t^{\mathrm{det}})$ on the adiabatic manifold. This is useful, in particular, to control the first exit-time $\tau_{\mathcal{D}_0}$ of y_t from the domain $\mathcal{D}_0$ in which the slow manifold is uniformly asymptotically stable.

As in the proof of Theorem 5.1.6, we use the change of variables

$$\begin{aligned} x_t &= \bar{x}(y_t^{\mathrm{det}} + \eta_t, \varepsilon) + \xi_t \ , \\ y_t &= y_t^{\mathrm{det}} + \eta_t \ , \end{aligned} \tag{5.1.115}$$

which produces (compare (5.1.32)) a nonlinear perturbation of the linear system

$$\begin{aligned} \mathrm{d}\xi_t^0 &= \frac{1}{\varepsilon} A(y_t^{\mathrm{det}}) \xi_t^0 \, \mathrm{d}t + \frac{\sigma}{\sqrt{\varepsilon}} F_0(y_t^{\mathrm{det}}) \, \mathrm{d}W_t \ , \\ \mathrm{d}\eta_t^0 &= \left[C(y_t^{\mathrm{det}}) \xi_t^0 + B(y_t^{\mathrm{det}}) \eta_t^0 \right] \mathrm{d}t + \sigma' G_0(y_t^{\mathrm{det}}) \, \mathrm{d}W_t \ . \end{aligned} \tag{5.1.116}$$

The coefficients are the matrices defined in (5.1.7) and (5.1.33). We can rewrite this system in compact form as

$$\mathrm{d}\zeta_t^0 = \mathcal{A}(y_t^{\mathrm{det}}) \zeta_t^0 \, \mathrm{d}t + \sigma \mathcal{F}_0(y_t^{\mathrm{det}}) \, \mathrm{d}W_t \ , \tag{5.1.117}$$

where $(\zeta^0)^{\mathrm{T}} = ((\xi^0)^{\mathrm{T}}, (\eta^0)^{\mathrm{T}})$ and

$$\mathcal{A}(y_t^{\mathrm{det}}) = \begin{pmatrix} \frac{1}{\varepsilon} A(y_t^{\mathrm{det}}) & 0 \\ C(y_t^{\mathrm{det}}) & B(y_t^{\mathrm{det}}) \end{pmatrix}, \quad \mathcal{F}_0(y_t^{\mathrm{det}}) = \begin{pmatrix} \frac{1}{\sqrt{\varepsilon}} F_0(y_t^{\mathrm{det}}) \\ \rho G_0(y_t^{\mathrm{det}}) \end{pmatrix} \ . \tag{5.1.118}$$

The solution of the linear SDE (5.1.116) is given by

$$\zeta_t^0 = \mathcal{U}(t)\zeta_0 + \sigma \int_0^t \mathcal{U}(t,s) \mathcal{F}_0(y_s^{\mathrm{det}}) \, \mathrm{d}W_s \ , \tag{5.1.119}$$

where $\mathcal{U}(t,s)$ denotes the principal solution of the deterministic homogeneous system $\dot{\zeta} = \mathcal{A}(y_t^{\mathrm{det}}) \zeta$. It can be written in the form

$$\mathcal{U}(t,s) = \begin{pmatrix} U(t,s) & 0 \\ S(t,s) & V(t,s) \end{pmatrix} \ . \tag{5.1.120}$$

Recall that $U(t,s)$ and $V(t,s)$ denote, respectively, the principal solutions of $\varepsilon \dot{\xi} = A(y_t^{\mathrm{det}}) \xi$ and $\dot{\eta} = B(y_t^{\mathrm{det}}) \eta$, while

$$S(t,s) = \int_s^t V(t,u) C(y_u^{\mathrm{det}}) U(u,s) \, \mathrm{d}u \ . \tag{5.1.121}$$

The Gaussian process ζ_t^0 has a covariance matrix of the form

$$\mathrm{Cov}(\zeta_t^0) = \sigma^2 \int_0^t \mathcal{U}(t,s)\mathcal{F}_0(y_s^{\mathrm{det}})\mathcal{F}_0(y_s^{\mathrm{det}})^{\mathrm{T}}\mathcal{U}(t,s)^{\mathrm{T}}\, ds$$
$$= \sigma^2 \begin{pmatrix} X(t) & Z(t) \\ Z(t)^{\mathrm{T}} & Y(t) \end{pmatrix}. \tag{5.1.122}$$

The matrices $X(t) \in \mathbb{R}^{n\times n}$, $Y(t) \in \mathbb{R}^{m\times m}$ and $Z(t) \in \mathbb{R}^{n\times m}$ are a particular solution of the following slow–fast system, which generalises (5.1.10):

$$\begin{aligned}
\varepsilon \dot{X} &= A(y)X + XA(y)^{\mathrm{T}} + F_0(y)F_0(y)^{\mathrm{T}}, \\
\varepsilon \dot{Z} &= A(y)Z + \varepsilon ZB(y)^{\mathrm{T}} + \varepsilon XC(y)^{\mathrm{T}} + \sqrt{\varepsilon}\rho F_0(y)G_0(y)^{\mathrm{T}}, \\
\dot{Y} &= B(y)Y + YB(y)^{\mathrm{T}} + C(y)Z + Z^{\mathrm{T}}C(y)^{\mathrm{T}} + \rho^2 G_0(y)G_0(y)^{\mathrm{T}}, \\
\dot{y} &= g(\bar{x}(y,\varepsilon), y).
\end{aligned} \tag{5.1.123}$$

This system admits a slow manifold given by

$$\begin{aligned}
X &= X^\star(y), \\
Z &= Z^\star(y) = -\sqrt{\varepsilon}\rho A(y)^{-1}F_0(y)G_0(y)^{\mathrm{T}} + \mathcal{O}(\varepsilon),
\end{aligned} \tag{5.1.124}$$

where $X^\star(y)$ is given by (5.1.14). It is straightforward to check that this manifold is uniformly asymptotically stable for sufficiently small ε, so that Fenichel's theorem yields the existence of an adiabatic manifold $X = \bar{X}(y,\varepsilon)$, $Z = \bar{Z}(y,\varepsilon)$, at a distance of order ε from the slow manifold. This manifold attracts nearby solutions of (5.1.123) exponentially fast, and thus asymptotically, the expectations of $\xi_t^0(\xi_t^0)^{\mathrm{T}}$ and $\xi_t^0(\eta_t^0)^{\mathrm{T}}$ will be close, respectively, to $\sigma^2 \bar{X}(y_t^{\mathrm{det}},\varepsilon)$ and $\sigma^2 \bar{Z}(y_t^{\mathrm{det}},\varepsilon)$.

In general, the matrix $Y(t)$ cannot be expected to approach some asymptotic value depending only on y_t^{det} and ε. In fact, if the deterministic orbit y_t^{det} is repelling, $\|Y(t)\|$ can grow exponentially fast. In order to measure this growth, we use again the functions

$$\chi^{(1)}(t) = \sup_{0 \leqslant s \leqslant t} \int_0^s \left(\sup_{u \leqslant v \leqslant s} \|V(s,v)\|\right) du, \tag{5.1.125}$$

$$\chi^{(2)}(t) = \sup_{0 \leqslant s \leqslant t} \int_0^s \left(\sup_{u \leqslant v \leqslant s} \|V(s,v)\|^2\right) du. \tag{5.1.126}$$

The solution of (5.1.123) with initial condition $Y(0) = Y_0$ satisfies

$$\begin{aligned}
Y(t;Y_0) = &\, V(t)Y_0 V(t)^{\mathrm{T}} \\
&+ \rho^2 \int_0^t V(t,s)G_0(y_s^{\mathrm{det}})G_0(y_s^{\mathrm{det}})^{\mathrm{T}}V(t,s)^{\mathrm{T}}\, ds \\
&+ \mathcal{O}((\varepsilon + \rho\sqrt{\varepsilon})\chi^{(2)}(t)).
\end{aligned} \tag{5.1.127}$$

We thus define an "asymptotic" covariance matrix $\bar{\bar{Z}}(t) = \bar{\bar{Z}}(t;Y_0,\varepsilon)$ by

$$\bar{\mathcal{Z}}(t; Y_0, \varepsilon) = \begin{pmatrix} \bar{X}(y_t^{\mathrm{det}}, \varepsilon) & \bar{Z}(y_t^{\mathrm{det}}, \varepsilon) \\ \bar{Z}(y_t^{\mathrm{det}}, \varepsilon)^{\mathrm{T}} & Y(t; Y_0) \end{pmatrix}, \tag{5.1.128}$$

and use $\bar{\mathcal{Z}}(t)^{-1}$ to characterise the ellipsoidal region in which $\zeta(t)$ is concentrated. The result is the following.

Theorem 5.1.18 (Concentration around a deterministic solution). *Assume that $\|\bar{X}(y_s^{\mathrm{det}}, \varepsilon)\|$ and $\|\bar{X}(y_s^{\mathrm{det}}, \varepsilon)^{-1})\|$ are uniformly bounded for $0 \leqslant s \leqslant t$ and that Y_0 has been chosen in such a way that $\|Y(s)^{-1}\| = \mathcal{O}(1/(\rho^2+\varepsilon))$ for $0 \leqslant s \leqslant t$. Fix an initial condition (x_0, y_0) with $y_0 \in \mathcal{D}_0$ and $x_0 = \bar{x}(y_0, \varepsilon)$, and let t be such that $y_s^{\mathrm{det}} \in \mathcal{D}_0$ for all $s \leqslant t$. Define*

$$R(t) = \|\bar{\mathcal{Z}}\|_{[0,t]} \Big[1 + \Big(1 + \|Y^{-1}\|_{[0,t]}^{1/2} \Big) \chi^{(1)}(t) + \chi^{(2)}(t) \Big]. \tag{5.1.129}$$

Let $\zeta_t^{\mathrm{T}} = (\xi_t^{\mathrm{T}}, \eta_t^{\mathrm{T}})$ denote the deviation of sample paths from the deterministic solution $(\bar{x}(y_t^{\mathrm{det}}, \varepsilon), y_t^{\mathrm{det}})$. Then there exist constants $\varepsilon_0, \Delta_0, h_0, L > 0$, independent of Y_0, y_0 and t, such that

$$\mathbb{P}^{x_0, y_0} \bigg\{ \sup_{0 \leqslant s \leqslant t \wedge \tau_{\mathcal{D}_0}} \langle \zeta_u, \bar{\mathcal{Z}}(u)^{-1} \zeta_u \rangle \geqslant h^2 \bigg\} \leqslant C_{n+m, \gamma, \Delta}(t, \varepsilon) \mathrm{e}^{-\kappa h^2 / 2\sigma^2} \tag{5.1.130}$$

holds, whenever $\varepsilon \leqslant \varepsilon_0$, $\Delta \leqslant \Delta_0$, $h \leqslant h_0 R(t)^{-1}$ and $0 < \gamma < 1$. Here

$$\kappa = \gamma \Big[1 - \mathcal{O}(\varepsilon + \Delta + h R(t)) \Big], \tag{5.1.131}$$

$$C_{n+m, \gamma, \Delta}(t, \varepsilon) = L \Big(1 + \frac{t}{\Delta \varepsilon} \Big) \Big[\Big(\frac{1}{1-\gamma} \Big)^{(n+m)/2} + \mathrm{e}^{(n+m)/4} \Big].$$

Proof. The proof differs from the one of Theorem 5.1.6 only in a few points on which we comment here. First we need bounds on the inverse of $\bar{\mathcal{Z}} = \bar{\mathcal{Z}}(u)$, which is given by

$$\bar{\mathcal{Z}}^{-1} = \begin{pmatrix} (\bar{X} - \bar{Z} Y^{-1} \bar{Z}^{\mathrm{T}})^{-1} & -\bar{X}^{-1} \bar{Z} (Y - \bar{Z}^{\mathrm{T}} \bar{X}^{-1} \bar{Z})^{-1} \\ -Y^{-1} \bar{Z}^{\mathrm{T}} (\bar{X} - \bar{Z} Y^{-1} \bar{Z}^{\mathrm{T}})^{-1} & (Y - \bar{Z}^{\mathrm{T}} \bar{X}^{-1} \bar{Z})^{-1} \end{pmatrix}. \tag{5.1.132}$$

Our assumptions imply that all entries are of order 1, except possibly the lower right one, which may reach order $1/(\rho^2 + \varepsilon)$. When extending the proof of Lemma 5.1.10, the functions $\Phi(t)$ and $\Psi(t)$ defined in (5.1.47) are replaced by

$$\widehat{\Phi}(t) = \int_0^t \mathrm{Tr} \big[\mathcal{J}(v)^{\mathrm{T}} \mathcal{U}(t,v)^{\mathrm{T}} \bar{\mathcal{Z}}(t)^{-1} \mathcal{U}(t,v) \mathcal{J}(v) \big] \, \mathrm{d}v,$$

$$\widehat{\Psi}(t) = \int_0^t \big\| \mathcal{J}(v)^{\mathrm{T}} \mathcal{U}(t,v)^{\mathrm{T}} \bar{\mathcal{Z}}(t)^{-1} \mathcal{U}(t,v) \mathcal{J}(v) \big\| \, \mathrm{d}v, \tag{5.1.133}$$

where

$$\mathcal{J}(v) = \frac{1}{\sqrt{2}M_1 h \|\bar{\mathcal{Z}}\|_\infty^{1/2}} \mathcal{F}_1(\zeta_v, v, \varepsilon) = \begin{pmatrix} \mathcal{O}(\frac{1}{\sqrt{\varepsilon}}) \\ \mathcal{O}(\rho) \end{pmatrix} \tag{5.1.134}$$

as long as $\langle \zeta_u, \bar{\mathcal{Z}}(u)^{-1}\zeta(u)\rangle \leqslant h^2$. Using the representations (5.1.120) of $\mathcal{U}$ and (5.1.132) of $\bar{\mathcal{Z}}^{-1}$ and expanding the matrix product, one obtains the relations

$$\widehat{\Phi}(t) \leqslant \Phi(t) + \rho^2 \int_0^t \mathrm{Tr}\big[V(t,v)^\mathrm{T} Y(t)^{-1} V(t,v)\big]\,dv$$
$$+ \mathcal{O}\big((n+m)(1+\chi^{(1)}(t)+\chi^{(2)}(t))\big),$$
$$\widehat{\Psi}(t) \leqslant \Psi(t) + \rho^2 \int_0^t \|V(t,v)^\mathrm{T} Y(t)^{-1} V(t,v)\|\,dv$$
$$+ \mathcal{O}\big(1+\chi^{(1)}(t)+\chi^{(2)}(t)\big). \tag{5.1.135}$$

Finally, to extend the proof of Proposition 5.1.11, one needs to show that the symmetric matrix $\mathcal{Q}(u)$, defined by

$$\mathcal{Q}(u)^2 = \mathcal{U}(u,s)^\mathrm{T}\bar{\mathcal{Z}}(u)^{-1}\mathcal{U}(u,s), \tag{5.1.136}$$

satisfies $\|\mathcal{Q}(u)\mathcal{Q}(t)^{-1}\| = 1 + \mathcal{O}(\Delta)$ for $u \in [t-\Delta\varepsilon, t]$. This follows from the estimate

$$\mathcal{U}(s,v)\mathcal{F}_0(y_v^{\mathrm{det}})\mathcal{F}_0(y_v^{\mathrm{det}})^\mathrm{T}\mathcal{U}(s,v)^\mathrm{T} = \begin{pmatrix} \mathcal{O}(1/\varepsilon) & \mathcal{O}(\Delta+\rho/\sqrt{\varepsilon}) \\ \mathcal{O}(\Delta+\rho/\sqrt{\varepsilon}) & \mathcal{O}(\Delta^2\varepsilon+\rho^2) \end{pmatrix}, \tag{5.1.137}$$

and the analogue of the differential equation (5.1.41). The remainder of the proof is similar. □

Concerning the implications of this result, let us first discuss timescales of order 1. Then the functions $\|\bar{\mathcal{Z}}\|_{[0,t]}$, $\chi^{(1)}(t)$ and $\chi^{(2)}(t)$ are at most of order 1, and $\|Y(t)^{-1}\|$ remains of the same order as $\|Y_0^{-1}\|$. The probability (5.1.130) becomes small as soon as $h \gg \sigma$. Because of the restriction $h \leqslant h_0 R(t)^{-1}$, the result is useful provided $\|Y^{-1}\|_{[0,t]} \ll \sigma^{-2}$. In order to obtain the optimal concentration result, we have to choose Y_0 according to two opposed criteria. On the one hand, we would like to take the smallest possible Y_0, in order to obtain the optimal domain of concentration $\langle \zeta_u, \bar{\mathcal{Z}}(u)^{-1}\zeta_u\rangle < h^2$. On the other hand, $\|Y_0^{-1}\|$ must not exceed certain bounds for Theorem 5.1.18 to be valid. Thus we require that

$$Y_0 > \big[\sigma^2 \vee (\rho^2 + \varepsilon)\big]1\!\!1_m \tag{5.1.138}$$

(in the sense of positive definite matrices). Because of the Gaussian decay of the probability (5.1.130) in σ/h, we can interpret the theorem by saying

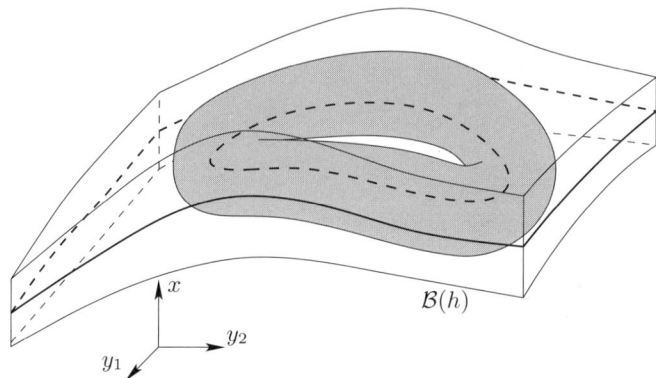

Fig. 5.3. Domain of concentration of sample paths around an asymptotically stable periodic orbit (*broken closed curve*), lying in a uniformly asymptotically stable adiabatic manifold.

that the typical spreading of paths in the y-direction is of order $\sigma(\rho + \sqrt{\varepsilon})$ if $\sigma < \rho + \sqrt{\varepsilon}$ and of order σ^2 if $\sigma > \rho + \sqrt{\varepsilon}$.

The term ρ is clearly due to the intensity $\sigma' = \rho\sigma$ of the noise acting on the slow variable. It prevails if $\rho > \sigma \vee \sqrt{\varepsilon}$. The term $\sqrt{\varepsilon}$ is due to the linear part of the coupling between slow and fast variables, while the behaviour in σ^2 observed when $\sigma > \rho + \sqrt{\varepsilon}$ can be traced back to the *nonlinear* coupling between slow and fast variables.

For longer timescales, the condition $h \leqslant h_0 R(t)^{-1}$ obliges us to take a larger Y_0, while $Y(t)$ typically grows with time. If the largest Lyapunov exponent of the deterministic orbit y_t^{det} is positive, this growth is exponential in time, so that the spreading of paths along the adiabatic manifold will reach order 1 in a time of order $\log|\sigma \vee (\rho^2 + \varepsilon)|$.

Example 5.1.19 (Asymptotically stable periodic orbit). Assume that y_t^{det} is an asymptotically stable periodic orbit with period T, entirely contained in $\mathcal{D}_0$ (and not too close to its boundary). Then all coefficients in (5.1.123) depend periodically on time, and, in particular, Floquet's theorem allows us to write

$$V(t) = P(t) e^{\Lambda t}, \qquad (5.1.139)$$

where $P(t)$ is a T-periodic matrix. The asymptotic stability of the orbit means that all eigenvalues but one of the monodromy matrix Λ have strictly negative real parts, the last eigenvalue, which corresponds to translations along the orbit, being 0. In that case, $\chi^{(1)}(t)$ and $\chi^{(2)}(t)$ grow only linearly with time, so that the spreading of paths in the y-direction remains small on timescales of order $1/(\sigma \vee (\rho^2 + \varepsilon))$.

In fact, we even expect this spreading to occur mainly along the periodic orbit, while the paths remain confined to a neighbourhood of the orbit on subexponential timescales. To see that this is true, we can use a new set of variables in the neighbourhood of the orbit. In order not to introduce too many new notations, we will replace y by (y, z), where $y \in \mathbb{R}^{m-1}$ describes the degrees of freedom transversal to the orbit, and $z \in \mathbb{R}$ parametrises the motion along the orbit. In fact, we can use an equal-time parametrisation of the orbit, so that $\dot{z} = 1$ on the orbit, i.e., we have $z_t^{\text{det}} = t \pmod{T}$. The SDE takes the form

$$
\begin{aligned}
\mathrm{d}x_t &= \frac{1}{\varepsilon} f(x_t, y_t, z_t) \, \mathrm{d}t + \frac{\sigma}{\sqrt{\varepsilon}} F(x_t, y_t, z_t) \, \mathrm{d}W_t \, , \\
\mathrm{d}y_t &= g(x_t, y_t, z_t) \, \mathrm{d}t + \sigma' G(x_t, y_t, z_t) \, \mathrm{d}W_t \, , \qquad (5.1.140) \\
\mathrm{d}z_t &= \bigl[1 + h(x_t, y_t, z_t)\bigr] \, \mathrm{d}t + \sigma' H(x_t, y_t, z_t) \, \mathrm{d}W_t \, ,
\end{aligned}
$$

where $h = \mathcal{O}(\|y_t\|^2 + \|x_t - x_t^{\text{det}}\|^2)$ and the characteristic multipliers associated with the periodic matrix $\partial_y g(x_t^{\text{det}}, 0, z_t^{\text{det}}, \varepsilon)$ are strictly smaller than one in modulus. As linear approximation of the dynamics of $\xi_t = x_t - x_t^{\text{det}}$ and $\eta_t = y_t - y_t^{\text{det}} = y_t$ we take

$$
\begin{aligned}
\mathrm{d}\xi_t^0 &= \frac{1}{\varepsilon} A(z_t^{\text{det}}) \xi_t^0 \, \mathrm{d}t + \frac{\sigma}{\sqrt{\varepsilon}} F_0(z_t^{\text{det}}) \, \mathrm{d}W_t \, , \\
\mathrm{d}\eta_t^0 &= \bigl[B(z_t^{\text{det}}) \eta_t^0 + C(z_t^{\text{det}}) \xi_t^0\bigr] \, \mathrm{d}t + \sigma' G_0(z_t^{\text{det}}) \, \mathrm{d}W_t \, , \qquad (5.1.141) \\
\mathrm{d}z_t^0 &= \mathrm{d}t + \sigma' H_0(z_t^{\text{det}}) \, \mathrm{d}W_t \, ,
\end{aligned}
$$

which depends periodically on time. One can again compute the covariance matrix of the Gaussian process $(\xi_t^0, \eta_t^0, z_t^0)$ as a function of the principal solutions U and V associated with A and B. In particular, the covariance matrix $Y(t)$ of η_t^0 still obeys the ODE

$$
\begin{aligned}
\dot{Y} &= B(z) Y + Y B(z)^{\mathrm{T}} + C(z) \bar{Z} + \bar{Z}^{\mathrm{T}} C(z)^{\mathrm{T}} + \rho^2 G_0(z) G_0(z)^{\mathrm{T}} \, , \\
\dot{z} &= 1 \, . \qquad (5.1.142)
\end{aligned}
$$

This is now a linear, inhomogeneous ODE with time-periodic coefficients. It is well known that such a system admits a unique periodic solution Y_t^{per}, which is of order $\rho^2 + \varepsilon$ since $\bar{Z}$ is of order $\rho\sqrt{\varepsilon} + \varepsilon$ and $\rho^2 G_0 G_0^{\mathrm{T}}$ is of order ρ^2. We can thus define an asymptotic covariance matrix $\bar{\mathcal{Z}}(t)$ of (ξ_t^0, η_t^0), which depends periodically on time. If $\zeta_t = (\xi_t, \eta_t)$, Theorem 5.1.18 shows that on timescales of order 1 (at least), the paths ζ_t are concentrated in a set of the form $\langle \zeta_t, \bar{\mathcal{Z}}(t)^{-1} \zeta_t \rangle < h^2$ (Fig. 5.3), while z_t remains h-close to z_t^{det}.

On longer timescales, the distribution of paths will be smeared out along the periodic orbit. However, the same line of reasoning as in Section 5.1.2, based on a comparison with different deterministic solutions on successive time intervals of order 1, can be used to show that ζ_t remains concentrated in the set $\langle \zeta_t, \bar{\mathcal{Z}}(t)^{-1} \zeta_t \rangle < h^2$ up to exponentially long timescales.

5.2 Periodic Orbits

In this section, we consider the effect of noise on slow–fast systems in which the fast system admits, for each value of the slow variable, an asymptotically stable periodic orbit. We have seen in Section 2.3.1 that solutions of the slow–fast system then track the slowly varying periodic orbit at a distance of order ε, while the dynamics of the slow variables is well-approximated by the system averaged over the fast motion along the orbit.

We will start, in Section 5.2.1, by considering the dynamics of the fast variables only, for frozen slow variables. We establish concentration and reduction results similar to those of the previous section. Then we pass to the general slow–fast case in Section 5.2.2.

5.2.1 Dynamics near a Fixed Periodic Orbit

We consider here an SDE of the form

$$\mathrm{d}x_s = f(x_s)\,\mathrm{d}s + \sigma F(x_s)\,\mathrm{d}W_s\;, \qquad (5.2.1)$$

where W_s is a k-dimensional Brownian motion, in the case where the deterministic system $\dot{x} = f(x)$ admits an asymptotically stable periodic orbit. More precisely, we will make the following assumptions.

Assumption 5.2.1 (Asymptotically stable periodic orbit).
- Domain and differentiability: *There are integers $n, k \geqslant 1$ such that $f \in \mathcal{C}^2(\mathcal{D}, \mathbb{R}^n)$, and $F \in \mathcal{C}^1(\mathcal{D}, \mathbb{R}^{n\times k})$, where $\mathcal{D}$ is an open subset of $\mathbb{R}^n$. We further assume that f, F and all their partial derivatives up to order 2, respectively 1, are uniformly bounded in norm in $\mathcal{D}$ by a constant M.*
- Periodic orbit: *There is a periodic function $\gamma^\star : \mathbb{R} \to \mathbb{R}^n$, of period T, such that*

$$\dot\gamma^\star(s) = f(\gamma^\star(s)) \qquad (5.2.2)$$

  for all $s \in \mathbb{R}$.
- Stability: *Let $A(s) = \partial_x f(\gamma^\star(s))$, and let $U(s)$ be the principal solution of $\dot\xi = A(s)\xi$. Then all eigenvalues but one of $U(T)$ are strictly smaller than 1 in modulus.*

In a neighbourhood of the periodic orbit, we may introduce a system of coordinates which separates the motions along the orbit, and transversal to it, setting for instance

$$x = \gamma^\star(T\theta) + r\;, \qquad (5.2.3)$$

where r is perpendicular to $\dot\gamma^\star(T\theta)$. In the deterministic case, drawing on the fact that $f(\gamma^\star(T\theta) + r) = \dot\gamma^\star(T\theta) + A(T\theta)r + \mathcal{O}(r^2)$, one obtains evolution equations of the form

$$\dot\theta = \frac{1}{T} + \mathcal{O}(\|r\|^2) , \qquad (5.2.4)$$
$$\dot r = A(T\theta)r + \mathcal{O}(\|r\|^2) .$$

Applying Itô's formula in the stochastic case yields a system of the form

$$\begin{aligned}
\mathrm{d}\theta_s &= \left[\frac{1}{T} + b_\theta(\theta_s, r_s; \sigma)\right] \mathrm{d}s + \sigma F_\theta(\theta_s, r_s)\, \mathrm{d}W_s , \\
\mathrm{d}r_s &= \left[A(T\theta_s)r_s + b_r(\theta_s, r_s; \sigma)\right] \mathrm{d}s + \sigma F_r(\theta_s, r_s)\, \mathrm{d}W_s .
\end{aligned} \qquad (5.2.5)$$

The functions b_θ and b_r contain terms of order r^2 and $n\sigma^2$. In principle, it would be possible to eliminate the terms of order $n\sigma^2$ in b_r, which are due to the second-order terms in Itô's formula, by a further change of variables. But this will not be necessary for our purposes.

As we expect r_t to remain small, it seems reasonable to approximate the dynamics of (5.2.5) by the linear equation

$$\mathrm{d}r_s^0 = A(T\theta_s^{\mathrm{det}})r_s^0\, \mathrm{d}s + \sigma F_r(\theta_s^{\mathrm{det}}, 0)\, \mathrm{d}W_s , \qquad (5.2.6)$$

where $\theta_s^{\mathrm{det}} = \theta_0 + s/T$. The process r_t^0 is Gaussian, with zero mean and covariance matrix $\sigma^2 X_s$, where X_s satisfies the linear ODE with periodic coefficients

$$\dot X = A(T\theta_s^{\mathrm{det}})X + X A(T\theta_s^{\mathrm{det}})^{\mathrm{T}} + F_r(\theta_s^{\mathrm{det}}, 0) F_r(\theta_s^{\mathrm{det}}, 0)^{\mathrm{T}} . \qquad (5.2.7)$$

This equation admits a unique periodic solution $X^{\mathrm{per}}(s)$, which describes the asymptotic covariance of r_s^0. We are thus led to introducing the set

$$\mathcal{B}(h) = \left\{(\theta, r) \colon \langle r, X^{\mathrm{per}}(T\theta)^{-1} r\rangle < h^2\right\} , \qquad (5.2.8)$$

which has the shape of a filled torus surrounding the periodic orbit. In doing so, we assumed that $X^{\mathrm{per}}(T\theta)$ is nonsingular for every θ. In the sequel, we will need to assume that in addition, $X^{\mathrm{per}}(T\theta)^{-1}$ is uniformly bounded in norm. A sufficient condition for this to hold true is that FF^{T} be positive definite along the orbit.

Theorem 5.2.2 (Stochastic motion near an asymptotically stable periodic orbit). *If $X^{\mathrm{per}}(T\theta)^{-1}$ is bounded in norm, then there exist constants $\varepsilon_0, \Delta_0, h_0, c_1, L > 0$ such that for all $\varepsilon \leqslant \varepsilon_0$, $\Delta \leqslant \Delta_0$, $h \leqslant h_0$, $\gamma \in (0,1)$, and all $s > 0$*

$$\mathbb{P}^{\theta_0, 0}\{\tau_{\mathcal{B}(h)} < s\} \leqslant C^+_{h/\sigma, n, \gamma, \Delta}(s)\, \mathrm{e}^{-\kappa_+ h^2/2\sigma^2} , \qquad (5.2.9)$$

where the exponent κ_+ is uniform in time and satisfies

$$\kappa_+ = \gamma\bigl[1 - c_1\bigl(h|\log h| + \Delta + \sigma^2/h^2\bigr)\bigr] , \qquad (5.2.10)$$

and the prefactor is given by

$$C^+_{h/\sigma, n, \gamma, \Delta}(s) = L\frac{(1+s)^2}{\Delta}\bigl[(1-\gamma)^{-n} + \mathrm{e}^{n/4}\bigr]\left(1 + \frac{h^2}{\sigma^2}\right) . \qquad (5.2.11)$$

Proof. The proof resembles the proof of Theorem 5.1.6 for $\varepsilon = 1$. One first bounds the probability that $\langle r_s, X^{\text{per}}(T\theta_s^{\text{det}})^{-1} r_s \rangle$ exceeds h^2, in a similar way as in Proposition 5.1.11. Bounding the probability of $\theta_s - \theta_s^{\text{det}}$ becoming larger than h_1 is much simpler as in Proposition 5.1.12, because the linear part of the equation for $\theta_s - \theta_s^{\text{det}}$ vanishes. As a consequence, the functions $\chi^{(i)}(s)$ grow only linearly with s. Another difference arises for the analogue of the endpoint estimate in Lemma 5.1.14, because the last error term in the exponent is of order $e^{-2K_0 T_0}/(1-\gamma)$ instead of $e^{-2K_0 T_0/\varepsilon}/(1-\gamma)$. When completing the proof by restarting the process at multiples of T_0, the optimal error term is obtained by choosing T_0 proportional to $|\log h|$, which yields an error of order $h|\log h|$ in the exponent. □

This result confirms that sample paths are likely to stay in $\mathcal{B}(h)$ up to times of order $e^{h^2/2\sigma^2}$. One is thus tempted to approximate the dynamics of the phase θ_s by the reduced equation

$$d\theta_s^0 = \frac{1}{T} ds + \sigma F_\theta(\theta_s^0, 0) dW_s . \qquad (5.2.12)$$

Note that since θ^0 lives on a circle, an examination of the Fokker–Planck equation with periodic boundary conditions shows that this equation admits a an invariant density of the form $1 + \mathcal{O}(\sigma^2)$. One expects that the distribution of θ_s^0 behaves approximately like a Gaussian with expectation s/T and variance of order $\sigma^2 s$ (which then has to be folded back to the circle). It is not difficult to adapt Theorem 5.1.17 to obtain the following estimates.

Theorem 5.2.3 (Reduction to phase dynamics). *There exist constants κ_1, κ_2, L such that, for any choice of h, h_1 at most up to order $1/s$,*

$$\mathbb{P}^{\theta_0, 0}\left\{ \sup_{0 \leqslant u \leqslant s \wedge \tau_{\mathcal{B}(h)}} \|\theta_u - \theta_u^0\| \geqslant h \right\} \leqslant L(1+s) \exp\left\{ -\frac{\kappa_1 h_1^2}{\sigma^2 (h^2 + h_1^2)s} \right\} , \qquad (5.2.13)$$

and

$$\mathbb{P}^{\theta_0}\left\{ \sup_{0 \leqslant u \leqslant s} \|\theta_u^0 - \theta_u^{\text{det}}\| \geqslant h_1 \right\} \leqslant L(1+s) \exp\left\{ -\frac{\kappa_2 h_1^2}{\sigma^2 s} \right\} . \qquad (5.2.14)$$

Proof. (5.2.13) follows from the fact that $\zeta_s = \theta_s - \theta_s^0$ satisfies an equation of the form

$$d\zeta_s = b_\theta(\theta_s^0 + \zeta_s, r_s) ds + \sigma\left[F_\theta(\theta_s^0 + \zeta_s, r_s) - F_\theta(\theta_s^0, r_s) \right] dW_s . \qquad (5.2.15)$$

For $s \leqslant \tau_{\mathcal{B}(h)}$, the diffusion term is of order $h + |\zeta_s|$, and so the result follows as in Proposition 5.1.12. The proof of (5.2.14) is similar (but much easier, since the equation for $\theta_s^0 - \theta_s^{\text{det}}$ contains no drift term). □

Estimate (5.2.14) shows that the spreading of the approximate phase θ_s^0 indeed grows like $\sigma\sqrt{s}$ around $\theta_s^{\det} = \theta_0 + s/T$. It thus takes a time of order $1/\sigma^2$ for sample paths to spread uniformly along the periodic orbit.[2]

Since the probability to leave $\tau_{\mathcal{B}(h)}$ before time s is small whenever $h \gg \sigma$, and the upper bound (5.2.13) becomes small as soon as $h_1^2 \gg \sigma^2 h^2 s \gg \sigma^4 s$, θ_s^0 approximates the real phase θ_s up to order $\sigma^2\sqrt{s}$, and is thus a better approximation of the real dynamics of θ_s than $\theta_s^{\det}$.

5.2.2 Dynamics near a Slowly Varying Periodic Orbit

We return now to slow–fast SDEs of the form

$$\begin{aligned} \mathrm{d}x_t &= \frac{1}{\varepsilon} f(x_t, y_t)\,\mathrm{d}t + \frac{\sigma}{\sqrt{\varepsilon}} F(x_t, y_t)\,\mathrm{d}W_t \;, \\ \mathrm{d}y_t &= g(x_t, y_t)\,\mathrm{d}t + \sigma' G(x_t, y_t)\,\mathrm{d}W_t \;, \end{aligned} \qquad (5.2.16)$$

where we assume that the deterministic system $x' = f(x,y)$ admits an asymptotically stable periodic orbit for each fixed y. More precisely, we shall make the following assumptions.

Assumption 5.2.4 (Slowly varying asymptotically stable periodic orbit).

- Domain and differentiability: *There are integers $n, m, k \geqslant 1$ such that $f \in \mathcal{C}^2(\mathcal{D}, \mathbb{R}^n)$, $g \in \mathcal{C}^2(\mathcal{D}, \mathbb{R}^m)$ and $F \in \mathcal{C}^1(\mathcal{D}, \mathbb{R}^{n\times k})$, $G \in \mathcal{C}^1(\mathcal{D}, \mathbb{R}^{m\times k})$, where $\mathcal{D}$ is an open subset of $\mathbb{R}^n \times \mathbb{R}^m$. We further assume that f, g, F, G and all their partial derivatives up to order 2, respectively 1, are uniformly bounded in norm in $\mathcal{D}$ by a constant M.*
- Periodic orbit: *There are an open subset $\mathcal{D}_0 \subset \mathbb{R}^m$, constants $T_2 > T_1 > 0$, and continuous functions $T : \mathcal{D}_0 \to [T_1, T_2]$, $\gamma^\star : \mathbb{R} \times \mathcal{D}_0 \to \mathbb{R}^n$, such that $\gamma^\star(t + T(y), y) = \gamma^\star(t, y)$ and*

$$\partial_t \gamma^\star(t, y) = f(\gamma^\star(t, y), y) \qquad (5.2.17)$$

for all (t, y) in $\mathbb{R} \times \mathcal{D}_0$.
- Stability: *Let $A(t, y) = \partial_x f(\gamma^\star(t, y), y)$ denote the linearisation of the fast vector field at the periodic orbit. Then all characteristic multipliers but one of the linear time-periodic equation $\xi' = A(t, y)\xi$, for fixed y, are strictly smaller than one in modulus.*

Recall from Section 2.3.2 that there exists an invariant tube tracking the family of periodic orbits, whose parametrisation we denote by $\bar{\Gamma}(\theta, y, \varepsilon) = \gamma^\star(T(y)\theta, y) + \mathcal{O}(\varepsilon)$. The change of variables $x = \bar{\Gamma}(\theta, y, \varepsilon) + r$, with r perpendicular to the orbit, yields a system of the form

---

[2] This fact is also compatible with spectral-theoretic results, which show that the generator of (5.2.12) has a spectral gap of order σ^2.

176   5 Multi-Dimensional Slow–Fast Systems

$$
\begin{aligned}
\mathrm{d}\theta_t &= \frac{1}{\varepsilon}\left[\frac{1}{T(y_t)} + b_\theta(\theta_t, r_t, y_t)\right]\mathrm{d}t + \frac{\sigma}{\sqrt{\varepsilon}}F_\theta(\theta_t, r_t, y_t)\,\mathrm{d}W_t\;, \\
\mathrm{d}r_t &= \frac{1}{\varepsilon}\bigl[\bar{A}(\theta_t, y_t)r_t + b_r(\theta_t, r_t, y_t)\bigr]\mathrm{d}t + \frac{\sigma}{\sqrt{\varepsilon}}F_r(\theta_t, r_t, y_t)\,\mathrm{d}W_t\;, \quad (5.2.18) \\
\mathrm{d}y_t &= \tilde{g}(\theta_t, r_t, y_t)\,\mathrm{d}t + \sigma'\widetilde{G}(\theta_t, r_t, y_t)\,\mathrm{d}W_t\;,
\end{aligned}
$$

where we have set $\bar{A}(\theta, y) = \partial_x f(\bar{\varGamma}(\theta, y, \varepsilon), y)$. The remainders $b_r(\theta, r, y)$ and $b_\theta(\theta, r, y)$ contain terms of order $\|r\|^2$ and terms of order $(n+m)\sigma^2$ stemming from Itô's formula. The functions $\tilde{g}$ and $\widetilde{G}$ are obtained by expressing g and G in the new variables.

The dynamics of r_t may be approximated by the linear equation

$$
\mathrm{d}r_t^0 = \frac{1}{\varepsilon}\bar{A}(\theta_t^{\mathrm{det}}, y_t^{\mathrm{det}})r_t^0\,\mathrm{d}t + \frac{\sigma}{\sqrt{\varepsilon}}F_r(\theta_t^{\mathrm{det}}, 0, y_t^{\mathrm{det}})\,\mathrm{d}W_t\;, \qquad (5.2.19)
$$

where $(\theta_t^{\mathrm{det}}, y_t^{\mathrm{det}})$ is a solution of the deterministic system, restricted to the invariant tube parametrised by $\bar{\varGamma}(\theta, y, \varepsilon)$. The process r_t^0 is Gaussian, with zero mean and covariance matrix $\sigma^2 X_t$, where X_t is a solution of the deterministic slow–fast system

$$
\begin{aligned}
\varepsilon\dot{X} &= \bar{A}(\theta, y)X + X\bar{A}(\theta, y)^{\mathrm{T}} + F(\theta, 0, y)F(\theta, 0, y)^{\mathrm{T}}\;, \\
\varepsilon\dot{\theta} &= \frac{1}{T(y)}\;, \qquad (5.2.20) \\
\dot{y} &= \tilde{g}(\theta, 0, y)\;.
\end{aligned}
$$

We already know from the previous section that for fixed y, this system admits a periodic solution $X^{\mathrm{per}}(\theta, y)$. This solution is asymptotically stable because the principal solution associated with $\bar{A}$ is contracting. We may thus conclude from Theorem 2.3.1 that (5.2.20) admits a solution of the form

$$
\bar{X}(\theta, y, \varepsilon) = X^{\mathrm{per}}(\theta, y) + \mathcal{O}(\varepsilon)\;. \qquad (5.2.21)
$$

We thus proceed to define the concentration domain of paths as

$$
\mathcal{B}(h) = \bigl\{(\theta, r, y)\colon \langle r, \bar{X}(\theta, y, \varepsilon)^{-1}r\rangle < h^2\bigr\}\;, \qquad (5.2.22)
$$

assuming again that $\bar{X}(\theta, y, \varepsilon)$ is invertible. This set is now a macaroni-shaped object, with thickness of order h, surrounding the invariant tube (Fig. 5.4). The following concentration result is proved exactly as Theorem 5.1.6.

Theorem 5.2.5 (Stochastic motion near a slowly varying asymptotically stable periodic orbit). *If $X^{\mathrm{per}}(\theta, y)^{-1}$ is bounded in norm, there exist constants $\varepsilon_0, \Delta_0, h_0, c, c_1, L > 0$ such that for all $\varepsilon \leqslant \varepsilon_0$, $\Delta \leqslant \Delta_0$, $h \leqslant h_0$ and all $\gamma \in (0, 1)$,*

$$
\mathbb{P}^{\theta_0, 0, y_0}\bigl\{\tau_{\mathcal{B}(h)} < t\bigr\} \leqslant C^+_{h/\sigma, n, m, \gamma, \Delta}(t, \varepsilon)\mathrm{e}^{-\kappa_+ h^2/2\sigma^2}\;, \qquad (5.2.23)
$$

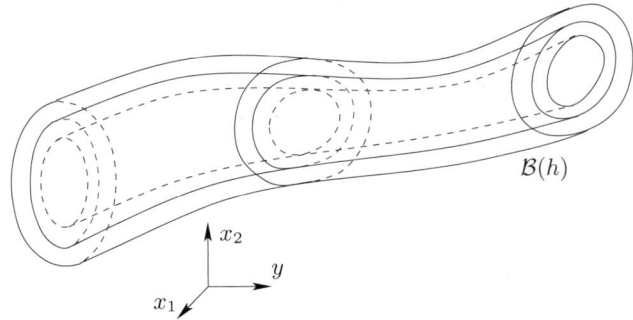

Fig. 5.4. Example of a macaroni-shaped domain of concentration of paths $\mathcal{B}(h)$ around a family of asymptotically stable periodic orbits, for one slow dimension and two fast dimensions.

where the exponent κ_+ is uniform in time and satisfies

$$\kappa_+ = \gamma\big[1 - c_1\big(h + \Delta + (n+m)\sigma^2/h^2 + \mathrm{e}^{-c/\varepsilon}/(1-\gamma)\big)\big]\,, \qquad (5.2.24)$$

and the prefactor is given by

$$C^+_{h/\sigma,n,m,\gamma,\Delta}(t,\varepsilon) = L\frac{(1+t)^2}{\Delta\varepsilon}\big[(1-\gamma)^{-n} + \mathrm{e}^{n/4} + \mathrm{e}^{m/4}\big]\bigg(1 + \frac{h^2}{\sigma^2}\bigg)\,. \qquad (5.2.25)$$

Remark 5.2.6. A natural question to ask is whether the dynamics of the full system (5.2.18) can be approximated by the dynamics of the reduced system

$$\begin{aligned}
\mathrm{d}\theta_t^0 &= \frac{1}{\varepsilon}\bigg[\frac{1}{T(y_t^0)} + b_\theta(\theta_t^0,0,y_t^0)\bigg]\mathrm{d}t + \frac{\sigma}{\sqrt{\varepsilon}}F_\theta(\theta_t^0,0,y_t^0)\,\mathrm{d}W_t\,,\\
\mathrm{d}y_t^0 &= \tilde{g}(\theta_t^0,0,y_t^0)\,\mathrm{d}t + \sigma'\widetilde{G}(\theta_t^0,0,y_t^0)\,\mathrm{d}W_t\,.
\end{aligned} \qquad (5.2.26)$$

It turns out, however, that due to the coupling with the slow variables via the period $T(y)$, the deviation $\theta_t - \theta_t^0$ can grow relatively fast (on the slow timescale). In the worst case, all information on the phase is lost after a time of order $\varepsilon|\log\sigma|$.

Remark 5.2.7. Another interesting question, which we do not pursue in detail here, is whether the dynamics of the slow variables can be approximated by a reduced equation. A natural candidate is the averaged reduced equation

$$\mathrm{d}\bar{y}_t^0 = \bar{g}(\bar{y}_t^0)\,\mathrm{d}t + \sigma'\overline{G}(\bar{y}_t^0)\,\mathrm{d}W_t\,, \qquad (5.2.27)$$

obtained by averaging $\tilde{g}$ and $\widetilde{G}$ over the angle θ. To analyse the accuracy of the approximation, one should start from Equation (5.2.18), and construct a change of variables decreasing the order in ε of the θ-dependent terms. One can then estimate deviations of the resulting process from the reduced process $\bar{y}_t^0$.

5.3 Bifurcations

In this section, we turn to the effect of noise on slow–fast systems admitting bifurcation points. We have already studied the effect of noise on the most generic one-dimensional bifurcations in Chapter 3, and so we concentrate here on specifically multidimensional effects. In Section 5.3.1, we consider situations in which the number q of bifurcating modes is smaller than the dimension n of the fast variable. We give results on the concentration of sample paths near invariant manifolds of dimension q, and on a possible reduction to a lower-dimensional equation, containing only q fast variables. In Section 5.3.2 we consider the effect of noise on dynamic Hopf bifurcations.

5.3.1 Concentration Results and Reduction

Let us consider a slow–fast SDE in the usual form

$$
\begin{aligned}
\mathrm{d}x_t &= \frac{1}{\varepsilon} f(x_t, y_t) \, \mathrm{d}t + \frac{\sigma}{\sqrt{\varepsilon}} F(x_t, y_t) \, \mathrm{d}W_t \; , \\
\mathrm{d}y_t &= g(x_t, y_t) \, \mathrm{d}t + \sigma' G(x_t, y_t) \, \mathrm{d}W_t \; ,
\end{aligned}
\tag{5.3.1}
$$

in the case where the associated system admits a bifurcation point. More precisely, we shall make the following assumptions.

Assumption 5.3.1 (Dynamic bifurcation).

- Domain and differentiability: *There are integers $n, m, k \geqslant 1$ such that $f \in \mathcal{C}^2(\mathcal{D}, \mathbb{R}^n)$, $g \in \mathcal{C}^2(\mathcal{D}, \mathbb{R}^m)$ and $F \in \mathcal{C}^1(\mathcal{D}, \mathbb{R}^{n \times k})$, $G \in \mathcal{C}^1(\mathcal{D}, \mathbb{R}^{m \times k})$, where $\mathcal{D}$ is an open subset of $\mathbb{R}^n \times \mathbb{R}^m$. We further assume that f, g, F, G and all their partial derivatives up to order 2, respectively 1, are uniformly bounded in norm in $\mathcal{D}$ by a constant M.*
- Bifurcation point: *There is a point $(\widehat{x}, \widehat{y}) \in \mathcal{D}$ such that $f(\widehat{x}, \widehat{y}) = 0$ and the Jacobian matrix $A = \partial_x f(\widehat{x}, \widehat{y})$ admits q eigenvalues on the imaginary axis, with $1 \leqslant q < n$. The $n - q$ other eigenvalues of A have strictly negative real parts.*

As in Section 2.2.1, we introduce coordinates $(x^-, z) \in \mathbb{R}^{n-q} \times \mathbb{R}^q$, with $x^- \in \mathbb{R}^{n-q}$ and $z \in \mathbb{R}^q$, in which the matrix $\partial_x f(\widehat{x}, \widehat{y})$ becomes block-diagonal, with a block $A^- \in \mathbb{R}^{(n-q) \times (n-q)}$ having eigenvalues in the left half-plane, and a block $A^0 \in \mathbb{R}^{q \times q}$ having eigenvalues on the imaginary axis. We call the components of z *bifurcating modes* and those of x^- *stable modes*. In the deterministic case $\sigma = 0$, we established in Section 2.2.1 the existence of a locally attracting invariant centre manifold

$$
\widehat{\mathcal{M}}_\varepsilon = \{(x^-, z, y) \colon x^- = \bar{x}^-(z, y, \varepsilon), (z, y) \in \mathcal{N}\} \tag{5.3.2}
$$

in a sufficiently small neighbourhood $\mathcal{N}$ of $(\widehat{z}, \widehat{y}) \in \mathbb{R}^q \times \mathbb{R}^m$.

We can rewrite the system (5.3.1) in (x^-, z, y)-coordinates as

$$dx_t^- = \frac{1}{\varepsilon} f^-(x_t^-, z_t, y_t) \, dt + \frac{\sigma}{\sqrt{\varepsilon}} F^-(x_t^-, z_t, y_t) \, dW_t \;,$$

$$dz_t = \frac{1}{\varepsilon} f^0(x_t^-, z_t, y_t) \, dt + \frac{\sigma}{\sqrt{\varepsilon}} F^0(x_t^-, z_t, y_t) \, dW_t \;, \qquad (5.3.3)$$

$$dy_t = g(x_t^-, z_t, y_t) \, dt + \sigma' G(x_t^-, z_t, y_t) \, dW_t \;.$$

Since x_t^- describes the stable modes, we expect it to remain close to the function $\bar{x}^-(z, y, \varepsilon)$ parameterising the invariant manifold $\widehat{\mathcal{M}}_\varepsilon$. We thus consider the deviation $\xi_t^- = x_t^- - \bar{x}^-(z_t, y_t, \varepsilon)$ of sample paths from $\widehat{\mathcal{M}}_\varepsilon$. It satisfies an SDE of the form

$$d\xi_t^- = \frac{1}{\varepsilon} \widehat{f}^-(\xi_t^-, z_t, y_t) \, dt + \frac{\sigma}{\sqrt{\varepsilon}} \widehat{F}^-(\xi_t^-, z_t, y_t) \, dW_t \;, \qquad (5.3.4)$$

where a straightforward computation, using Itô's formula, shows that

$$\widehat{f}^-(\xi^-, z, y) = f^-(\bar{x}^-(z, y, \varepsilon) + \xi^-, z, y) \qquad (5.3.5)$$
$$- \partial_z \bar{x}^-(z, y, \varepsilon) f^0(\bar{x}^-(z, y, \varepsilon) + \xi^-, z, y)$$
$$- \varepsilon \partial_y \bar{x}^-(z, y, \varepsilon) g(\bar{x}^-(z, y, \varepsilon) + \xi^-, z, y) + \mathcal{O}((m+q)\sigma^2) \;,$$

the last term being due to the second-order term in Itô's formula. The invariance of $\widehat{\mathcal{M}}_\varepsilon$ implies that all terms in (5.3.5) except for the last one vanish for $\xi^- = 0$, compare (2.2.4). The linearisation of $\widehat{f}^-$ at $\xi^- = 0$ can be approximated by the matrix

$$A^-(z, y, \varepsilon) = \partial_{x^-} f^-(\bar{x}^-(z, y, \varepsilon), z, y) - \partial_z \bar{x}^-(z, y, \varepsilon) \partial_{x^-} f^0(\bar{x}^-(z, y, \varepsilon), z, y)$$
$$- \varepsilon \partial_y \bar{x}^-(z, y, \varepsilon) \partial_{x^-} g(\bar{x}^-(z, y, \varepsilon), z, y) \;. \qquad (5.3.6)$$

Since $A^-(\widehat{z}, \widehat{y}, 0) = \partial_{x^-} f^-(\widehat{x}^-, \widehat{z}, \widehat{y}) = A^-$, the eigenvalues of $A^-(z, y, \varepsilon)$ have negative real parts, bounded away from zero, provided we take $\mathcal{N}$ and ε small enough. In the sequel, we shall no longer indicate the ε-dependence of A^-.

We now approximate the dynamics of (ξ_t^-, z_t, y_t) by the linear system

$$d\xi_t^0 = \frac{1}{\varepsilon} A^-(z_t^{\text{det}}, y_t^{\text{det}}) \xi_t^0 \, dt + \frac{\sigma}{\sqrt{\varepsilon}} F_0^-(z_t^{\text{det}}, y_t^{\text{det}}) \, dW_t \;,$$

$$dz_t^{\text{det}} = \frac{1}{\varepsilon} f^0(\bar{x}^-(z_t^{\text{det}}, y_t^{\text{det}}, \varepsilon), z_t^{\text{det}}, y_t^{\text{det}}) \, dt \;, \qquad (5.3.7)$$

$$dy_t^{\text{det}} = g(\bar{x}^-(z_t^{\text{det}}, y_t^{\text{det}}, \varepsilon), z_t^{\text{det}}, y_t^{\text{det}}) \, dt \;,$$

where $F_0^-(z, y) = \widehat{F}^-(0, z, y)$ is the value of the diffusion coefficient on the invariant manifold. The process ξ_t^0 is Gaussian with zero mean and covariance matrix X_t^-, where X_t^- obeys the slow–fast ODE

$$\varepsilon \dot{X}^- = A^-(z, y) X^- + X^- A^-(z, y)^{\text{T}} + F_0^-(z, y) F_0^-(z, y)^{\text{T}} \;,$$
$$\varepsilon \dot{z} = f^0(\bar{x}^-(z, y, \varepsilon), z, y) \;, \qquad (5.3.8)$$
$$\dot{y} = g(\bar{x}^-(z, y, \varepsilon), z, y) \;.$$

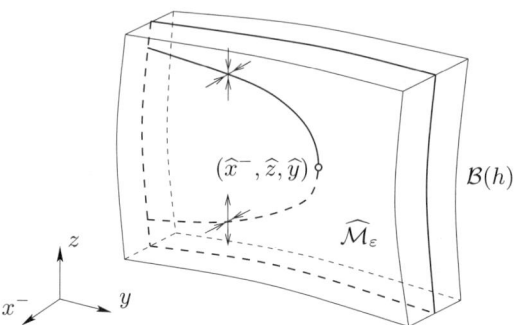

Fig. 5.5. Example of a bifurcation point, for one slow and two fast variables. The bifurcation shown here is a saddle–node bifurcation, with arrows indicating the stability of the equilibrium branches. Sample paths are concentrated in a set $\mathcal{B}(h)$ surrounding the deterministic invariant manifold $\widehat{\mathcal{M}}_\varepsilon$.

Since all eigenvalues of $A^-(z,y)$ have negative real parts, this system admits a locally attracting invariant manifold $X^- = \bar{X}^-(z,y,\varepsilon)$ for $(z,y) \in \mathcal{N}$ and ε small enough. We use this manifold to define the domain of concentration of paths

$$\mathcal{B}^-(h) = \{(x^-, z, y) \colon (z,y) \in \mathcal{N},$$
$$\langle x^- - \bar{x}^-(z,y,\varepsilon), \bar{X}^-(z,y,\varepsilon)^{-1}(x^- - \bar{x}^-(z,y,\varepsilon))\rangle < h^2\}. \quad (5.3.9)$$

In the sequel, we will need the stopping times

$$\tau_{\mathcal{B}_-(h)} = \inf\{t > 0 \colon (x_t^-, z_t, y_t) \notin \mathcal{B}^-(h)\}, \quad (5.3.10)$$
$$\tau_\mathcal{N} = \inf\{t > 0 \colon (z_t, y_t) \notin \mathcal{N}\}. \quad (5.3.11)$$

Then the following result shows that sample paths are concentrated in $\mathcal{B}^-(h)$ as long as (z_t, y_t) remains in $\mathcal{N}$.

Theorem 5.3.2 (Multidimensional bifurcation). *Assume that the norms $\|\bar{X}^-(z,y,\varepsilon)\|$ and $\|\bar{X}^-(z,y,\varepsilon)^{-1}\|$ are uniformly bounded in $\mathcal{N}$. Choose a deterministic initial condition $(z_0, y_0) \in \mathcal{N}$, $x_0^- = \bar{x}^-(z_0, y_0, \varepsilon)$. Then, there exist constants $h_0, \Delta_0, L > 0$ and $\nu \in (0,1]$ such that for all $h \leqslant h_0$, all $\Delta \leqslant \Delta_0$ and all $0 < \gamma < 1$,*

$$\mathbb{P}^{x_0^-, z_0, y_0}\{\tau_{\mathcal{B}^-(h)} < t \wedge \tau_\mathcal{N}\} \leqslant C_{h/\sigma, n, m, q, \gamma, \Delta}(t, \varepsilon) e^{-\kappa h^2 / 2\sigma^2}, \quad (5.3.12)$$

provided $\varepsilon|\log(h(1-\gamma))| \leqslant 1$. Here the exponent satisfies

$$\kappa = \gamma\bigl[1 - \mathcal{O}(\Delta) - \mathcal{O}(h^\nu(1-\gamma)^{1-\nu}|\log(h(1-\gamma))|)\bigr], \quad (5.3.13)$$

and the prefactor is given by

5.3 Bifurcations

$$C_{h/\sigma,n,m,q,\gamma,\Delta}(t,\varepsilon) = L\left(1 + \frac{t}{\Delta\varepsilon}\right)\left(1 + \frac{t}{\varepsilon}\right)\left(1 + \frac{h^2}{\sigma^2}\right) \tag{5.3.14}$$
$$\times \left[(1-\gamma)^{-(n-q)} + e^{(n-q)/4} + e^{m/4} + e^{q/4}\right].$$

Proof. Given constants $h_y, h_z > 0$, we introduce stopping times

$$\tau_y = \inf\{s > 0 \colon \|y_s - y_s^{\text{det}}\| \geqslant h_y\}, \tag{5.3.15}$$
$$\tau_z = \inf\{s > 0 \colon \|z_s - z_s^{\text{det}}\| \geqslant h_z\}. \tag{5.3.16}$$

Using similar ideas as in Proposition 5.1.11, one starts by proving that for ξ_0^- satisfying $\langle \xi_0^-, \overline{X}^-(y_0, z_0, \varepsilon)^{-1}\xi_0^- \rangle \leqslant h^2$, one has

$$\mathbb{P}^{\xi_0^-, z_0, y_0}\left\{\sup_{0 \leqslant s \leqslant t \wedge \tau_{\mathcal{B}^-(h)} \wedge \tau_z} \|y_s - y_s^{\text{det}}\| \geqslant h_y\right\}$$
$$\leqslant 2\left[\frac{t}{\Delta\varepsilon}\right]e^{m/4}\exp\left\{-\kappa_0 \frac{h_y^2(1 - \mathcal{O}(\Delta\varepsilon))}{\sigma^2(\rho^2 + \varepsilon)\chi^{(2)}(t)}\right.$$
$$\left.\times \left[1 - \mathcal{O}\left(\chi^{(1)}(t)h_y\left(1 + \frac{h^2}{h_y^2} + \frac{h_z^2}{h_y^2} + (m+q)\frac{\sigma^2}{h_y^2}\right)\right)\right]\right\}, \tag{5.3.17}$$

where the $\chi^{(i)}(t)$ are the quantities introduced in (5.1.66) and (5.1.67), which measure the growth of perturbations in the y-direction. Similarly,

$$\mathbb{P}^{\xi_0^-, z_0, y_0}\left\{\sup_{0 \leqslant s \leqslant t \wedge \tau_{\mathcal{B}^-(h)} \wedge \tau_y} \|z_s - z_s^{\text{det}}\| \geqslant h_z\right\}$$
$$\leqslant 2\left[\frac{t}{\Delta\varepsilon}\right]e^{q/4}\exp\left\{-\kappa_0 \frac{\varepsilon h_z^2(1 - \mathcal{O}(\Delta\varepsilon))}{\sigma^2 \chi_z^{(2)}(t)}\right.$$
$$\left.\times \left[1 - \mathcal{O}\left(\chi_z^{(1)}(t)h_z\left(1 + \frac{h^2}{h_z^2} + \frac{h_y^2}{h_z^2} + (m+q)\frac{\sigma^2}{h_z^2}\right)\right)\right]\right\}, \tag{5.3.18}$$

where the $\chi_z^{(i)}(t)$ are defined analogously as the $\chi^{(i)}(t)$, but for perturbations in the z-direction.

Let $T > 0$ be a constant that we will specify below. Choosing h_y proportional to h, h_z proportional to $(1 + \chi_z^{(2)}(T)/\varepsilon)^{1/2}h$, and proceeding as in Propositions 5.1.11 and 5.1.13, one obtains that for $\langle \xi_0^-, \overline{X}^-(y_0, z_0, \varepsilon)^{-1}\xi_0^- \rangle \leqslant \alpha^2 h^2$,

$$\mathbb{P}^{\xi_0^-, z_0, y_0}\left\{\sup_{0 \leqslant s \leqslant T \wedge \tau_N} \langle \xi_s^-, \overline{X}^-(y_s, z_s, \varepsilon)^{-1}\xi_s^- \rangle \geqslant h^2\right\}$$
$$\leqslant C_{h/\sigma,n,m,q,\gamma,\Delta}(T,\varepsilon)e^{-\kappa^+(\alpha)h^2/2\sigma^2} \tag{5.3.19}$$

holds with an exponent

$$\kappa^+(\alpha) = \gamma\left[1 - \alpha^2 - \mathcal{O}(\Delta) - \mathcal{O}\left(\left(1 + \frac{\chi_z^{(2)}(T)}{\varepsilon}\right)h\right)\right]. \tag{5.3.20}$$

The probability that $\langle \xi_T^-, \bar{X}^-(y_T, z_T, \varepsilon)^{-1} \xi_T^- \rangle$ exceeds h satisfies a similar bound as in Lemma 5.1.14, with an exponent of the form

$$\kappa' = \gamma \left[1 - \mathcal{O}(\Delta) - \mathcal{O}\left(\left(1 + \frac{\chi_z^{(2)}(T)}{\varepsilon} \right) h \right) - \mathcal{O}\left(\frac{\mathrm{e}^{-2K_0 T/\varepsilon}}{1-\gamma} \right) \right]. \quad (5.3.21)$$

The main difference with the end of the proof of Theorem 5.1.6 lies in the choice of the time T at which we restart the process when considering longer timescales. It has to be such that both error terms $\chi_z^{(2)}(T) h/\varepsilon$ and $\mathrm{e}^{-2K_0 T/\varepsilon}/(1-\gamma)$ in the exponents are small. In the worst case, perturbations in the z-direction grow like $\mathrm{e}^{K_+ t/\varepsilon}$ for some $K_+ > 0$, and then $\chi_z^{(2)}(T)$ is of order $T \mathrm{e}^{2K_+ T/\varepsilon}$. The optimal choice of T is then close to $\varepsilon \theta$, where

$$\mathrm{e}^{-\theta} = \left[h(1-\gamma) \right]^{1/(2(K_0 + K_+))}, \quad (5.3.22)$$

and this yields the error term $h^{\nu}(1-\gamma)^{1-\nu} |\log(h(1-\gamma))|$ in the exponent (5.3.13), with $\nu = K_0/(K_0 + K_+)$. $\square$

Theorem 5.3.2 shows that sample paths tend to concentrate in a neighbourhood of order σ of the invariant manifold $\widehat{\mathcal{M}}_\varepsilon$. This suggests to approximate the original system by its projection

$$\begin{aligned}
\mathrm{d}z_t^0 &= \frac{1}{\varepsilon} f^0(\bar{x}^-(z_t^0, y_t^0, \varepsilon), z_t^0, y_t^0) \, \mathrm{d}t + \frac{\sigma}{\sqrt{\varepsilon}} F^0(\bar{x}^-(z_t^0, y_t^0, \varepsilon), z_t^0, y_t^0) \, \mathrm{d}W_t, \\
\mathrm{d}y_t^0 &= g(\bar{x}^-(z_t^0, y_t^0, \varepsilon), z_t^0, y_t^0) \, \mathrm{d}t + \sigma' G(\bar{x}^-(z_t^0, y_t^0, \varepsilon), z_t^0, y_t^0) \, \mathrm{d}W_t,
\end{aligned} \quad (5.3.23)$$

called the *reduced stochastic system*. Being of smaller dimension, this system is in general easier to analyse than the original equation. In particular, if $\sigma' = 0$ and z is one-dimensional (that is, $q = 1$), we recover equations of the type studied in Chapter 3.

In order to quantify the deviations of $\zeta_t^0 = (z_t^0, y_t^0)$ from (z_t, y_t), we can proceed similarly as in Section 5.1.3. However, we would like to avoid having to compare with the deterministic solution, and instead compare both processes directly. To do this, we need to take into account the fact that ζ_t^0 is a random process. For a fixed initial condition $\zeta_0^0 \in \mathcal{N}$, we define the (random) matrices

$$B(\zeta_t^0, \varepsilon) = \begin{pmatrix} \partial_z f^0 & \partial_y f^0 \\ \varepsilon \partial_z g & \varepsilon \partial_y g \end{pmatrix} \bigg|_{x^- = \bar{x}^-(z_t^0, y_t^0, \varepsilon), z = z_t^0, y = y_t^0}, \quad (5.3.24)$$

$$C(\zeta_t^0, \varepsilon) = \begin{pmatrix} \partial_{x^-} f^0 \\ \varepsilon \partial_{x^-} g \end{pmatrix} \bigg|_{x^- = \bar{x}^-(z_t^0, y_t^0, \varepsilon), z = z_t^0, y = y_t^0}. \quad (5.3.25)$$

Observe that $C((\hat{z}, \hat{y}), 0) = 0$ because of our choice of coordinates, so that $\|C(\zeta_t^0, \varepsilon)\|$ will be small in a neighbourhood of the origin. We denote, for each realisation $\zeta^0(\omega)$, by $\mathcal{V}_\omega$ the principal solution[3] of

---

[3] Note that we may assume that almost all realisations $\zeta^0(\omega)$ are continuous.

$$\mathrm{d}\zeta_t(\omega) = \frac{1}{\varepsilon} B(\zeta_t^0(\omega), \varepsilon) \zeta_t(\omega) \, \mathrm{d}t \; . \tag{5.3.26}$$

We need to assume the existence of deterministic functions $\vartheta(t,s)$, $\vartheta_C(t,s)$, and a stopping time $\tau \leqslant \tau_{\mathcal{B}^-(h)}$ such that

$$\|\mathcal{V}_\omega(t,s)\| \leqslant \vartheta(t,s) \; , \qquad \|\mathcal{V}_\omega(t,s) C(\zeta_s^0(\omega), \varepsilon)\| \leqslant \vartheta_C(t,s) \tag{5.3.27}$$

hold for all $s \leqslant t \leqslant \tau(\omega)$ and (almost) all paths $(\zeta_u^0(\omega))_{u \geqslant 0}$ of (5.3.23). Then we define

$$\begin{aligned}\chi^{(i)}(t) &= \sup_{0 \leqslant s \leqslant t} \frac{1}{\varepsilon} \int_0^s \vartheta(s,u)^i \, \mathrm{d}u \; , \\ \chi_C^{(i)}(t) &= \sup_{0 \leqslant s \leqslant t} \frac{1}{\varepsilon} \int_0^s \Bigl(\sup_{u \leqslant v \leqslant s} \vartheta_C(s,v)^i\Bigr) \mathrm{d}u\end{aligned} \tag{5.3.28}$$

for $i = 1, 2$, and the following result holds.

Theorem 5.3.3 (Reduction near bifurcation point). *Assume that there exist constants $\Delta, \vartheta_0 > 0$ (of order 1) such that $\vartheta(s,u) \leqslant \vartheta_0$ and $\vartheta_C(s,u) \leqslant \vartheta_0$ whenever $0 < s - u \leqslant \Delta \varepsilon$. Then there are constants $h_0, \kappa_0, L > 0$ such that for all $h \leqslant h_0 [\chi^{(1)}(t) \vee \chi_C^{(1)}(t)]^{-1}$ and all initial conditions $(x_0^-, z_0^0, y_0^0) \in \mathcal{B}^-(h)$,*

$$\mathbb{P}^{x_0^-, z_0^0, y_0^0}\Bigl\{\sup_{0 \leqslant s \leqslant t \wedge \tau} \|(z_s, y_s) - (z_s^0, y_s^0)\| \geqslant h\Bigr\}$$
$$\leqslant C_{m,q}(t,\varepsilon) \exp\Bigl\{-\kappa_0 \frac{h^2}{2\sigma^2} \frac{1}{\chi_C^{(2)}(t) + h \chi_C^{(1)}(t) + h^2 \chi^{(2)}(t)}\Bigr\} \; , \tag{5.3.29}$$

where

$$C_{m,q}(t,\varepsilon) = L\Bigl(1 + \frac{t}{\varepsilon}\Bigr) e^{(m+q)/4} \; . \tag{5.3.30}$$

Proof. The deviation $\zeta_t = (z_t - z_t^0, y_t - y_t^0)$ satisfies an SDE of the form

$$\begin{aligned}\mathrm{d}\xi_t^- &= \frac{1}{\varepsilon}\bigl[A^-(\zeta_t^0)\xi_t^- + b(\xi_t^-, \zeta_t, \zeta_t^0)\bigr] \mathrm{d}t + \frac{\sigma}{\sqrt{\varepsilon}} \widetilde{F}(\xi_t^-, \zeta_t, \zeta_t^0) \, \mathrm{d}W_t \; , \\ \mathrm{d}\zeta_t &= \frac{1}{\varepsilon}\bigl[C(\zeta_t^0)\xi_t^- + B(\zeta_t^0)\zeta_t + c(\xi_t^-, \zeta_t, \zeta_t^0)\bigr] \mathrm{d}t + \frac{\sigma}{\sqrt{\varepsilon}} \widetilde{\mathcal{G}}(\xi_t^-, \zeta_t, \zeta_t^0) \, \mathrm{d}W_t \; ,\end{aligned} \tag{5.3.31}$$

where $\|b\|$ is of order $\|\xi^-\|^2 + \|\zeta\|^2 + (m+q)\sigma^2$, $\|c\|$ is of order $\|\xi^-\|^2 + \|\zeta\|^2$ and $\|\widetilde{\mathcal{G}}\|$ is of order $\|\xi^-\| + \|\zeta\|$, while $\|\widetilde{F}\|$ is bounded.

For a given continuous sample path $\{\zeta_t^0(\omega)\}_{t \geqslant 0}$, we denote by U_ω and $\mathcal{V}_\omega$ the principal solutions of $\varepsilon \dot{\xi}^- = A^-(\zeta_t^0(\omega))\xi^-$ and $\varepsilon \dot{\zeta} = B(\zeta_t^0(\omega))\zeta$. If we further define

$$\mathcal{S}_\omega(t,s) = \frac{1}{\varepsilon} \int_s^t \mathcal{V}_\omega(t,u) C(\zeta_u^0(\omega)) U_\omega(u,s) \, \mathrm{d}u \; , \tag{5.3.32}$$

we can write the solution of (5.3.31) as

$$
\begin{aligned}
\zeta_t(\omega) &= \frac{\sigma}{\sqrt{\varepsilon}} \int_0^t \mathcal{V}_\omega(t,s) \widetilde{\mathcal{G}}(\xi_s^-(\omega), \zeta_s(\omega), \zeta_s^0(\omega)) \, \mathrm{d}W_s(\omega) \\
&+ \frac{\sigma}{\sqrt{\varepsilon}} \int_0^t \mathcal{S}_\omega(t,s) \widetilde{F}(\xi_s^-(\omega), \zeta_s(\omega), \zeta_s^0(\omega)) \, \mathrm{d}W_s(\omega) \\
&+ \frac{1}{\varepsilon} \int_0^t \mathcal{V}_\omega(t,s) c(\xi_s^-(\omega), \zeta_s(\omega), \zeta_s^0(\omega)) \, \mathrm{d}s \\
&+ \frac{1}{\varepsilon} \int_0^t \mathcal{S}_\omega(t,s) b(\xi_s^-(\omega), \zeta_s(\omega), \zeta_s^0(\omega)) \, \mathrm{d}s \, .
\end{aligned}
\qquad (5.3.33)
$$

Concerning the first two summands in (5.3.33), note that the identities

$$
\begin{aligned}
\mathcal{V}_\omega(t,s) &= \mathcal{V}_\omega(t,0) \mathcal{V}_\omega(s,0)^{-1}, \\
\mathcal{S}_\omega(t,s) &= \mathcal{S}_\omega(t,0) \mathcal{U}_\omega(s,0)^{-1} + \mathcal{V}_\omega(t,0) \mathcal{S}_\omega(s,0)^{-1}
\end{aligned}
\qquad (5.3.34)
$$

allow to rewrite the stochastic integrals in such a way that the integrands are adapted with respect to the filtration generated by $\{W_s\}_{s \geq 0}$. Now the remainder of the proof follows closely the proof of Proposition 5.1.12, using the relations (5.3.27). □

Theorem 5.3.3 shows that typical solutions of the reduced system (5.3.23) approximate solutions of the initial system (5.3.5) up to order $\sigma \chi_C^{(2)}(t)^{1/2} + \sigma^2 \chi_C^{(1)}(t)$, as long as $\chi^{(1)}(t) \ll 1/\sigma$. Checking the validity of Condition (5.3.27) for a reasonable stopping time τ is, of course, not straightforward, but it depends only on the dynamics of the reduced system, which is usually easier to analyse.

Example 5.3.4 (Pitchfork bifurcation). Assume the reduced equation has the form

$$
\begin{aligned}
\mathrm{d}z_t^0 &= \frac{1}{\varepsilon} [y_t^0 z_t^0 - (z_t^0)^3] \, \mathrm{d}t + \frac{\sigma}{\sqrt{\varepsilon}} \, \mathrm{d}W_t, \\
\mathrm{d}y_t^0 &= \mathrm{d}t,
\end{aligned}
\qquad (5.3.35)
$$

i.e., there is a pitchfork bifurcation at the origin. We fix an initial time $t_0 < 0$ and choose an initial condition (z_0, y_0) with $y_0 = t_0$, so that $y_t^0 = t$. In Section 3.4 we proved that if $\sigma \leq \sqrt{\varepsilon}$, the paths $\{z_s\}_{s \geq t_0}$ are concentrated, up to time $\sqrt{\varepsilon}$, in a strip of width of order $\sigma/(|y^0|^{1/2} \vee \varepsilon^{1/4})$ around the corresponding deterministic solution.

Using for τ the first-exit time from a set of this form, one finds that $\chi_C^{(2)}(\sqrt{\varepsilon})$ is of order $\sqrt{\varepsilon} + \sigma^2/\varepsilon$ and that $\chi_C^{(1)}(\sqrt{\varepsilon})$ is of order $1 + \sigma/\varepsilon^{3/4}$. Thus, up to time $\sqrt{\varepsilon}$, the typical spreading of z_s around reduced solutions z_s^0 is at most of order $\sigma \varepsilon^{1/4} + \sigma^2/\sqrt{\varepsilon}$, which is smaller than the spreading of z_s^0 around a deterministic solution. Hence the reduced system provides a good approximation to the full system up to time $\sqrt{\varepsilon}$.

From then on, however, $x_C^{(2)}(t)$ grows like $e^{t^2/\varepsilon}$ until the paths leave a neighbourhood of the unstable equilibrium $z = 0$, which typically occurs at a time of order $\sqrt{\varepsilon|\log\sigma|}$. Thus the spreading is too fast for the reduced system to provide a good approximation to the dynamics. This shows that Theorem 5.3.3 is not quite sufficient to reduce the problem to a one-dimensional one, and a more detailed description has to be used for the region of instability.

5.3.2 Hopf Bifurcation

In this section, we consider the case where the fast system admits a Hopf bifurcation point. In order to keep the discussion reasonably simple, rather than considering the most general case, we will restrict our attention to situations in which

- the diffusion coefficient for the fast variable depends only on the slow variable,
- there is no noise term acting on the slow variable,
- the slow variable is one-dimensional, while the fast variable and the Brownian motion are two-dimensional.

We shall thus consider slow–fast SDEs of the form

$$\mathrm{d}x_t = \frac{1}{\varepsilon}f(x_t, y_t)\,\mathrm{d}t + \frac{\sigma}{\sqrt{\varepsilon}}F(y_t)\,\mathrm{d}W_t\;, \qquad (5.3.36)$$
$$\mathrm{d}y_t = g(x_t, y_t)\,\mathrm{d}t\;,$$

under the following assumptions.

Assumption 5.3.5 (Hopf bifurcation).

- Domain and differentiability: *There is an open set $\mathcal{D} \subset \mathbb{R}^2 \times \mathbb{R}$ and an open interval $I \subset \mathbb{R}$ such that $f : \mathcal{D} \to \mathbb{R}^2$, $g : \mathcal{D} \to \mathbb{R}$ and $F : I \to \mathbb{R}^{2\times 2}$ are real-analytic, and uniformly bounded in norm by a constant M.*
- Slow manifold: *There is a function $x^\star : I \to \mathbb{R}^2$ such that $(x^\star(y), y) \in \mathcal{D}$ and $f(x^\star(y), y) = 0$ for all $y \in I$.*
- Hopf bifurcation: *The Jacobian matrix $A^\star(y) = \partial_x f(x^\star(y), y)$ has complex conjugate eigenvalues $a^\star(y) \pm \mathrm{i}\omega^\star(y)$. There is a $y_0 \in I$ such that $a^\star(y)$ has the same sign as $y - y_0$, and $\mathrm{d}_y a^\star(y_0)$ is strictly positive. The imaginary part $\omega^\star(y)$ is bounded away from 0 in I. Finally, $g(0, y) > 0$ for $y \in I$.*
- Non-degeneracy of noise term: *$F(y)F(y)^{\mathrm{T}}$ is positive definite for all $y \in I$.*

Neishtadt's theorem (Theorem 2.2.12) shows that in the deterministic case $\sigma = 0$, (5.3.36) admits a solution tracking the slow manifold at a distance of order ε, provided we make I small enough. Performing a translation of the fast variables to this particular solution, we can achieve that the drift term f vanishes in $x = 0$. The system can thus be written in the form

186 5 Multi-Dimensional Slow–Fast Systems

$$\mathrm{d}x_t = \frac{1}{\varepsilon}\bigl[A(y_t)x_t + b(x_t, y_t)\bigr]\,\mathrm{d}t + \frac{\sigma}{\sqrt{\varepsilon}}F(y_t)\,\mathrm{d}W_t\;, \tag{5.3.37}$$
$$\mathrm{d}y_t = g(x_t, y_t)\,\mathrm{d}t\;,$$

where $A(y)$ is ε-close to $A^\star(y)$, and $b(x,y)$ is of order $\|x\|^2$. We denote the eigenvalues of $A(y)$ by $a(y) \pm \mathrm{i}\omega(y)$.

We still have the possibility of further simplifying the system by performing an appropriate linear transformation. We may choose this transformation either in such a way that the diffusion coefficient becomes the identity matrix, or in such a way that the linear part of the drift term assumes some canonical form. We will opt for the second possibility.

Lemma 5.3.6. *There exists a nonsingular linear transformation* $x = S(y, \varepsilon)\tilde{x}$, *casting the system (5.3.37) into the form*

$$\mathrm{d}\tilde{x}_t = \frac{1}{\varepsilon}\bigl[\widetilde{A}(y_t)\tilde{x}_t + \tilde{b}(\tilde{x}_t, y_t)\bigr]\,\mathrm{d}t + \frac{\sigma}{\sqrt{\varepsilon}}\widetilde{F}(y_t)\,\mathrm{d}W_t\;, \tag{5.3.38}$$
$$\mathrm{d}y_t = \tilde{g}(\tilde{x}_t, y_t)\,\mathrm{d}t\;,$$

where $\widetilde{A}(y)$ *has the canonical form*

$$\widetilde{A}(y) = \begin{pmatrix} \tilde{a}(y) & -\tilde{\omega}(y) \\ \tilde{\omega}(y) & \tilde{a}(y) \end{pmatrix}, \tag{5.3.39}$$

with $\tilde{a}(y) = a(y) + \mathcal{O}(\varepsilon)$ *and* $\tilde{\omega}(y) = \omega(y) + \mathcal{O}(\varepsilon)$.

Proof. For any invertible matrix $S(y)$, the transformation $x = S(y)\tilde{x}$ yields a system of the form (5.3.38) with

$$\widetilde{A}(y) = S(y)^{-1}A(y)S(y) - \varepsilon S(y)^{-1}S'(y)\tilde{g}(\tilde{x}, y)\;. \tag{5.3.40}$$

It is always possible to choose $S(y)$ in such a way that $S^{-1}AS = \begin{pmatrix} a & -\omega \\ \omega & a \end{pmatrix}$ is in canonical form. We may thus assume that $A(y)$ is already in canonical form, up to corrections of order ε, but with different correction terms depending on the matrix element (e.g., the diagonal elements differ slightly). It then remains to eliminate these differences. This can be done by constructing $S = \begin{pmatrix} 1 & s_2 \\ s_1 & 1 \end{pmatrix}$ and $\widetilde{A}$ in such a way that the system

$$\varepsilon \dot{S} = AS - S\widetilde{A} \tag{5.3.41}$$

admits a bounded solution. Writing out this system in components, one obtains two algebraic and two closed differential equations, which can then be analysed by the methods we previously developed. □

In the sequel, we shall drop the tildes in (5.3.38), which amounts to considering the system (5.3.37) with $A(y)$ in canonical form. We may further assume, by translating y if necessary, that $a(y)$ vanishes for $y = 0$.

5.3 Bifurcations

Consider now the linear approximation

$$dx_t^0 = \frac{1}{\varepsilon} A(y_t^{\text{det}}) x_t^0 \, dt + \frac{\sigma}{\sqrt{\varepsilon}} F(y_t^{\text{det}}) \, dW_t \,, \tag{5.3.42}$$

$$dy_t^{\text{det}} = g(0, y_t^{\text{det}}) \, dt \,.$$

The Gaussian process x_t^0 has zero expectation and covariance matrix $\sigma^2 X(t)$, where $X(t)$ satisfies the slow–fast ODE

$$\varepsilon \dot X = A(y) X + X A(y)^{\text{T}} + F(y) F(y)^{\text{T}} \,, \tag{5.3.43}$$

$$\dot y = g(0, y) \,.$$

For negative y, bounded away from zero, we know from Section 5.1.1 that this system admits an invariant manifold $X = \bar X(y, \varepsilon)$, where the matrix elements of $\bar X(y, \varepsilon)$ remain of order 1. The question now is to know what happens as y_t approaches zero, where the ODE (5.3.43) admits a bifurcation point.

Lemma 5.3.7. *The system (5.3.43) admits a solution $\bar X(y_t, \varepsilon)$ whose eigenvalues grow like $1/|y_t|$ for $y_t < -\sqrt{\varepsilon}$, and stay of order $1/\sqrt{\varepsilon}$ for $|y_t| \leqslant \sqrt{\varepsilon}$. For $y_t > \sqrt{\varepsilon}$, the eigenvalues of $\bar X(y_t, \varepsilon)$ grow exponentially fast in y_t^2/ε.*

Proof. Writing

$$X = \begin{pmatrix} x_1 & x_2 \\ x_2 & x_3 \end{pmatrix}, \quad F(y) F(y)^{\text{T}} = \begin{pmatrix} d_1(y) & d_2(y) \\ d_2(y) & d_3(y) \end{pmatrix}, \tag{5.3.44}$$

and substituting into (5.3.43), one obtains the system

$$\varepsilon \dot x_1 = 2a(y) x_1 + 2\omega(y) x_2 + d_1(y) \,,$$
$$\varepsilon \dot x_2 = 2a(y) x_2 + \omega(y)(x_3 - x_1) + d_2(y) \,, \tag{5.3.45}$$
$$\varepsilon \dot x_3 = 2a(y) x_3 - 2\omega(y) x_2 + d_3(y) \,.$$

In particular, the trace of X satisfies the equation

$$\varepsilon \frac{d}{dt} \operatorname{Tr} X = 2a(y) \operatorname{Tr} X + \operatorname{Tr}\bigl(F(y) F(y)^{\text{T}}\bigr) \,. \tag{5.3.46}$$

By our non-degeneracy assumption, the last term on the right-hand side is positive, bounded away from zero. Thus $\operatorname{Tr} X$ behaves just as $\zeta(t)$ in Lemma 3.4.2. It remains to prove that the eigenvalues of X behave like its trace. The eigenvalues can be expressed in terms of $\operatorname{Tr} X$, x_2 and $z = x_3 - x_1$ as

$$\frac{1}{2}\Bigl[\operatorname{Tr} X \pm \sqrt{z^2 + 4x_2^2}\,\Bigr] \,. \tag{5.3.47}$$

The variables x_2 and z satisfy the system

$$\varepsilon \begin{pmatrix} \dot x_2 \\ \dot z \end{pmatrix} = \begin{pmatrix} 2a(y) & \omega(y) \\ -4\omega(y) & 2a(y) \end{pmatrix} \begin{pmatrix} x_2 \\ z \end{pmatrix} + \begin{pmatrix} d_2(y) \\ d_3(y) - d_1(y) \end{pmatrix} \,. \tag{5.3.48}$$

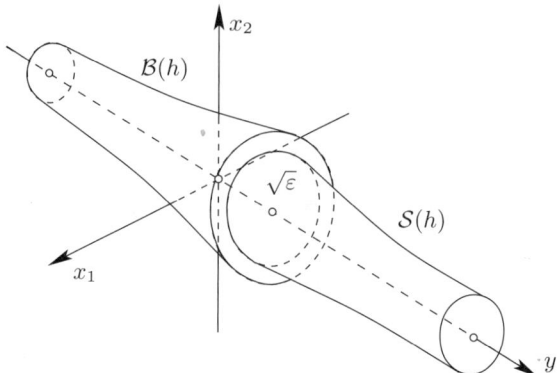

Fig. 5.6. Concentration sets of paths near a dynamic Hopf bifurcation point.

The square matrix has eigenvalues $2(a(y) \pm i\omega(y))$, and is thus invertible near $y = 0$. Equation (5.3.48) describes a particular (linear) case of a dynamic Hopf bifurcation. Neishtadt's theorem can thus be applied to show that both variables remain of order 1 in a neighbourhood of $y = 0$, which completes the proof. □

This lemma shows that the domain

$$\mathcal{B}(h) = \{(x,y) \in \mathcal{D}: y \in I, \langle x, \overline{X}(y,\varepsilon)^{-1} x \rangle < h^2\} \qquad (5.3.49)$$

has a diameter of order h for negative y, bounded away from zero, which then grows and reaches order $h/\varepsilon^{1/4}$ for $|y| \leqslant \sqrt{\varepsilon}$ (Fig. 5.6). For larger y, the diameter of $\mathcal{B}(h)$ grows exponentially fast. Our aim is now to compare the first-exit time $\tau_{\mathcal{B}(h)}$ of sample paths from $\mathcal{B}(h)$ with the first time

$$\bar{\tau}(y) = \inf\{t > 0: y_t > y\} \qquad (5.3.50)$$

at which y_t reaches a given value y. The following result shows that sample paths are concentrated in $\mathcal{B}(h)$ as long as $y_t \leqslant \sqrt{\varepsilon}$, provided $\sigma \leqslant \sqrt{\varepsilon}$.

Theorem 5.3.8 (Stochastic Hopf bifurcation – stable phase). *Fix an initial condition $(0, y_0) \in \mathcal{B}(h)$ with $y_0 \leqslant \sqrt{\varepsilon}$. Then there exist constants $\varepsilon_0, \Delta_0, h_0, c_1, L > 0$ such that for all $\varepsilon \leqslant \varepsilon_0$, all $\Delta \leqslant \Delta_0$, all $\gamma \in (0,1)$ and all $h \leqslant h_0 \sqrt{\varepsilon}$,*

$$\mathbb{P}^{0,y_0}\{\tau_{\mathcal{B}(h)} < \bar{\tau}(\sqrt{\varepsilon})\} \leqslant \frac{L}{\Delta \varepsilon} \frac{1}{1-\gamma} e^{-\kappa_+ h^2/2\sigma^2}, \qquad (5.3.51)$$

where the exponent κ_+ satisfies

$$\kappa_+ = \gamma \bigl[1 - c_1(\Delta + h^2/\varepsilon)\bigr]. \qquad (5.3.52)$$

Proof. The proof resembles the proof of Proposition 5.1.13. The main difference lies in the estimate of the effect of nonlinear terms, see in particular

Equation (5.1.65) in Proposition 5.1.11. On the one hand, the cubic nature of the nonlinearity causes h to appear with a higher power. On the other hand, the norm of $\overline{X}(y,\varepsilon)$ and the term $\Theta(t)$ grow as y_t approaches zero. The effect of this growth can be controlled by replacing the bounds $K_\pm$ by bounds $K_\pm(t,s)$, valid on the particular time interval under consideration. This yields the error term h^2/ε. □

Consider now the dynamics after y_t has reached $\sqrt{\varepsilon}$, still assuming that $\sigma \leqslant \sqrt{\varepsilon}$. We now expect sample paths to leave the vicinity of the equilibrium branch at $x=0$ exponentially fast. The escape is diffusion-dominated in a set of the form

$$\mathcal{S}(h) = \{(x,y)\colon y > \sqrt{\varepsilon},\ y \in I,\ \|x\| \leqslant h\rho(y)\}, \qquad (5.3.53)$$

where the appropriate definition of $\rho(y)$ turns out to be

$$\rho(y) = \sqrt{\frac{\operatorname{Tr}(F(y)F(y)^T)}{2a(y)}}. \qquad (5.3.54)$$

Then one can prove the following analogue of Theorem 3.2.2 and Theorem 3.4.4:

Theorem 5.3.9 (Stochastic Hopf bifurcation – diffusion-dominated escape). *Let $\mu > 0$, and set $C_\mu = (2+\mu)^{-(1+\mu/2)}$. Then, for any h and any initial condition $(x_0,y_0) \in \mathcal{S}(h)$ such that $\sigma < h < (y_0^2 C_\mu \sigma^{1+\mu})^{1/(3+\mu)}$, and any $y \in I$ with $y \geqslant y_0 \vee \sqrt{\varepsilon}$,*

$$\mathbb{P}^{x_0,y_0}\{\tau_{\mathcal{S}(h)} \geqslant \bar{\tau}(y)\} \leqslant \left(\frac{h}{\sigma}\right)^{2\mu} \exp\left\{-\kappa_\mu \frac{\alpha(y,y_0)}{\varepsilon}\right\}, \qquad (5.3.55)$$

where $\alpha(y,y_0) = \int_{y_0}^{y} a(z)\,dz$, and the exponent κ_μ is given by

$$\kappa_\mu = \frac{2\mu}{1+\mu}\left[1 - \mathcal{O}\!\left(\varepsilon\frac{1+\mu}{\mu}\right) - \mathcal{O}\!\left(\frac{1}{\mu \log(1+h/\sigma)}\right)\right]. \qquad (5.3.56)$$

Proof. As in Proposition 3.2.5 and Proposition 3.2.6, we introduce a partition of $[y_0,y]$ by setting $\alpha(y_k,y_{k-1}) = \varepsilon\Delta$. On each interval of the partition, we approximate x_t by a Gaussian process $x_t^{k,0}$, obeying the linear equation (5.3.42) with initial condition $(x_{\bar{\tau}(y_{k-1})}, y_{k-1})$. The main difference with the proof of Proposition 3.2.6 lies in the estimation of the term $P_{k,0}$, bounding the probability that the linear approximation $x_t^{k,0}$ remains small (compare (3.2.26) and (3.2.40)). A simple endpoint estimate shows that

$$\begin{aligned}
P_{k,0}(x,H_0) &\leqslant \mathbb{P}^{x,y_{k-1}}\{\|x_{\bar{\tau}(y_k)}^{k,0}\| < H_0\rho(y_k)\} \\
&\leqslant \frac{\pi H_0^2 \rho(y_k)^2}{2\pi\sqrt{\det \operatorname{Cov}(x_{\bar{\tau}(y_k)}^{k,0})}}.
\end{aligned} \qquad (5.3.57)$$

Using the same representation of the eigenvalues of the covariance matrix as in Lemma 5.3.7, we get

$$\sqrt{\det \operatorname{Cov}\bigl(x^{k,0}_{\bar\tau(y_k)}\bigr)} \geq \frac{\sigma^2}{2} \rho(y_k)^2 \bigl[e^{2\Delta} - 1\bigr] \inf_{y_{k-1} \leq z \leq y_k} \frac{\rho(z)^2}{\rho(y_k)^2}$$
$$\times \biggl[1 - \mathcal{O}\biggl(\frac{1}{\rho(y_k)^4 e^{4\Delta}}\biggr)\biggr]. \qquad (5.3.58)$$

The remainder of the proof is similar to the proof of Proposition 3.2.6. □

The probability in (5.3.55) becomes small as soon as y is such that $\alpha(y, y_0) \gg \varepsilon(1+\mu)\log(h/\sigma)$. Since $\alpha(y, y_0)$ grows quadratically with y, we can conclude that sample paths are likely to leave the domain $\mathcal{S}(h)$ after a time of order $\sqrt{\varepsilon \log(h/\sigma)}$.

To complete the discussion, we should show that sample paths leave a neighbourhood of order $\sqrt{y}$ of the equilibrium branch as soon as y reaches order $\sqrt{\varepsilon|\log \sigma|}$, and then, if the Hopf bifurcation is supercritical, approach the periodic orbit originating in the bifurcation. This analysis having not yet been worked out in detail, we shall limit ourselves to giving an idea of how one could proceed.

Using the Itô formula to pass to polar coordinates, one obtains a system of the form

$$dr_t = \frac{1}{\varepsilon}\bigl[a(y_t)r_t + b_r(r_t, \theta_t, y_t)\bigr] dt + \frac{\sigma}{\sqrt{\varepsilon}} F_r(\theta_t, y_t) dW_t,$$
$$d\theta_t = \frac{1}{\varepsilon}\bigl[\omega(y_t) + b_\theta(r_t, \theta_t, y_t)\bigr] dt + \frac{\sigma}{\sqrt{\varepsilon}} \frac{F_\theta(\theta_t, y_t)}{r_t} dW_t, \qquad (5.3.59)$$

where b_r contains terms of order r^2 and σ^2/r, and b_θ contains terms of order r and σ^2/r^2, while F_r and F_θ are of order 1. Note, in particular, that outside $\mathcal{S}(\sigma)$, the terms of order σ^2/r and σ^2/r^2, which are due to the second-oder term in Itô's formula, become negligible with respect to the leading part of the corresponding drift term.

In the analysis, we are mainly interested in the dynamics of r_t. As the motion of θ_t occurs on a faster timescale as the motion of r_t for small y_t, we expect the system (5.3.59) to be well-approximated by its averaged version

$$d\bar r_t = \frac{1}{\varepsilon}\bigl[a(y_t)\bar r_t + \bar b_r(\bar r_t, y_t)\bigr] dt + \frac{\sigma}{\sqrt{\varepsilon}} \bar F_r(y_t) dW_t. \qquad (5.3.60)$$

Now this equation is similar to the equation describing a dynamic pitchfork bifurcation, studied in Section 3.4.

Bibliographic Comments

The results in Sections 5.1 and 5.3.1 have been developed in [BG03], while the results of the other sections are presented here for the first time.

There exist various related results on reduction of slow–fast stochastic differential equations. The reduction to low-dimensional effective equations near bifurcation points of stochastic ODEs or PDEs has been considered, e.g., in [Blö03, BH04]. The effect of noise on dynamic Hopf bifurcations has been addressed in [SRT04] in the context of bursting in neural dynamics (see also Section 6.3.2). The questions concerning averaging, encountered in particular in situations with periodic orbits, are related to stochastic averaging, discussed in particular in [Kif01a, Kif01b, Kif03, BK04].

6
Applications

In this last chapter, we illustrate the methods developed in Chapters 3 and 5, by applying them to a few concrete examples. This will show what kind of information can be obtained from the sample-path approach, and also which aspects of the theory still need to be developed further.

In order to concentrate on physical implications, the examples we present have been arranged by field of application, rather than by the type of mathematical problem involved. In fact, it turns out that the same type of equation, admitting an S-shaped slow manifold, appears in very different fields.[1]

- In Section 6.1, we consider two important examples of second-order differential equations perturbed by noise: The overdamped Langevin equation, for which we examine the accuracy of the Smoluchowski approximation, and the van der Pol oscillator forced by noise.
- In Section 6.2, we describe some simple climate models, which aim at establishing connections between the North-Atlantic thermohaline circulation and warming events during Ice Ages. Here noise, modelling the influence of unresolved degrees of freedom, may trigger transitions between climate regimes, with far-reaching consequences.
- Section 6.3 reviews some of the basic models for action-potential generation in neurons, and discusses the phenomena of excitability and periodic bursting.
- Finally, in Section 6.4 we examine the effect of noise on hysteresis in a mean-field model of ferromagnets, and on the slow–fast dynamics of Josephson junctions.

---

[1]This may be due, in part, to the fact that such bistable situations are in some sense generic, especially the presence of saddle–node bifurcation points. It may also be related to a widespread (and generally rewarding) tendency, when modelling a specific system, to seek inspiration from similar phenomena in other fields.

6.1 Nonlinear Oscillators

6.1.1 The Overdamped Langevin Equation

Langevin discussed the motion of a particle subject to viscous drag and white noise [Lan08] (see [LG97] for a translation and comments), in order to provide an alternative derivation of Einstein's formula for the mean square displacement of a Brownian particle [Ein05].

More generally, consider a particle of unit mass moving in a potential $U(x)$, subject to a viscous drag with coefficient γ and white noise of intensity σ_0. Its dynamics is governed by the SDE

$$\frac{d^2 x_s}{ds^2} + \gamma \frac{dx_s}{ds} + \nabla U(x_s) = \sigma_0 \frac{dW_s}{ds}, \qquad (6.1.1)$$

also called Langevin equation by extension. For large damping, one often considers this equation to be equivalent to the first-order equation

$$\frac{dx_s}{ds} = -\frac{1}{\gamma} \nabla U(x_s) + \frac{\sigma_0}{\gamma} \frac{dW_s}{ds}. \qquad (6.1.2)$$

The associated forward Kolmogorov equation is also called *Smoluchowski equation*. Can we quantify how well the second-order system is approximated by the first-order system?

As we have seen in Example 2.1.3, for large damping γ, and without the noise term, one can write (6.1.1) in slow–fast form, using $t = s/\gamma$ as slow time. Incorporating the noise term one arrives at the SDE

$$\begin{aligned} dx_t &= \frac{1}{\varepsilon}(y_t - x_t)\, dt\,, \\ dy_t &= -\nabla U(x_t)\, dt + \sigma\, dW_t\,, \end{aligned} \qquad (6.1.3)$$

where $\sigma = \sigma_0 \gamma^{-1/2}$ and $\varepsilon = 1/\gamma^2$. This system is almost in the standard form (5.0.1) considered in Chapter 5, the only particularity being that there is no noise acting on the fast variables. Thus we cannot directly apply Theorem 5.1.6 on concentration and Theorem 5.1.17 on reduction, which require the noise term acting on the slow variables to be dominated by a noise term acting on the fast ones. However, owing to the relatively simple form of the system, it is still possible to derive analogous concentration and reduction results.

We showed in Example 2.1.9 that in the deterministic case, the system (6.1.3) admits an invariant manifold $x = \bar{x}(y, \varepsilon)$, where $\bar{x}(y, \varepsilon)$ satisfies the PDE

$$\varepsilon \partial_y \bar{x}(y, \varepsilon) \nabla U(\bar{x}(y, \varepsilon)) = \bar{x}(y, \varepsilon) - y\,, \qquad (6.1.4)$$

and admits an expansion of the form

$$\bar{x}(y, \varepsilon) = y + \varepsilon \nabla U(y) + \mathcal{O}(\varepsilon^2)\,. \qquad (6.1.5)$$

Thus if we can show that sample paths of the stochastic process are concentrated near the invariant manifold, solutions of the original system should be well-approximated by those of the reduced system

$$dy_t^0 = -\nabla U(\bar{x}(y_t^0, \varepsilon)) \, dt + \sigma \, dW_t \,. \tag{6.1.6}$$

In the limit $\varepsilon \to 0$, i.e., $\gamma \to \infty$, this system becomes

$$dy_t^0 = -\nabla U(y_t^0) \, dt + \sigma \, dW_t \,, \tag{6.1.7}$$

which validates (6.1.2) since $x = y$ on the slow manifold.

The following result estimates the probability of sample paths deviating from the invariant manifold. For simplicity, we shall assume that second-order derivatives of the potential are uniformly bounded in $\mathbb{R}^n$. The result is easily extended to situations where the derivatives are only bounded in an open subset $\mathcal{D} \subset \mathbb{R}^n$, by introducing an appropriate first-exit time.

Proposition 6.1.1. *Assume that $U : \mathbb{R}^n \to \mathbb{R}$ is twice continuously differentiable, and all its derivatives up to order 2 are uniformly bounded. Then there exist $h_0, \Delta_0, c_1 > 0$ such that for any $\rho \in (0,1)$, $\Delta \leqslant \Delta_0$ and $h \leqslant h_0$,*

$$\mathbb{P}^{\bar{x}(y_0,\varepsilon),y_0}\left\{\sup_{0 \leqslant s \leqslant t} \|x_s - \bar{x}(y_s, \varepsilon)\| > h\right\} \leqslant C_{n,\rho,\Delta}(t,\varepsilon) \exp\left\{-\kappa \frac{h^2}{\varepsilon\sigma^2}\right\}, \tag{6.1.8}$$

where

$$\kappa = \rho\bigl[1 - c_1(\Delta + n\varepsilon(\varepsilon\sigma^2/h^2)h + \varepsilon)\bigr]\,, \tag{6.1.9}$$

and the prefactor is given by

$$C_{n,\rho,\Delta}(t,\varepsilon) = \left(1 + \frac{t}{\Delta\varepsilon}\right)\bigl[(1-\rho)^{-n/2} + e^{n/4}\bigr]\,. \tag{6.1.10}$$

Proof. The deviation $\xi_t = x_t - \bar{x}(y_t, \varepsilon)$ satisfies an SDE of the form

$$d\xi_t = \frac{1}{\varepsilon}\bigl[-\xi_t + b(\xi_t, y_t)\bigr]\,dt + \sigma\bigl[1 + \varepsilon G_1(y_t)\bigr]\,dW_t\,, \tag{6.1.11}$$

where $b(\xi, y)$ contains terms of order $\varepsilon\|\xi\|$ and $n\varepsilon^2\sigma^2$ (the latter stemming from second-order terms in Itô's formula), and $G_1(y)$ is bounded. The solution can be represented as

$$\xi_t = \sigma \int_0^t e^{-(t-s)/\varepsilon}\,dW_s$$
$$+ \frac{1}{\varepsilon}\int_0^t e^{-(t-s)/\varepsilon}b(\xi_s, y_s)\,ds + \varepsilon\sigma\int_0^t e^{-(t-s)/\varepsilon}G_1(y_s)\,dW_s\,, \tag{6.1.12}$$

where the first term on the right-hand side is an Ornstein–Uhlenbeck process with asymptotic covariance $(\varepsilon\sigma^2/2)\mathbb{1}$. The probability of this term becoming large can be estimated by splitting the time interval and then proceeding as in Lemma 5.1.8. The remainder of the proof follows along the lines of the proof of Proposition 5.1.13. Since the domain of concentration considered has a constant diameter, and the linear term in (6.1.11) does not depend on y, there is however no need to control the deviation of y_t from its deterministic counterpart. Also, it is not necessary to restart the process after times of order 1, as in the general case. □

The proposition shows that x_t is unlikely to deviate from $\bar{x}(y_t, \varepsilon)$ by much more than $\sigma \varepsilon^{1/2} = \sigma_0 \gamma^{-3/2}$. Note that this is smaller than in the standard case, where the typical spreading of paths has order σ, which is due to the fact that the noise only acts indirectly on the fast variables via the coupling with the slow ones.

Since x_t is likely to remain close to $\bar{x}(y_t, \varepsilon)$, the reduced system (6.1.6) should give, as expected, a good approximation of the slow dynamics. Estimating the deviation of the slow variables directly shows that there exists a constant L such that

$$\|y_t - y_t^0\| \leq L\left[h^2 \chi^{(1)}(t) + (h^2 + n\varepsilon\sigma^2)\chi_C^{(1)}(t)\right] \qquad (6.1.13)$$

holds, whenever $t < \tau_{\mathcal{B}(h)}$ is such that $\chi^{(1)}(t)$ and $\chi_C^{(1)}(t)$ are at most of order $1/h$, where $\chi^{(1)}(t)$ and $\chi_C^{(1)}(t)$ are the analogues of the quantities defined in (5.3.28), measuring the divergence of neighbouring solutions of the reduced deterministic system. Thus, we can conclude that the Smoluchowski approximation is reliable with a precision of order $\sigma^2\varepsilon[\chi^{(1)}(t) + (n+1)\chi_C^{(1)}(t)]$ up to these times t.

6.1.2 The van der Pol Oscillator

Van der Pol introduced the equation now bearing his name in order to describe the dynamics of a triode [vdP20, vdP27], and was also the first to use the term relaxation oscillations [vdP26]. See, for instance, [PRK01] for a brief historical account of the oscillator's discovery.

The van der Pol oscillator can be realised by the equivalent electric RCL circuit depicted in Fig. 6.1. Its particularity is the nonlinear resistor denoted $r(v)$, whose resistance decreases when the voltage increases. Kirchhoff's laws imply that the voltage at the inductance, at the capacitor as well as at the nonlinear resistor is given by

$$v = Ri + u = R(i_1 + i_2 + i_3) + u, \qquad (6.1.14)$$

so that

$$\begin{aligned}\frac{dv}{dt} &= R\frac{d}{dt}(i_1 + i_2 + i_3) + \frac{du}{dt} \\ &= R\left(\frac{v}{L} + C\frac{d^2v}{dt^2} + \frac{d}{dt}\frac{v}{r(v)}\right) + \frac{du}{dt}.\end{aligned} \qquad (6.1.15)$$

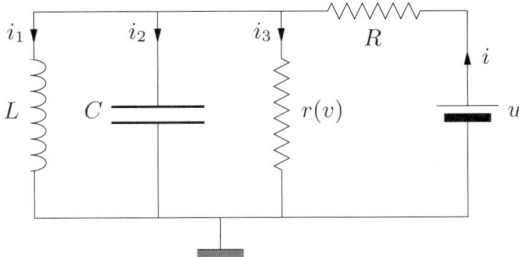

Fig. 6.1. Van der Pol's oscillating circuit includes an inductance L, a capacitor C and two resistors, one of them with a resistance depending on the voltage.

This second-order differential equation can be rewritten as

$$\frac{d^2v}{dt^2} + \frac{1}{RC}\frac{d}{dt}\left(\frac{Rv}{r(v)} - v\right) + \omega_0^2 v = -\frac{1}{RC}\frac{du}{dt}, \qquad (6.1.16)$$

where $\omega_0^2 = 1/LC$. Scaling time by a factor ω_0 eliminates the factor ω_0^2 from the equation. The nonlinear resistance $r(v)$ is a decreasing function of v, and one usually assumes that $r(v)$ is proportional to R/v^2, where the proportionality factor may be chosen equal to $1/3$ by rescaling v. This yields the dimensionless equation

$$\frac{d^2x}{dt^2} + \gamma\frac{d}{dt}\left(\frac{1}{3}x^3 - x\right) + x = -\gamma\frac{du}{dt}, \qquad (6.1.17)$$

with $\gamma = \omega_0/RC$. For large damping γ (cf. Example 2.1.6), this equation is equivalent to the slow–fast first-order system

$$\varepsilon \dot{x} = y + x - \frac{x^3}{3}, \qquad (6.1.18)$$
$$\dot{y} = -x - \dot{u},$$

where $\varepsilon = 1/\gamma^2$, and we have scaled time by a factor γ. The slow manifold of this equation consists of three parts, connected by two saddle–node bifurcation points, and the system displays relaxation oscillations (cf. Example 2.2.3).

Assume now that this system is perturbed by noise. If the noise term were to act on the fast variable x, we would be in the situation considered in Section 3.3, where we studied the saddle–node bifurcation with noise. In particular, for noise intensities $\sigma > \sqrt{\varepsilon}$, sample paths would be likely to switch between stable slow manifolds already at a time of order $\sigma^{4/3}$ before reaching the bifurcation point. This would decrease the size of typical cycles.

Another situation arises when, say, the voltage u of the power supply is random and described by white noise. Consider the SDE

$$dx_t = \frac{1}{\varepsilon}\left(y_t + x_t - \frac{x_t^3}{3}\right)dt, \qquad (6.1.19)$$
$$dy_t = -x_t\,dt + \sigma\,dW_t.$$

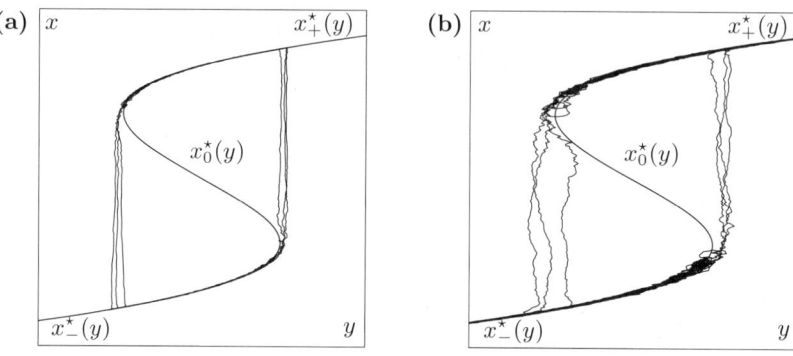

Fig. 6.2. Sample paths of the van der Pol Equation (6.1.19) with noise added on the slow variable y, for $\varepsilon = 0.001$ and **(a)** $\sigma = 0.2$, **(b)** $\sigma = 0.8$.

The interesting part of the dynamics takes place near the saddle–node bifurcation points. Let (x_c, y_c) denote one of them, say, $(x_c, y_c) = (1, -\frac{2}{3})$. The local dynamics of $(\tilde{x}_t, \tilde{y}_t) = (x_t - x_c, -(y_t - y_c))$ near this bifurcation point is a small perturbation

$$\begin{aligned} d\tilde{x}_t &= \frac{1}{\varepsilon}\bigl(-\tilde{y}_t - \tilde{x}_t^2\bigr)\,dt \;, \\ d\tilde{y}_t &= dt + \sigma\,dW_t \end{aligned} \qquad (6.1.20)$$

of the normal form. Recall that in the deterministic case $\sigma = 0$, there is an adiabatic solution $\bar{x}(\tilde{y}_t, \varepsilon)$, tracking the slow manifold $x^\star(y) = |y|^{1/2}$ at a distance increasing like $\varepsilon/|\tilde{y}_t|$ for $\tilde{y}_t \leqslant -\varepsilon^{2/3}$. The noise-induced deviation $\xi_t = \tilde{x}_t - \bar{x}(\tilde{y}_t, \varepsilon)$ from this solution satisfies the SDE

$$d\xi_t = \frac{1}{\varepsilon}\bigl[a(\tilde{y}_t)\xi_t - \xi_t^2 - \tfrac{1}{2}\sigma^2\varepsilon\,\partial_{yy}\bar{x}(\tilde{y}_t, \varepsilon)\bigr]\,dt + \sigma g(\tilde{y}_t)\,dW_t \;, \qquad (6.1.21)$$

where the linearised drift term

$$a(y) = -2\bar{x}(y, \varepsilon) \qquad (6.1.22)$$

behaves like $-|y|^{1/2}$ for $y \leqslant -\varepsilon^{2/3}$, and the effective noise intensity

$$g(y) = -\partial_y \bar{x}(y, \varepsilon) \qquad (6.1.23)$$

behaves like $|y|^{-1/2}$ for $y \leqslant -\varepsilon^{2/3}$. The linear approximation

$$d\xi_t^0 = \frac{1}{\varepsilon}a(\tilde{y}_t^{\mathrm{det}})\xi_t^0\,dt + \sigma g(\tilde{y}_t^{\mathrm{det}})\,dW_t \qquad (6.1.24)$$

turns out to have a variance growing like $\sigma^2\varepsilon/|\tilde{y}_t^{\mathrm{det}}|^{3/2}$, so that the typical spreading of sample paths around the adiabatic solution in the fast direction is of order $\sigma\varepsilon^{1/2}/|\tilde{y}_t|^{3/4}$. There are thus two situations to be considered:

- If $\sigma < \varepsilon^{1/3}$, the maximal spreading of paths near the bifurcation point is simply of order σ, and thus smaller than the distance between adiabatic solutions tracking the stable and unstable equilibrium branches, which has order $\varepsilon^{1/3}$. As a consequence, sample paths are unlikely to make a transition to the other stable slow manifold before passing the bifurcation point (Fig. 6.2a).
- If $\sigma > \varepsilon^{1/3}$, the typical spreading of paths becomes comparable to the distance between stable and unstable adiabatic solutions for y_t of order $-\sigma^{4/5}\varepsilon^{2/5}$, and transitions to the other stable slow manifold become possible (Fig. 6.2b).

The situation is thus similar to the situation of noise acting on the fast variable, only the threshold noise intensity necessary for early transitions to happen is much higher, due to the fact that noise only enters indirectly via the coupling with to the slow variables.

6.2 Simple Climate Models

Describing the evolution of the Earth's climate over time spans of several millennia seems an impossible task. Nonetheless, numerous models have been developed, which try to capture the dynamics of the more relevant quantities, for instance atmosphere and ocean temperatures averaged over long time intervals and large volumes (for an overview of various climate models, see, for instance, [Olb01]). Among these models, one distinguishes the following:

- General Circulation Models (GCMs), which are discretised versions of partial differential equations governing the atmospheric and oceanic dynamics, and including the effect of land masses, ice sheets, etc.
- Earth Models of Intermediate Complexity (EMICs), which concentrate on certain parts of the climate system, and use a more coarse-grained description of the remainder.
- Simple conceptual models, such as box models, whose variables are quantities averaged over very large volumes, and whose dynamics are based on global conservation laws.

While the first two types of models can only be analysed numerically, models of the third type are usually chosen in such a way that they are accessible to analytic methods, and can be used to gain some insight into the basic mechanisms governing the climate system. Still, even the most refined GCMs have limited resolution, with high-frequency and short-wavelength modes being neglected.

Several means exist to include the effect of unresolved degrees of freedom in the model. A first method, called parametrisation, assumes that the unresolved degrees of freedom can be expressed as a function of the resolved ones (like fast variables are enslaved by the slow ones on a stable slow manifold of a slow–fast system). The parametrisation is then chosen on more or less empirical grounds. This procedure is analogous to the one used in the kinetic theory of gases. A second method consists in averaging the equations for the resolved degrees of freedom over the unresolved ones, using, if possible, an invariant measure of the unresolved system in the averaging process.

A third approach, which was – in the context of climate models – proposed by Hasselmann in 1976 ([Has76], see also [Arn01]), is to model the effect of unresolved degrees of freedom by a noise term. This approach has not yet been rigorously justified, though some partial results exist (e.g. [Kha66, Kif01a, Kif01b, Kif03, BK04, JGB+03]); in particular, deviations from the averaged equations can often be proved to obey a central limit theorem. Anyhow, the stochastic approach has by now been used in numerous case studies, and has the advantage of providing a plausible model for certain rapid transition phenomena observed in the climate system. The examples presented here have partly been discussed in [BG02c].

6.2.1 The North-Atlantic Thermohaline Circulation

The dynamics of the Earth's oceans is characterised by a complicated system of surface, intermediate-depth and bottom currents, driven by thermal and haline (i.e., salinity) differences. In particular, the North-Atlantic thermohaline circulation (THC) transports enormous quantities of heat from the tropics as far North as the Barents Sea, causing the comparatively mild climate of Western and Northern Europe. The northward Gulf Stream is compensated by southward deep currents of cold water, originating for the most part from deep-water-formation regions located in the Greenland and Labrador Seas.

It is believed, however, that the situation has not always been like that in the past, and that during long time spans, the THC was locked in a stable state with far less heat transported to the North (see for instance [Cro75, Rah95, Mar00]). This on-and-off switching of the THC might be responsible for the repeated occurrence of minor Ice Ages during the last several ten thousand years, the so-called Dansgaard–Oeschger events.

A simple model for oceanic circulation exhibiting bistability is Stommel's box model [Sto61], where the North Atlantic is represented by two boxes, a low-latitude box with temperature T_1 and salinity S_1, and a high-latitude box with temperature T_2 and salinity S_2 (Fig. 6.3). Here we will follow the presentation in [Ces94], where the intrinsic dynamics of salinity and of temperature are not modelled in the same way. The differences $\Delta T = T_1 - T_2$ and $\Delta S = S_1 - S_2$ are assumed to evolve in time s according to the equations

$$\frac{\mathrm{d}}{\mathrm{d}s}\Delta T = -\frac{1}{\tau_r}(\Delta T - \theta) - Q(\Delta \rho)\Delta T \;,$$
$$\frac{\mathrm{d}}{\mathrm{d}s}\Delta S = \frac{F}{H}S_0 - Q(\Delta \rho)\Delta S \;. \qquad (6.2.1)$$

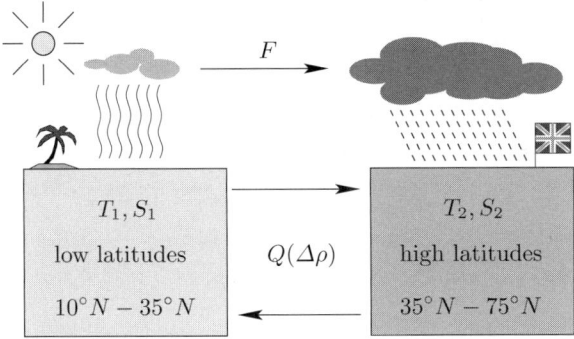

Fig. 6.3. Stommel's box model for the North-Atlantic thermohaline circulation. Differences in insolation and an atmospheric freshwater flux induce differences in temperature and salinity between low- and high-latitude boxes. The resulting density difference drives a northward surface current and a southward bottom current.

Here τ_r is the relaxation time of ΔT to its reference value θ, S_0 is a reference salinity, and H is the depth of the model ocean. F is the freshwater flux, modelling imbalances between evaporation (which dominates at low latitudes) and precipitation (which dominates at high latitudes). The dynamics of ΔT and ΔS are coupled via the density difference $\Delta \rho$, approximated by the linearised equation of state

$$\Delta \rho = \alpha_S \Delta S - \alpha_T \Delta T , \qquad (6.2.2)$$

which induces an exchange of mass $Q(\Delta \rho)$ between the boxes. We will use here Cessi's model [Ces94] for Q,

$$Q(\Delta \rho) = \frac{1}{\tau_d} + \frac{q}{V} \Delta \rho^2 , \qquad (6.2.3)$$

where τ_d is the diffusion timescale, q the Poiseuille transport coefficient and V the volume of the box. Stommel uses a different relation, with $\Delta \rho^2$ replaced by $|\Delta \rho|$, but we will not make this choice here because it leads to a singularity (and thus adds some technical difficulties which do not lie at the heart of our discussion here).

Using the dimensionless variables $x = \Delta T/\theta$ and $y = \alpha_S \Delta S/(\alpha_T \theta)$, and rescaling time by a factor τ_d, (6.2.1) can be rewritten as

$$\begin{aligned} \varepsilon \dot{x} &= -(x-1) - \varepsilon x \left[1 + \eta^2 (x-y)^2\right] , \\ \dot{y} &= \mu - y \left[1 + \eta^2 (x-y)^2\right] , \end{aligned} \qquad (6.2.4)$$

where $\varepsilon = \tau_r/\tau_d$, $\eta^2 = \tau_d (\alpha_T \theta)^2 q/V$, and μ is proportional to the freshwater flux F, with a factor $\alpha_S S_0 \tau_d/(\alpha_T \theta H)$. Cessi uses the estimates $\eta^2 \simeq 7.5$, $\tau_r \simeq 25$ days and $\tau_d \simeq 219$ years. This yields $\varepsilon \simeq 3 \times 10^{-4}$, implying that (6.2.4) is a slow–fast system.

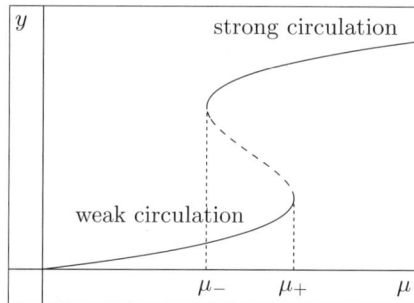

Fig. 6.4. Bifurcation diagram of the reduced equation, shown here for $\eta^2 = 7.5$. For freshwater fluxes in an interval $\mu \in (\mu_-, \mu_+)$, there are two possible stable states for the thermohaline circulation. The upper branch corresponds to larger salinity differences, and thus to a more important mass (and heat) transport to the North Atlantic.

As already pointed out in Example 2.1.5, the slow–fast system (6.2.4) admits a slow manifold of the form $x^*(y) = 1 + \mathcal{O}(\varepsilon)$, which is uniformly asymptotically stable (for y in a bounded set). By Fenichel's theorem, there exists an invariant manifold $x = \bar{x}(y, \varepsilon) = 1 + \mathcal{O}(\varepsilon)$, on which the flow is governed by the reduced equation

$$\dot{y} = \bar{g}(y, \varepsilon), \quad \text{with} \quad \bar{g}(y, \varepsilon) = \mu - y\big[1 + \eta^2(\bar{x}(y, \varepsilon) - y)^2\big]. \quad (6.2.5)$$

Note that an affine transformation brings $\bar{g}(y, 0) = \mu - y\big[1 + \eta^2(1-y)^2\big]$ into a more familiar cubic form, namely

$$g(u, 0) := \eta \bar{g}\left(\frac{2}{3} + \frac{u}{\eta}, 0\right) = \eta\left[\mu - \frac{2}{3} - \frac{2}{27}\eta^2\right] + \left[\frac{1}{3}\eta^2 - 1\right]u - u^3. \quad (6.2.6)$$

Thus $\bar{g}(y, 0)$ admits three roots if $\eta^2 > 3$ and μ belongs to an interval centred in (μ_-, μ_+), where

$$\mu_\pm = \frac{2}{3} + \frac{2}{27}\eta^2 \pm \frac{2}{3\sqrt{3}\eta}\left[\frac{1}{3}\eta^2 - 1\right]^{3/2}, \quad (6.2.7)$$

and one root if $\mu \notin [\mu_-, \mu_+]$. For $\mu = \mu_\pm$, the system admits two saddle–node bifurcation points. In other words, Stommel's box model displays bistability for freshwater fluxes in a certain range. If the freshwater flux, i.e., μ, is too small, then there is only one equilibrium point, corresponding to a small salinity difference, and thus to a small mass exchange. If μ is too large, then there also is only one equilibrium, but this time it corresponds to large ΔS and Q.

Let us now turn to random perturbations of the equations (6.2.4), in which the additional noise term will enable transitions between the stable states. We consider the situation where the freshwater flux is perturbed by white noise, modelling weather fluctuations which are basically uncorrelated, while the temperature dynamics is perturbed by an Ornstein–Uhlenbeck process.[2] The resulting equations read

$$\begin{aligned} dx_t &= \frac{1}{\varepsilon}\left[-(x_t-1) - \varepsilon x_t\left(1+\eta^2(x_t-y_t)^2\right)\right]dt + d\xi_t, \\ dy_t &= \left[\mu - y_t\left(1+\eta^2(x_t-y_t)^2\right)\right]dt + \sigma_2\,dW_t^2, \end{aligned} \qquad (6.2.8)$$

where the Ornstein–Uhlenbeck process ξ_t is given by

$$d\xi_t = -\frac{\gamma}{\varepsilon}\xi_t\,dt + \frac{\sigma_1}{\sqrt{\varepsilon}}\,dW_t^1. \qquad (6.2.9)$$

Note that we chose to adapt the noise intensity and relaxation time to the fast timescale. This system can be rewritten in canonical form for fast variables (x,ξ) and one slow variable y, with a fast drift term

$$f(x,\xi,y) = \begin{pmatrix} -(x-1) - \varepsilon x\left(1+\eta^2(x-y)^2\right) - \gamma\xi \\ -\gamma\xi \end{pmatrix}, \qquad (6.2.10)$$

a slow drift term

$$g(x,y) = \mu - y\left(1+\eta^2(x-y)^2\right), \qquad (6.2.11)$$

and diffusion matrices

$$F = \begin{pmatrix} 1 & 0 \\ 1 & 0 \end{pmatrix}, \qquad G = \begin{pmatrix} 0 & 1 \end{pmatrix}. \qquad (6.2.12)$$

The linearisation of the fast drift coefficient around the invariant manifold has the form $A(x,\xi)^{\mathrm{T}}$, with

$$A = \begin{pmatrix} -1 & -\gamma \\ 0 & -\gamma \end{pmatrix} + \mathcal{O}(\varepsilon), \qquad (6.2.13)$$

and a direct computation shows that the solution of the Lyapunov equation $AX + XA^{\mathrm{T}} + FF^{\mathrm{T}} = 0$ is given by

$$X^\star = \begin{pmatrix} \dfrac{1}{2(1+\gamma)} & \dfrac{1}{2(1+\gamma)} \\ \dfrac{1}{2(1+\gamma)} & \dfrac{1}{2\gamma} \end{pmatrix} + \mathcal{O}(\varepsilon). \qquad (6.2.14)$$

Recall that the asymptotic covariance of the process linearised around the slow manifold is of the form $\sigma_1^2(X^\star + \mathcal{O}(\varepsilon))$. The term $\sigma_1^2/2\gamma$ is simply the asymptotic variance of the Ornstein–Uhlenbeck process ξ_t. The other matrix elements describe the variance of $x_t - 1$, and the covariance of $x_t - 1$ and ξ_t

---

[2] A different scaling of noise intensities arises when the perturbation acts on the mass exchange term, as for instance in [Mon02].

in the linear approximation. Theorem 5.1.6 then shows that sample paths of the *nonlinear* system are concentrated in a set of the form

$$\mathcal{B}(h) = \left\{ (x,\xi,y) \colon (x - \bar{x}(y,\varepsilon),\xi)\bar{X}(y,\varepsilon)^{-1}(x - \bar{x}(y,\varepsilon),\xi)^{\mathrm{T}} < h^2 \right\}, \quad (6.2.15)$$

where $\bar{X}(y,\varepsilon) = X^\star(y) + \mathcal{O}(\varepsilon)$, whenever $h \gg \sigma_1$. We see in particular that larger damping γ yields a smaller spreading of typical sample paths.

The reduced stochastic equation has the form

$$\mathrm{d}y_t^0 = \left[\mu - y_t^0\bigl(1 + \eta^2(1 - y_t^0)^2\bigr) + \mathcal{O}(\varepsilon) \right] \mathrm{d}t + \sigma_2 \, \mathrm{d}W_t^2, \quad (6.2.16)$$

and describes, for appropriate μ, a bistable one-dimensional system with noise. There will thus be rare transitions between the two stable states, unless the freshwater flux μ is close to one of the two thresholds $\mu_\pm$, and the probability of transitions (in one direction) increases. The closer μ gets to its lower critical value μ_-, the more likely transitions from warm to cold climate in Northern Europe become.

6.2.2 Ice Ages and Dansgaard–Oeschger Events

We already mentioned in Chapter 4 that the discovery of stochastic resonance originated in an attempt to model the regular occurrence of major Ice Ages by a stochastic differential equation. The climate during major Ice Ages, however, has not been uniformly cold. The analysis of ice cores from the Greenland ice sheet has revealed the existence of repeated sudden warming events, in a time span between roughly 80 000 and 10 000 years before present (Fig. 6.5). These temperature increases are called *Dansgaard–Oeschger (DO) events*. A remarkable aspect of DO events is that they appear to be governed by a 1470-year cycle: Although they do not occur periodically, the time spans between DO events are close to multiples of 1470 years [DJC93, Sch02b, Sch02a, Rah03].

One of the hypotheses that have been put forward to explain this cycle is that DO events are caused by periodic changes in the freshwater flux driving the North-Atlantic thermohaline circulation. Such a periodic change can have various origins (cf. [TGST03]):

- *Feedback:* The atmospheric temperature gradient between low and high latitude, which depends on the intensity of the THC, influences the northward moisture transport (see for instance [PS02]).
- *External periodic forcing*: The system is driven by a periodic change of some external factor, e.g., the insolation (Milankovitch factors).
- *Internal periodic forcing:* The climate system itself displays a periodic oscillation, due to its intrinsic dynamics, for instance as a consequence of a Hopf bifurcation (see, e.g., [SP99]).

To describe the first situation, i.e., feedback, consider the following modification of (6.2.8):

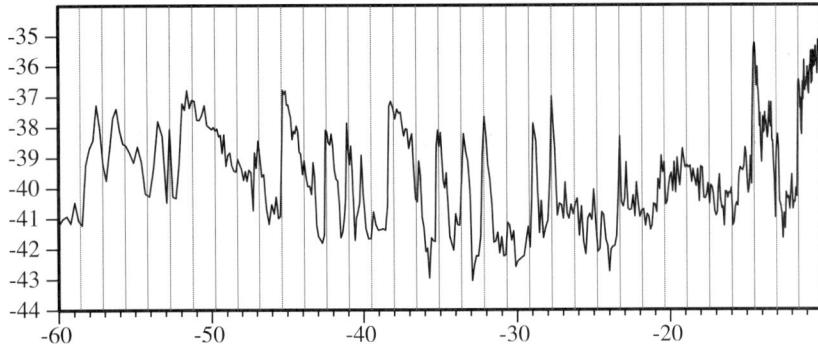

Fig. 6.5. Oxygen isotope (δ^{18}O) record from the GISP2 Greenland ice core drilling [GS97]. The isotope concentration gives an indication of the average temperature in Greenland, shown here for a period between 60 000 and 10 000 years before present. The vertical lines are equally spaced by 1470 years. Dansgaard–Oeschger (DO) events are represented by rapid increases in the temperature. (Data provided by the National Snow and Ice Data Center, University of Colorado at Boulder, and the WDC-A for Paleoclimatology, National Geophysical Data Center, Boulder, Colorado. http://arcss.colorado.edu/data/gisp_grip.)

$$\begin{aligned} \mathrm{d}x_t &= \frac{1}{\varepsilon_0}\bigl[-(x_t-1) - \varepsilon_0 x_t\bigl(1+\eta^2(x_t-y_t)^2\bigr)\bigr]\,\mathrm{d}t + \frac{\sigma}{\sqrt{\varepsilon_0}}\,\mathrm{d}W_t^0\,, \\ \mathrm{d}y_t &= \bigl[z_t - y_t\bigl(1+\eta^2(x_t-y_t)^2\bigr)\bigr]\,\mathrm{d}t + \sigma_1\,\mathrm{d}W_t^1\,, \quad (6.2.17) \\ \mathrm{d}z_t &= \varepsilon h(x_t,y_t,z_t)\,\mathrm{d}t + \sqrt{\varepsilon}\sigma_2\,\mathrm{d}W_t^2\,, \end{aligned}$$

where we have chosen a white-noise perturbation of x for simplicity. The new feature is that the freshwater flux z_t changes according to some, as yet unspecified, function h of all variables. We also introduced three well-separated timescales for the variations of temperature, salinity and freshwater flux, whose ratios are measured by small parameters ε_0 and ε.

By Theorem 5.1.17, we may consider, at least up to some timescale, the stochastic system projected onto the adiabatic manifold $x = \bar{x}(y,z,\varepsilon_0)$, where $\bar{x}(y,z,\varepsilon_0) = 1 + \mathcal{O}(\varepsilon_0)$. Speeding up time by a factor ε, we arrive at the system

$$\begin{aligned} \mathrm{d}y_t &= \frac{1}{\varepsilon}\bigl[z_t - y_t\bigl(1+\eta^2(1-y_t)^2\bigr) + \mathcal{O}(\varepsilon_0)\bigr]\,\mathrm{d}t + \frac{\sigma_1}{\sqrt{\varepsilon}}\,\mathrm{d}W_t^1\,, \\ \mathrm{d}z_t &= h(1+\mathcal{O}(\varepsilon_0),y_t,z_t)\,\mathrm{d}t + \sigma_2\,\mathrm{d}W_t^2\,. \end{aligned} \quad (6.2.18)$$

The drift term for the (now) fast variable y vanishes on the S-shaped curve of equation $z = y(1+\eta^2(1-y)^2) + \mathcal{O}(\varepsilon_0)$, which contains two saddle–node bifurcation points. In fact, up to an affine transformation, (6.2.18) is equivalent to the van der Pol oscillator, at least as far as the fast dynamics is concerned. If the function h is positive in a neighbourhood of the upper stable equilibrium branch, and negative in a neighbourhood of the lower stable branch,

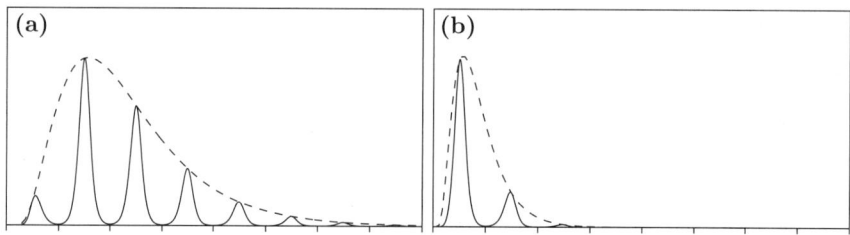

Fig. 6.6. Examples of residence-time distributions in the cold stadial (a) and warm interstadial state (b), for a period $T = 7$ and values of the Kramers times chosen to fit the observations.

the system will display relaxation oscillations (compare Section 6.1.2). For sufficiently large noise intensity, Theorem 3.3.4 shows that transitions from one stable state to the other one are likely to occur *before* the saddle–node bifurcation is reached, that is, before the freshwater flux reaches its critical value μ_+ or μ_-.

In order to describe the situations with external or internal periodic forcing, rather than feedback, we take, instead of an additional dynamic variable z, a time-periodic freshwater flux $\mu(t)$ (see, e.g., [VBAC$^+$01, GR02]). This yields a reduced equation

$$\mathrm{d}y_t = \frac{1}{\varepsilon}\bigl[\mu(t) - y_t\bigl(1 + \eta^2(1 - y_t)^2\bigr) + \mathcal{O}(\varepsilon_0)\bigr]\,\mathrm{d}t + \frac{\sigma_1}{\sqrt{\varepsilon}}\,\mathrm{d}W_t^1\;. \qquad (6.2.19)$$

The adiabatic parameter ε is given by the ratio of the diffusion time scale τ_d (219 years) and the forcing period of 1470 years, that is, $\varepsilon \simeq 0.15$. The behaviour depends essentially on the range of the freshwater flux $\mu(t)$. If it remains within the interval (μ_-, μ_+) in which the frozen system is bistable, we are in a situation where stochastic resonance can occur.[3]

In Section 4.1.4 we computed the residence-time distribution for a symmetric bistable system subject to a periodic forcing. The only new feature in the asymmetric case is that the first-passage-time distributions from the left to the right equilibrium state and from the right to the left equilibrium state can differ, due in particular to different values of the quasipotential. The general expression for the densities of the residence-time distributions in the two equilibrium states is

$$q_\pm(t) = \frac{1}{\mathcal{N}_\pm}\,\frac{\mathrm{e}^{-t/T_\mathrm{K}^\pm(\sigma)}}{T_\mathrm{K}^\pm(\sigma)}\,f_\mathrm{trans}^\pm(t)Q_{\lambda T}^{\Delta\pm}(t/T)\;. \qquad (6.2.20)$$

Here $\mathcal{N}_\pm$ is the normalisation making $q_\pm(t)$ a probability density. The $T_\mathrm{K}^\pm(\sigma)$ are the Kramers times for leaving the respective equilibrium state, $f_\mathrm{trans}^\pm(t)$

---

[3] Note that SR has been proposed as a mechanism at work in other related fields, e.g., in geodynamo models [LSCDM05].

are transient terms approaching 1 in a time of order $|\log \sigma|$, λ is the Lyapunov exponent of the unstable orbit separating the basins of attraction of the two stable states, and T is the period of the forcing. The two quantities Δ_+ and $\Delta_- = T - \Delta_+$ are phase shifts depending on the location of the most probable exit paths, and the $Q^\Delta_{\lambda T}$ are universal periodic modulations given by

$$Q^\Delta_{\lambda T}(t) = \frac{\lambda T}{2} \sum_{k=-\infty}^{\infty} \frac{1}{\cosh^2(\lambda T(t + \Delta/T - k))} \, . \quad (6.2.21)$$

The difference between the residence-time distributions $q_\pm(t)$ in the two equilibrium states is mainly due to the different values of the Kramers times $T^\pm_K(\sigma)$, which control the decay of the peaks. It is possible to find parameter values reproducing the behaviour seen in Fig. 6.5, where the system typically spends only half a period in the warm state, while in can spend several periods in the cold state.

Note that both for a freshwater flux influenced by feedback or by external periodic forcing, we started with a three-timescale system, that we reduced in two successive steps. The climate system is characterised by a whole hierarchy of further timescales, with different periodic behaviours influencing each other (see, for instance, [SPT02, KST04]). Singular perturbation theory allows these timescales to be treated successively, provided they are sufficiently well separated.

6.3 Neural Dynamics

Neurons communicate by generating action potentials, that is, abrupt changes in the electrical potential across the cell's membrane. These potential changes then propagate away from the cell body along axons, essentially without changing their shape. It is commonly admitted in neuroscience that the shape of these spikes does not carry any information; rather, it is the frequency of successively emitted spikes that is responsible for information transmission between neurons.

Much effort has gone into the mathematical modelling of the generation of action potentials. These are built up by flows of ions, mainly sodium, calcium and potassium, through the cell membrane. The fast activation of ion channels triggers a relatively slower process of inactivation, responsible for the ultimate decay of the spike.

One of the first mathematical models capturing the dynamics of action-potential generation is a four-dimensional ODE introduced by Hodgkin and Huxley in 1952 [HH52], which is based on an equivalent electric circuit.[4] The

---
[4] Another class of neural models are the integrate-and-fire models, also based on an equivalent electric circuit. In these models, firing occurs upon the membrane potential reaching a threshold, and the membrane potential is reset afterwards. We will not discuss this class of models here.

evolution of the action potential v is modelled by the equation

$$C\dot{v} = G_{\mathrm{Na}} m^3 h (V_{\mathrm{Na}} - v) + G_{\mathrm{K}} n^4 (V_{\mathrm{K}} - v) + G_{\mathrm{L}} (V_{\mathrm{L}} - v) + i(t) , \quad (6.3.1)$$

where the G_x, $x \in \{\mathrm{Na}, \mathrm{K}, \mathrm{L}\}$, are constant conductivities, the V_x are constant reference potentials, and $i(t)$ is an external current. The dynamic variables m, h and n control, respectively, sodium activation, sodium inactivation and potassium activation. They obey linear ODEs

$$\begin{aligned} \tau_m(v) \, \dot{m} &= m^\star(v) - m , \\ \tau_h(v) \, \dot{h} &= h^\star(v) - h , \\ \tau_n(v) \, \dot{n} &= n^\star(v) - n , \end{aligned} \quad (6.3.2)$$

in which the v-dependence of relaxation times $\tau_x(v)$, $x \in \{m, h, n\}$, and equilibrium values $x^\star(v)$ are fitted to experimental data. The equilibrium values are monotonous in v and approach constants in both limits $v \to \pm\infty$.

This model was later simplified to yield various two-dimensional approximations. These are based, firstly, on the experimental observation that $\tau_m(v)$ is much shorter than $\tau_h(v)$ and $\tau_n(v)$. The system can thus be reduced to the slow manifold $m = m^\star(v)$. Secondly, one observes that $h + n$ is approximately constant, so that it suffices to keep one of these variable. This yields a two-dimensional approximation of general dimensionless form

$$\begin{aligned} \varepsilon \dot{x} &= f(x, y) + i(t) , \\ \dot{y} &= g(x, y) , \end{aligned} \quad (6.3.3)$$

where x is proportional to the action potential v, and y is proportional to n. The parameter ε turns out to be (moderately) small because the relaxation time of v is somewhat shorter than $\tau_h(v)$ and $\tau_n(v)$. For instance, the Morris–Lecar model [ML81], which was devised for giant barnacle (*Balanus nubilus*) muscle fibres, assumes that

$$\begin{aligned} f(x, y) &= c_1 m^\star(x)(1 - x) + c_2 y (V_2 - x) + c_3 (V_3 - x) , \\ g(x, y) &= (w^\star(x) - y) \cosh\left(\frac{x - x_3}{x_4}\right) , \end{aligned} \quad (6.3.4)$$

where the c_i, V_i and x_i are constants, and

$$\begin{aligned} m^\star(x) &= \frac{1}{2}\left[1 + \tanh\left(\frac{x - x_1}{x_2}\right)\right] , \\ w^\star(x) &= \frac{1}{2}\left[1 + \tanh\left(\frac{x - x_3}{x_4}\right)\right] . \end{aligned} \quad (6.3.5)$$

The function $f(x, y)$ is also often fitted to experimental data rather than deduced from the Hodgkin–Huxley equations. In particular, the Fitzhugh–Nagumo model [Fit61, NAY62] postulates

$$f(x,y) = x - x^3 + y,$$
$$g(x,y) = \alpha - \beta x - \gamma y, \qquad (6.3.6)$$

yielding a generalisation of the van der Pol equations.

Among the typical behaviours of action potentials, the following are of particular interest to the mathematical modeller.

1. *Excitability:* A weak perturbation of a system at rest (called *quiescent*) causes the generation of a spike, i.e., an abrupt increase of the action potential, at the end of which the system returns to rest.
2. *Bursting:* The neuron activity alternates between quiescent phases and phases of repetitive spiking, during which the period and amplitude of spikes are slowly modulated.

An extensive classification of bifurcation scenarios in ODEs leading to these and other spiking behaviours can be found in [Izh00]. Among the possible perturbations causing an excitable system to spike, one may consider stochastic perturbations. See, for instance, [Tuc89] for an overview of stochastic models in neuroscience, and [Lon93, PM04] for studies of specific noise-induced phenomena.

6.3.1 Excitability

A dynamical system is called *excitable* if it admits an asymptotically stable equilibrium point, but is such that a small change in a parameter would make the point unstable, while creating a stable periodic orbit. In other words, an excitable system operates close to a bifurcation point, and is thus sensitive to small perturbations away from the stable equilibrium.

One distinguishes between two types of excitability. In Type I, the period of the stable periodic orbit diverges as the bifurcation parameter approaches its threshold value. In Type II excitability, orbits of finite period appear as soon as the bifurcation parameter crosses its threshold. The type of excitability displayed by a specific system depends on the nature of the involved bifurcation.

A typical bifurcation yielding Type I excitability is the saddle–node-to-invariant-circle bifurcation. It occurs in the Morris–Lecar model for certain parameter values. The slow manifold $f(x,y) = 0$ is typically S-shaped, consisting of one unstable and two stable branches. These branches meet at saddle–node bifurcation points. Assume the nullcline $g(x,y) = 0$ intersects the unstable branch twice and one of the stable branches once (Fig. 6.7a). The intersection with the stable branch is a stable equilibrium point of the equations (usually a node), while the other intersections are unstable, at least one of them being of saddle type. Assume further that both unstable manifolds of the saddle closest to the node are attracted by the node. When the nullcline $g(x,y) = 0$ moves towards the saddle–node bifurcation point, one of the two unstable manifolds shrinks until the saddle and the node meet and then vanish. In the process, the unstable manifolds have been replaced by a periodic orbit.

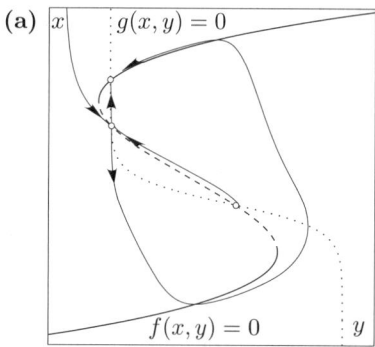 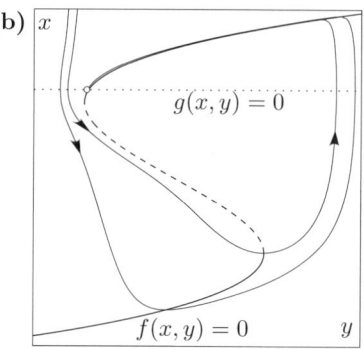

Fig. 6.7. Typical examples of excitable slow–fast systems. The intersections of nullclines $g(x,y) = 0$ (*dotted curves*) and stable (*full heavy curves*) or unstable (*broken heavy curves*) branches of the slow manifold $f(x,y) = 0$ are equilibrium points, marked by small circles. **(a)** For Type I excitability, both unstable manifolds of the saddle point are attracted by a close-by node. **(b)** For Type II excitability, certain trajectories pass close to a stable equilibrium point, but then make a large excursion before converging to this point. In both cases, a small displacement of the nullcline can produce a periodic orbit.

It is possible to describe this bifurcation by an equivalent one-dimensional equation for a variable θ, measuring the position on the unstable manifolds or periodic orbit, respectively. A nonlinear transformation of space and time brings the equation for θ into canonical form

$$\dot\theta = (1 - \cos\theta) + (1 + \cos\theta)\lambda\,, \qquad (6.3.7)$$

where λ is the bifurcation parameter (see, for instance, [Izh00, Theorem 1, p. 1180]). Note that the right-hand side derives from a "washboard potential"

$$U(\theta) = -(1+\lambda)\theta + (1-\lambda)\sin\theta\,. \qquad (6.3.8)$$

For $\lambda > 0$, the right-hand side of (6.3.7) is always positive, so that θ increases monotonously, meaning that, in this case, there is a periodic orbit, and $U(\theta)$ is a monotonously decreasing function of the position θ. For $\lambda < 0$, pairs of stable and unstable equilibrium points appear, meaning that $U(\theta)$ admits pairs of minima and maxima. In that case, there are thus potential barriers, placed periodically, of height of order $|\lambda|^{1/2}$ (for small $|\lambda|$).

Adding noise to the equation (6.3.7) makes it possible for solutions to overcome the potential barriers, the expected time between transitions being exponentially large in $|\lambda|^{1/2}/\sigma^2$ for noise intensity σ. Each time a potential barrier is crossed, the system relaxes to the bottom of the next potential well. In the original variables, all potential wells correspond to the unique stable equilibrium point, and the barrier crossings and subsequent relaxation events correspond to spiking behaviour.

6.3 Neural Dynamics

One can determine the relation of the fictive bifurcation parameter λ in (6.3.7) to the actual parameters in the Morris–Lecar model in the following way. Assume that the nullcline $g(x, y) = 0$ lies at a horizontal distance δ from the saddle–node bifurcation point (x_c, y_c) of the slow manifold. The local dynamics near this bifurcation point is a small perturbation of the normal form

$$\varepsilon \dot{\tilde{x}} = -\tilde{y} - \tilde{x}^2 \,,$$
$$\dot{\tilde{y}} = -\delta - \tilde{y} \,, \qquad (6.3.9)$$

where $\tilde{x} = x - x_c$ and $\tilde{y} = -(y - y_c)$. The equilibrium points of the normal form are located at $\tilde{x} = \pm \delta^{1/2}$, $\tilde{y} = -\delta$. If noise of intensity $\sigma/\sqrt{\varepsilon}$ is added to the fast variable $\tilde{x}$, the effective dynamics is governed by the equation

$$d\tilde{x}_t = \frac{1}{\varepsilon}\left[\delta - \tilde{x}_t^2\right] dt + \frac{\sigma}{\sqrt{\varepsilon}} dW_t \,. \qquad (6.3.10)$$

The drift term $\delta - \tilde{x}_t^2$ derives from a potential with a barrier of height proportional to $\delta^{3/2}$. In analogy with the results of Section 4.2.2, we conclude that spikes are rare if $\sigma \ll \delta^{3/4}$, and frequent if $\sigma \gg \delta^{3/4}$.

A typical bifurcation yielding Type II excitability is the Hopf bifurcation. It occurs, e.g., in the (unforced) Fitzhugh–Nagumo equations

$$\varepsilon \dot{x} = x - x^3 + y \,,$$
$$\dot{y} = \alpha - \beta x - \gamma y \,. \qquad (6.3.11)$$

The slow manifold is given by the S-shaped curve of equation $y = x^3 - x$. The dynamics depends crucially on the location of the nullcline $\beta x + \gamma y = \alpha$. If it only intersects the unstable part of the slow manifold, then the system performs relaxation oscillations like the van der Pol oscillator. If, however, the nullcline intersects one of the stable branches of the slow manifold (Fig. 6.7b), this intersection is a stable equilibrium point (of focus type if the intersection is close to the saddle–node point on the slow manifold). Though it may not be obvious from the picture, when the nullcline moves below the saddle–node point, the system undergoes a Hopf bifurcation, and a periodic orbit of rapidly growing diameter appears. The most important fact is that before the bifurcation, there can be trajectories closely missing the stable equilibrium, and making a large excursion before actually reaching the equilibrium point. A small perturbation can induce another excursion, corresponding to a spike.

Consider for instance the case $\alpha = x_c + \delta$, $\beta = 1$, $\gamma = 0$, where $(x_c, y_c) = (1/\sqrt{3}, -2/3\sqrt{3})$, is the location of the saddle–node bifurcation point, and δ is a small parameter. The local dynamics near the bifurcation point can be described by the normal form

$$\varepsilon \dot{\tilde{x}} = -\tilde{y} - \tilde{x}^2 \,,$$
$$\dot{\tilde{y}} = \tilde{x} - \tilde{\delta} \,, \qquad (6.3.12)$$

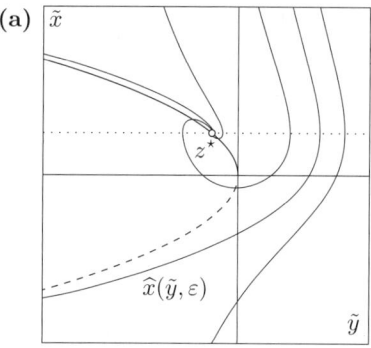 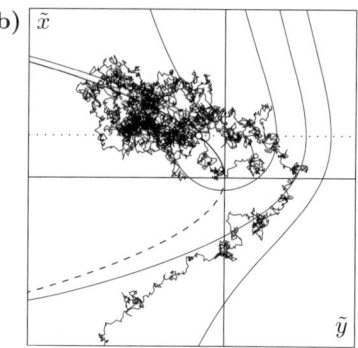

Fig. 6.8. Dynamics of the normal form (6.3.12) describing the dynamics, near a saddle–node bifurcation point, of a system displaying Type II excitability. **(a)** Without noise, trajectories starting above and to the left of the particular solution $\widehat{x}(\tilde{y},\varepsilon)$, which delimits spiking and non-spiking behaviour, are attracted by the equilibrium point $z^\star$. Those starting below and to the right escape a neighbourhood of the bifurcation point. In the original system (6.3.11), they ultimately converge to $z^\star$, having made a large excursion which corresponds to a spike. **(b)** Noise can cause sample paths to leave the vicinity of $z^\star$, cross $\widehat{x}(\tilde{y},\varepsilon)$ and thus produce a spike.

where $\tilde{x} = \sqrt{3}(x - x_c)$, $\tilde{y} = -\sqrt{3}(y - y_c)$, and $\tilde{\delta} = \sqrt{3}\delta$. Trajectories tracking the stable slow manifold $\tilde{x} = \sqrt{-\tilde{y}}$ converge to a stable equilibrium point $z^\star = (\tilde{\delta}, -\tilde{\delta}^2)$. A special rôle is played by the trajectory of equation $\tilde{x} = \widehat{x}(\tilde{y},\varepsilon)$ tracking the *unstable* slow manifold $\tilde{x} = -\sqrt{-\tilde{y}}$. It separates the (local) basins of attraction of $z^\star$ and the orbits escaping to negative $\tilde{x}$ (Fig. 6.8a). In the global system (6.3.11), of course, these orbits ultimately return to the equilibrium point, but only after having followed the lower stable branch of the slow manifold. We can thus consider the trajectory $\widehat{x}(\tilde{y},\varepsilon)$ as delimiting spiking and non-spiking behaviour.

Adding noise to the system results in sample paths making excursions away from $z^\star$. Occasionally, a sample path will reach and cross $\widehat{x}(\tilde{y},\varepsilon)$, and the system will spike (Fig. 6.8b). Shape and duration of spikes will be close to their deterministic value, but the time between spikes will be random, as it is related to the activation process of going from $z^\star$ to $\widehat{x}(\tilde{y},\varepsilon)$.

6.3.2 Bursting

A neuron is said to exhibit *periodic bursting* behaviour if it switches periodically between quiescent phases, and phases of repetitive spiking. Often the shape and frequency of the spikes are slowly modulated, on a timescale which is much longer than the timescale of individual spike generation. In this case, the system can be modelled by a slow–fast ODE of the form

$$\varepsilon \dot{x} = f(x, u)\,,$$
$$\dot{u} = g(x, u)\,, \qquad (6.3.13)$$

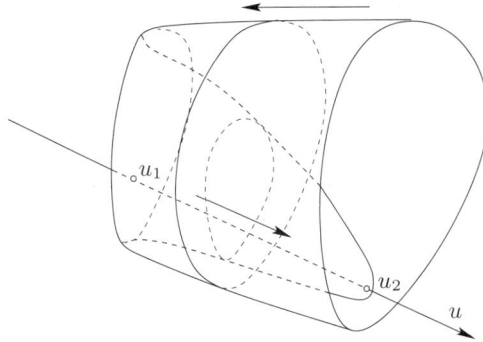

Fig. 6.9. Bifurcation scenario leading to elliptic bursting. For $u_1 < u < u_2$, the system admits a stable equilibrium point, corresponding to a quiescent neuron, and a stable limit cycle, corresponding to spiking. The slow dynamics increases u near the quiescent state, driving the system to a Hopf bifurcation point. Near the limit cycle, it decreases u, driving the system to a saddle–node bifurcation point of periodic orbits. As a result, the system alternates between quiescent and spiking phases.

where x now stands for the dynamic variables associated with action-potential generation, and u describes slowly changing parameters of the system, such as the current $i(t)$.

Periodic bursting can occur when the associated system $\frac{d}{ds}x_s = f(x_s, u)$ admits a stable equilibrium point and a limit cycle for certain values of u. A typical example is shown in Fig. 6.9, where $x \in \mathbb{R}^2$ and $u \in \mathbb{R}$:

- For $u < u_1$, all orbits converge to a globally asymptotically stable equilibrium point $x^\star(u)$.
- At $u = u_1$, a pair of periodic orbits of opposite stability is created in a saddle–node bifurcation of periodic orbits.
- For $u_1 < u < u_2$, the system admits a stable equilibrium point $x^\star(u)$, an unstable and a stable periodic orbit.
- At $u = u_2$, the unstable periodic orbit collapses on the equilibrium point in a subcritical Hopf bifurcation.
- For $u > u_2$, the system admits a limit cycle and an unstable equilibrium point.

Assume now that the slow vector field $g(x, u)$ is positive near $x^\star(u)$, and negative near the stable periodic orbit. A solution starting near $x^\star(u_0)$ for some $u_0 < u_1$ will track the equilibrium branch $x^\star(u)$ adiabatically: The neuron is in a quiescent state. Owing to the phenomenon of bifurcation delay it keeps tracking $x^\star(u)$ for some time after passing the Hopf bifurcation point at $u = u_2$. Ultimately, however, the solution will depart from $x^\star(u)$ and approach the periodic orbit: The neuron enters the spiking phase. Since g is negative near the limit cycle, u decreases while the trajectory spirals around the set of periodic orbits, yielding the slow modulation of the spikes. Finally, as u reaches u_1, the periodic orbit disappears, and the neuron returns to its quiescent state. Then the whole cycle starts over again.

Adding a noise term of small intensity $\sigma/\sqrt{\varepsilon}$ to the fast variables has two main effects:

- The bifurcation delay occurring after the passage through the Hopf bifurcation point is decreased from order 1 to order $\sqrt{\varepsilon|\log\sigma|}$, shortening the quiescent phase.
- For sufficiently strong noise, there is a chance that sample paths cross the unstable orbit and approach the quiescent state, some time before the saddle–node bifurcation of the periodic orbits.

The overall effect of noise is thus to shorten both the quiescent and the spiking phases. The main reason for this shortening is the fact that noise decreases the Hopf bifurcation delay. For $\sigma \gg \sqrt{\varepsilon}$, it even becomes possible for sample paths to escape from the quiescent state through the unstable orbit, some time before the Hopf bifurcation point is reached.

6.4 Models from Solid-State Physics

Stochastic models appear in various fields of solid-state physics, e.g. in lasers, which present many phenomena similar to neurons (see, for instance, [HN79]). In this section, we discuss stochastic models occurring in two other fields of solid-state physics, namely in mean-field models for ferromagnets, and in the dynamics of Josephson junction.

6.4.1 Ferromagnets and Hysteresis

For a long time, hysteresis in ferromagnets was considered a purely static phenomenon. As a consequence, it has been modelled by various integral operators relating the "output" of the system to its "input", for operators not depending on the speed of variation of the input (see for instance [May91] and [MNZ93] for reviews).

This situation changed drastically in 1990, when Rao and coauthors published a numerical study of the effect of the input's frequency on shape and area of hysteresis cycles [RKP90]. They proposed in particular that the area $\mathcal{A}$ of a hysteresis cycle, which measures the energy dissipation per period, should obey a scaling law of the form

$$\mathcal{A} \simeq A^\alpha \varepsilon^\beta \qquad (6.4.1)$$

for small amplitude A and frequency ε of the periodic input (e.g. the magnetic field), and some model-dependent exponents α and β. This work triggered a substantial amount of numerical, experimental and theoretical studies, aiming at establishing the validity of the scaling law (6.4.1) for various systems, a problem which has become known as *dynamical hysteresis*.

6.4 Models from Solid-State Physics

The first model investigated in [RKP90] is a Langevin partial differential equation for the spatially extended, N-component order parameter (e.g., the magnetisation), in a $(\Phi^2)^2$-potential with $O(N)$-symmetry, in the limit $N \to \infty$. Their numerical experiments suggested that (6.4.1) holds with $\alpha \simeq 2/3$ and $\beta \simeq 1/3$. Various theoretical arguments [DT92, SD93, ZZ95] indicate that the scaling law should be valid, but with $\alpha = \beta = 1/2$.

The second model considered in [RKP90] is an Ising model with Monte-Carlo dynamics. Here the situation is not so clear. Different numerical simulations (for instance [LP90, AC95, ZZL95]) suggested scaling laws with widely different exponents. More careful simulations [SRN98], however, showed that the behaviour of hysteresis cycles depends in a complicated way on the mechanism of magnetisation reversal, and no universal scaling law of the form (6.4.1) should be expected. Though much progress has been made in recent years on the mechanism of magnetisation reversal for quenched field (see [SS98], for instance, and [dH04] for a recent review), only few rigorous results on hysteresis in the Ising model are available so far.

A third kind of models for which scaling laws of hysteresis cycles have been investigated, belong to the mean field class, and include the Curie–Weiss model. In the limit of infinite system size (see, for instance, [Mar79]), the dynamics of the magnetisation can be described by the equation

$$\varepsilon \dot{x} = -x + \tanh\bigl(\beta(x + \lambda(t))\bigr), \qquad \lambda(t) = A\cos t, \qquad (6.4.2)$$

sometimes called *Suzuki–Kubo (SK) equation*. The right-hand side vanishes on the slow manifold $\lambda = \beta^{-1} \tanh^{-1} x - x$, which is S-shaped for $\beta > 1$. There are two saddle–node bifurcation points at $\pm(x_c, \lambda_c)$, where

$$x_c = \sqrt{1 - \beta^{-1}}, \qquad \lambda_c = \beta^{-1} \tanh^{-1}\sqrt{1-\beta^{-1}} - \sqrt{1-\beta^{-1}}. \qquad (6.4.3)$$

The equation (6.4.2) was examined numerically by Tomé and de Oliveira in [TdO90], where it was shown that the behaviour changes drastically when the amplitude of the forcing crosses a threshold close to $|\lambda_c|$, a phenomenon the authors termed "dynamic phase transition".

We have seen several examples of periodically forced bistable systems, such as the *Ginzburg–Landau (GL) equation*

$$\varepsilon \dot{x} = x - x^3 + \lambda(t), \qquad \lambda(t) = A\cos t, \qquad (6.4.4)$$

whose slow manifolds contain saddle–node bifurcation points. When passing such bifurcation points, the solutions react after a delay of order $\varepsilon^{2/3}$ by jumping to the other stable slow manifold. As a result, they converge to a cycle in the (λ, x)-plane, enclosing an area which obeys the scaling law

$$\mathcal{A} \simeq \mathcal{A}_0 + \varepsilon^{2/3} \qquad (6.4.5)$$

for sufficiently large driving amplitude A. Here $\mathcal{A}_0$ is a constant independent of ε, called the *static hysteresis area*. The same relation holds for (6.4.2), with a different $\mathcal{A}_0$.

As pointed out in [Rao92], the difference between the scaling laws (6.4.1) and (6.4.5) can be attributed to the existence of a potential barrier for the one-dimensional order parameter, which is absent in higher dimensions. The deterministic equation, however, neglects both thermal fluctuations and the finite system size, whose effects may be modelled by additive white noise (see for instance [Mar77, Mar79]). Noise, however, may help to overcome the potential barrier and thus change the scaling law.

Let us thus consider SDEs of the form

$$\mathrm{d}x_t = \frac{1}{\varepsilon} f(x_t, t) \,\mathrm{d}t + \frac{\sigma}{\sqrt{\varepsilon}} \,\mathrm{d}W_t , \qquad (6.4.6)$$

where $f(x,t)$ is either equal to $x - x^3 + \lambda(t)$, in the case of the GL equation, or to $-x + \tanh(\beta(x + \lambda(t)))$, for the SK equation. In each case, we assume that the periodic driving has the form $\lambda(t) = A\cos t$. Let $A_c = |\lambda_c|$ be the critical driving amplitude allowing to reach the saddle–node bifurcation point, that is, $A_c = 2/3\sqrt{3}$ for the GL equations, and $A_c = \sqrt{1 - \beta^{-1}} - \beta^{-1}\tanh^{-1}\sqrt{1 - \beta^{-1}}$ for the SK equations. We introduce a parameter $\delta = A - A_c$, measuring the "imperfection of the bifurcation" (cf. Section 4.2.1): For $\delta = 0$, the system effectively sees a transcritical bifurcation point. For $\delta < 0$, it encounters an avoided transcritical bifurcation, while for $\delta > 0$, it successively passes two saddle–node bifurcation points, separated by a distance of order $\delta^{1/2}$.

We already analysed the deterministic case for $A < A_c$ in Section 4.2.1, see Proposition 4.2.2. The case $A > A_c$ can be analysed by very similar means. One obtains the existence of constants $\gamma_1 > \gamma_0 > 0$ such that

- If $\delta = A - A_c \leqslant \gamma_0 \varepsilon$, the deterministic equation $\varepsilon \dot{x} = f(x,t)$ has exactly two stable periodic solutions $x_t^{\mathrm{per},+}$ and $x_t^{\mathrm{per},-}$, and one unstable periodic solution $x_t^{\mathrm{per},0}$. These solutions track, respectively, the two stable equilibrium branches and the unstable equilibrium branch, at a distance not larger than $\mathcal{O}(\varepsilon|\delta|^{-1/2} \wedge \sqrt{\varepsilon})$, and enclose an area

$$\mathcal{A}(\varepsilon) \asymp \varepsilon A . \qquad (6.4.7)$$

All solutions which do not start on $x_t^{\mathrm{per},0}$ are attracted either by $x_t^{\mathrm{per},+}$ or by $x_t^{\mathrm{per},-}$.

- If $\delta = A - A_c \geqslant \gamma_1 \varepsilon$, the deterministic equation $\varepsilon\dot{x} = f(x,t)$ admits exactly one periodic solution x_t^{per}, tracking the upper stable branch for decreasing λ, and the lower stable branch for increasing λ. This solution is stable and encloses an area $\mathcal{A}(\varepsilon)$ satisfying

$$\mathcal{A}(\varepsilon) - \mathcal{A}_0 \asymp \varepsilon^{2/3} \delta^{1/3} . \qquad (6.4.8)$$

For $\gamma_0 \varepsilon < \delta < \gamma_1 \varepsilon$, the situation is more complicated, as there may be more than three coexisting periodic orbits.

We return now to the study of Equation (6.4.6) with positive noise intensity σ. For $A < A_c$, we know from Theorem 4.2.3 and Theorem 4.2.4 that

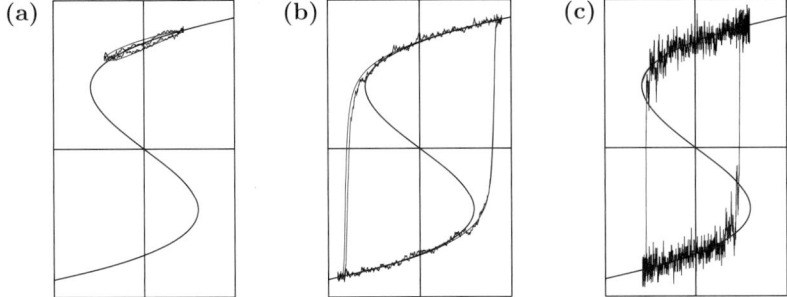

Fig. 6.10. Hysteresis cycles in the three parameter regimes. (**a**) *Small-amplitude regime:* If $\delta = A - A_c \leqslant \gamma_0 \varepsilon$ and $\sigma \ll (|\delta| \vee \varepsilon)^{3/4}$, typical cycles enclose an area of order εA. (**b**) *Large-amplitude regime:* If $\delta = A - A_c \geqslant \gamma_1 \varepsilon$ and $\sigma \ll \varepsilon^{1/2} \delta^{1/4}$, typical cycles enclose an area of order $\mathcal{A}_0 + \varepsilon^{2/3} \delta^{1/3}$. (**c**) *Strong-noise regime:* In cases of large noise intensity, typical cycles enclose an area of order $\mathcal{A}_0 - \sigma^{4/3}$, which is smaller than the static hysteresis area $\mathcal{A}_0$.

sample paths are likely to stay close to the same stable equilibrium branch if $\sigma \ll \sigma_c = (|\delta| \vee \varepsilon)^{3/4}$ (Fig. 6.10a). They are likely to switch from one branch to the other if $\sigma \gg \sigma_c$ (Fig. 6.10c). The switching typically occurs at a time of order $\sigma^{2/3}$ before $|\lambda(t)| = A$, that is, when $A - |\lambda(t)| \asymp \sigma^{4/3}$.

We also know from Theorem 3.3.3 and Theorem 3.3.4 that for $A - A_c$ of order one, a transition from late jumps (Fig. 6.10b) to early crossings occurs when σ becomes larger than $\sigma_c = \varepsilon^{1/2}$. It is not difficult to extend this result to all $A > A_c$, in which case one obtains a threshold noise intensity of the form $\sigma_c = \varepsilon^{1/2}(\delta \vee \varepsilon)^{1/4}$. Again, for $\sigma \gg \sigma_c$, transitions already occur as soon as $A - |\lambda(t)| \asymp \sigma^{4/3}$.

Since the area of hysteresis cycles measures the energy dissipation per period, it is of interest to determine its dependence on the parameters ε, δ and σ. This turns out to be possible, thanks to our relatively precise knowledge of the sample paths' behaviour. In the stochastic case, the hysteresis area is itself a random variable, that can be defined as

$$\mathcal{A}(\varepsilon, \delta, \sigma; \omega) = -\int_{-1/2}^{1/2} x_t(\omega) \lambda'(t) \, dt \ . \tag{6.4.9}$$

A key observation is that the deviation of the area from its deterministic counterpart is given by

$$\mathcal{A}(\varepsilon, \delta, \sigma; \omega) - \mathcal{A}(\varepsilon, \delta, 0) = -\int_{-1/2}^{1/2} (x_t(\omega) - x_t^{\text{det}}) \lambda'(t) \, dt \ , \tag{6.4.10}$$

which is accessible because of the information we have on the behaviour of $x_t - x_t^{\text{det}}$. Without entering into technical details, which can be found in [BG02b], we may summarise the results as follows. There are three main parameter regimes to be considered (Fig. 6.11), see also [BG02a].

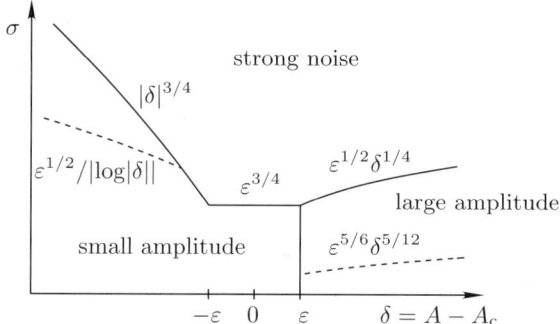

Fig. 6.11. Parameter regimes for dynamical hysteresis, shown, for fixed ε, in the plane $(\delta = A - A_c, \sigma)$. For noise intensities σ below a threshold depending on δ and ε, the behaviour resembles the deterministic behaviour. For noise intensities above this threshold, typical hysteresis cycles shrink by an amount depending on σ. Broken curves subdivide the three main regimes into smaller zones, in which we obtain estimates of variable degree of precision.

- *Small-amplitude regime:* $\delta = A - A_c \leqslant \gamma_0 \varepsilon$ and $\sigma \leqslant (|\delta| \vee \varepsilon)^{3/4}$ (Fig. 6.10a). The random area $\mathcal{A}(\varepsilon, \delta, \sigma)$ is concentrated around its deterministic limit $\mathcal{A}(\varepsilon, \delta, 0) \asymp A\varepsilon$. More precisely, one can show that
  - if $\delta \geqslant -\varepsilon$ or $\sigma \leqslant \varepsilon^{1/2}/|\log|\delta||$, the distribution of $\mathcal{A}(\varepsilon, \delta, \sigma)$ is close to a normal distribution, with variance of order $\sigma^2 \varepsilon$;
  - if $\delta < -\varepsilon$ and $\sigma > \varepsilon^{1/2}/|\log|\delta||$, the distribution of $\mathcal{A}(\varepsilon, \delta, \sigma)$ is not necessarily close to a Gaussian, but its expectation differs from $\mathcal{A}(\varepsilon, \delta, 0)$ by at most $\mathcal{O}(\sigma^2 |\log|\delta||)$, and the standard deviation of $\mathcal{A}(\varepsilon, \delta, \sigma)$ is of the same order at most. Finally, the tails of the distribution of $\mathcal{A}(\varepsilon, \delta, \sigma)$ decrease exponentially with an exponent of order $\sigma^2 |\log|\delta||$.
- *Large-amplitude regime:* $\delta = A - A_c \geqslant \gamma_1 \varepsilon$ and $\sigma \leqslant \varepsilon^{1/2} \delta^{1/4}$ (Fig. 6.10b). The random area $\mathcal{A}(\varepsilon, \delta, \sigma)$ is concentrated around its deterministic counterpart $\mathcal{A}(\varepsilon, \delta, 0)$, which satisfies $\mathcal{A}(\varepsilon, \delta, 0) - \mathcal{A}_0 \asymp \varepsilon^{2/3} \delta^{1/3}$. More precisely, one can show that
  - if $\sigma \leqslant \varepsilon^{5/6} \delta^{5/12}$, the distribution of $\mathcal{A}(\varepsilon, \delta, \sigma)$ is close to a normal distribution, with variance of order $\sigma^2 \varepsilon^{1/3} \delta^{1/6}$;
  - if $\varepsilon^{5/6} \delta^{5/12} < \sigma \leqslant \varepsilon^{1/2} \delta^{1/4}$, the distribution of $\mathcal{A}(\varepsilon, \delta, \sigma)$ is not necessarily close to a Gaussian, but it is concentrated in an interval of size $\varepsilon^{2/3} \delta^{1/3}$ around $\mathcal{A}(\varepsilon, \delta, 0)$.
- *Strong-noise regime:* Either $\delta \leqslant \varepsilon$ and $\sigma > (|\delta| \vee \varepsilon)^{3/4}$, or $\delta > \varepsilon$ and $\sigma > \varepsilon^{1/2} \delta^{1/4}$ (Fig. 6.10c).
  The random area $\mathcal{A}(\varepsilon, \delta, \sigma)$ is concentrated around a deterministic reference value $\widehat{\mathcal{A}}$, satisfying $\widehat{\mathcal{A}} - \mathcal{A}_0 \asymp -\sigma^{4/3}$. More precisely, one can show that
  - if either $\delta \leqslant \varepsilon$ or $\sigma > \delta^{3/4}$, the tail probability $\mathbb{P}\{\mathcal{A}(\varepsilon, \delta, \sigma) - \widehat{\mathcal{A}} > H\}$ decays exponentially, with an exponent of order $H/\sigma^2 |\log \sigma|$; while the

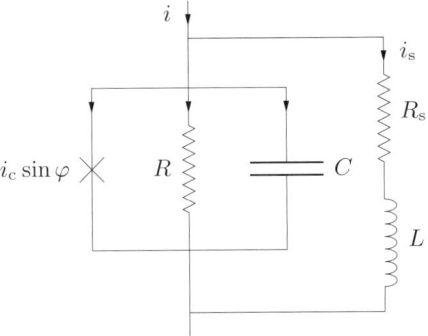

Fig. 6.12. The resistive-capacitive-inductively shunted model of a Josephson junction. The ideal junction is traversed by a current $i_c \sin\varphi$. The other elements of the equivalent circuit model imperfections in the junction and its wiring.

  tail probability $\mathbb{P}\{\mathcal{A}(\varepsilon,\delta,\sigma) - \widehat{\mathcal{A}} < -H\}$ decays exponentially with an exponent of order $H^{3/2}/\sigma^2$;
- if $\delta > \varepsilon$ and $\sigma \leqslant \delta^{3/4}$, the tail probability $\mathbb{P}\{\mathcal{A}(\varepsilon,\delta,\sigma) - \widehat{\mathcal{A}} > H\}$ decays exponentially, with an exponent of order $H/\sigma^2 |\log(\sigma^{4/3}\delta^{-1/2})|$; while the tail probability $\mathbb{P}\{\mathcal{A}(\varepsilon,\delta,\sigma) - \widehat{\mathcal{A}} < -H\}$ decays exponentially with an exponent of order $H^{3/2}/\sigma^2$.

The most interesting aspect of these estimates is that noise of sufficient intensity decreases the hysteresis area, and thus *reduces* the energy dissipation per period.

6.4.2 Josephson Junctions

Josephson junctions are devices made of two pieces of superconductor, which present nontrivial current–voltage characteristics. The state of the junction is described by the so-called Josephson phase φ, which corresponds to the phase difference of the complex wave functions of Cooper pairs in the two superconductors. An ideal junction can function in two states: In the superconducting state, the Josephson phase φ is constant, and a current proportional to $\sin\varphi$ flows through the junction without any resistance. In the resistive state, the phase φ is not constant, and a voltage proportional to the derivative $\dot\varphi$ appears across the junction.

In practice, Josephson junctions do not behave in this idealised way, but show some parasite resistance, capacitance and inductance. We consider here the resistive-capacitive-inductively shunted junction (RCLSJ) model, which is based on the equivalent circuit of Fig. 6.12. The ideal Josephson junction is traversed by a current $i_c \sin\varphi$, where i_c is a constant. The voltage v at the junction is given by the phase-voltage relation $(\hbar/2e)(\mathrm{d}\varphi/\mathrm{d}t)$. Kirchhoff's laws thus yield the equations

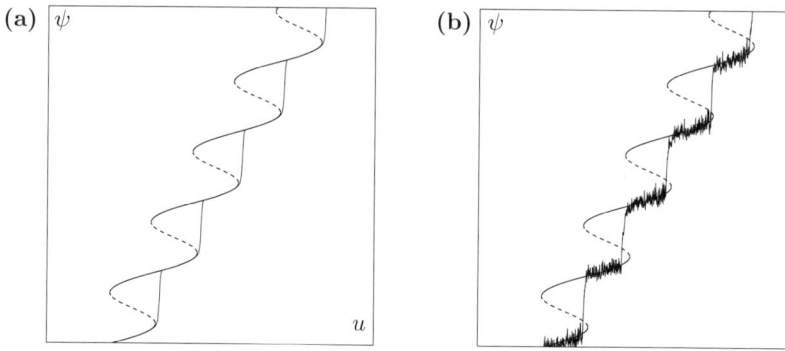

Fig. 6.13. Solutions of equations (6.4.16), describing the dynamics of a Josephson junction in the overdamped regime, for $\hat{\alpha} = 5$, $J = 1.5$ and $\varepsilon = 0.01$, (a) without noise, (b) with noise of intensity $\sigma = 0.5$ added to the fast variable.

$$i = C\frac{dv}{dt} + \frac{v}{R} + i_c \sin\varphi + i_s,$$
$$v = \frac{\hbar}{2e}\frac{d\varphi}{dt} = L\frac{di_s}{dt} + i_s R_s, \quad (6.4.11)$$

for the dynamic variables, i.e., the Josephson phase φ and the current i_s (cf. [NP03]). Passing to dimensionless variables, and rescaling time, this system can be rewritten in the form

$$\beta\gamma^2\ddot{\varphi} + \gamma\dot{\varphi} + \sin\varphi = J - I_s,$$
$$\dot{\varphi} = \alpha\dot{I}_s + I_s. \quad (6.4.12)$$

Here $J = i/i_c$ and $I_s = i_s/i_c$ are dimensionless currents, $\alpha = 2ei_c L/\hbar$ represents a dimensionless inductance, $\beta = 2ei_c R^2 C/\hbar$ is called *Stewart–McCumber parameter*, its inverse measures the dissipation, and $\gamma = R_s/R$. Introducing the variable

$$u = J - I_s + \frac{1+\gamma}{\alpha}\varphi + \frac{\beta\gamma^2}{\alpha}\dot{\varphi} \quad (6.4.13)$$

allows the system (6.4.12) to be rewritten in the form

$$\beta\gamma^2\ddot{\varphi} + \left(1 + \frac{\beta\gamma}{\alpha}\right)\gamma\dot{\varphi} + \frac{1+\gamma}{\alpha}\varphi + \sin\varphi = u,$$
$$\alpha\dot{u} = J - \sin\varphi. \quad (6.4.14)$$

Note that $\dot{u}$ vanishes if and only if all the current flows through the ideal junction. The parameters α and γ measure, respectively, the relaxation times of u and of φ. If $\gamma \ll \alpha$, (6.4.14) is a slow–fast system. Scaling time by a factor α and setting $\varepsilon = \gamma/\alpha$, we can rewrite it as

$$\beta\varepsilon\dot{\varphi} = \psi - (1 + \beta\varepsilon)\varphi,$$
$$\varepsilon\dot{\psi} = u - \hat{\alpha}^{-1}\varphi - \sin\varphi, \quad (6.4.15)$$
$$\dot{u} = J - \sin\varphi,$$

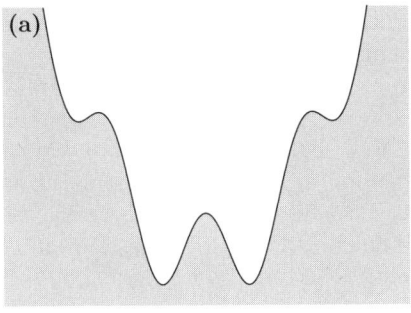 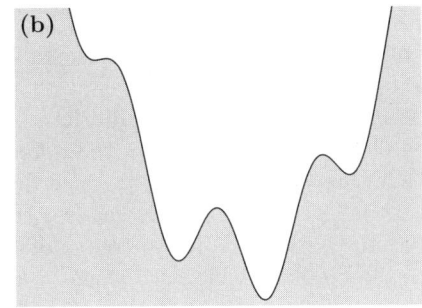

Fig. 6.14. The potential $\psi \mapsto U(\psi, u)$, for $\widehat{\alpha} = 10$, and **(a)** $u = \pi/\widehat{\alpha} \simeq 0.31$, and **(b)** $u = 0.46$, slightly before the leftmost well disappears in a saddle–node bifurcation.

where $\widehat{\alpha}^{-1} = \alpha^{-1} + \varepsilon$. The slow manifold is thus given by the two equations $\psi = (1 + \beta\varepsilon)\varphi$ and $u = \widehat{\alpha}^{-1}\varphi + \sin\varphi$. If $\widehat{\alpha} > 1$, it admits several stable and unstable branches, connected by saddle–node bifurcation points.

The simplest situation to analyse is the strong-damping case $\beta \ll 1$. Then the variable φ is an order of magnitude faster than ψ, and converges to $\psi + \mathcal{O}(\beta)$. One can thus replace (6.4.15) by the reduced system (written for simplicity for $\beta = 0$)

$$\varepsilon\dot{\psi} = u - \widehat{\alpha}^{-1}\psi - \sin\psi = -\frac{\partial}{\partial\psi}U(\psi, u),$$

$$\dot{u} = J - \sin\psi, \qquad (6.4.16)$$

where we have introduced a potential

$$U(\psi, u) = \frac{(\psi - \widehat{\alpha}u)^2}{2\widehat{\alpha}} - \cos\psi. \qquad (6.4.17)$$

The dynamics depends crucially on the value of the current J. If $|J| < 1$, the system has at least one asymptotically stable equilibrium point, which corresponds to a superconducting state since $\dot{\varphi} = 0$. If $|J| > 1$, then u varies monotonously, and ψ tries to track the slow manifold. If $\widehat{\alpha} > 1$, the phase repeatedly jumps from one stable branch of the slow manifold to the next one, each time it passes a bifurcation point (Fig. 6.13a).

Adding noise to the system allows the sample paths to jump from one piece of stable slow manifold to the next one, already some time before reaching the saddle–node bifurcation point (Fig. 6.13b). This is the same effect as discussed for stochastic resonance in Chapter 4 and can be analysed by the same methods.

For weak damping, the dynamics can be more complicated. The associated fast system then describes the underdamped motion of a particle in the potential (6.4.17), and inertial effects may cause the particle to cross several potential valleys in a row. This can be seen by analysing the behaviour of the unstable manifold of a saddle, when approaching a saddle–node bifurcation. While for large damping the unstable manifold is always attracted by the next potential minimum, for smaller damping it may bridge several potential wells before reaching an equilibrium. When weak noise is added to the system, sample paths escaping from the potential well slightly before the bifurcation occurs are likely to track the unstable manifold of the saddle, ending up in the same potential well as the deterministic system would have. Again, the methods developed in the course of the book will allow for a description of the sample-path behaviour when overcoming the potential barrier.

A

A Brief Introduction to Stochastic Differential Equations

A.1 Brownian Motion

Brownian motion can be constructed as a scaling limit of a symmetric random walk,

$$W_t = \lim_{n \to \infty} \frac{1}{\sqrt{n}} \sum_{i=1}^{\lfloor nt \rfloor} X_i , \qquad (A.1.1)$$

where the X_i are independent, identically distributed real-valued random variables of zero mean and unit variance, and the limit is to be understood as convergence in distribution. An equivalent definition of Brownian motion is the following one.

Definition A.1.1 (Brownian motion). *A one-dimensional standard Brownian motion, or Wiener process, is a stochastic process $\{W_t\}_{t \geq 0}$, i.e., a collection of real-valued random variables, satisfying the three conditions*

1. $W_0 = 0$;
2. Independent increments: *For all $t > s \geq 0$, the increment $W_t - W_s$ is independent of $\{W_u\}_{0 \leq u \leq s}$;*
3. Gaussian increments: *For all $t > s \geq 0$, the increment $W_t - W_s$ is normally distributed with zero mean and variance $t - s$.*

Using, for instance, Kolmogorov's continuity theorem, one shows that $\{W_t\}_{t \geq 0}$ admits a continuous version. Thus we may assume that the sample paths $t \mapsto W_t(\omega)$ are continuous, and use $(\Omega, \mathcal{F}, \mathbb{P})$ as probability space, where $\Omega = \mathcal{C}_0([0, \infty), \mathbb{R})$ is the space of continuous functions $f : [0, \infty) \to \mathbb{R}$ with $f(0) = 0$. However, the sample paths of a Brownian motion are almost surely nowhere differentiable.

In the sequel we will also deal with Brownian motion starting at an arbitrary time t_0 in some point x_0. Such a Brownian motion is easily constructed from standard Brownian motion $\{W_t\}_{t \geq 0}$ by considering $\{x_0 + W_{t_0+t}\}_{t \geq 0}$.

Whenever not starting in $(t_0, x_0) = (0,0)$, we indicate this by writing $\mathbb{P}^{t_0,x_0}$ instead of $\mathbb{P} = \mathbb{P}^{0,0}$.

The following properties follow immediately from the definition A.1.1:

- *Markov property:* $\{W_t\}_{t\geqslant 0}$ is a Markov process, i.e.,

$$\mathbb{P}\{W_{t+s} \in A \mid W_u, \ u \leqslant t\} = \mathbb{P}\{W_{t+s} \in A \mid W_t\}, \qquad (A.1.2)$$

for any Borel set A and all $t, s \geqslant 0$, and the transition probabilities are Gaussian:

$$\mathbb{P}\{W_{t+s} \in A \mid W_t = x\} = \mathbb{P}^{t,x}\{W_{t+s} \in A\} = \int_A \frac{\mathrm{e}^{-(y-x)^2/2s}}{\sqrt{2\pi s}}\,\mathrm{d}y \quad (A.1.3)$$

for any Borel set A, and all $t, s \geqslant 0$.
- *Scaling property:* For any $c > 0$, $\{cW_{t/c^2}\}_{t\geqslant 0}$ is a standard Brownian motion.
- *Gaussian process:* $\{W_t\}_{t\geqslant 0}$ is a Gaussian process (i.e., all its finite-dimensional marginals are Gaussian random variables), of zero mean, and covariance matrix

$$\mathrm{Cov}\{W_t, W_s\} := \mathbb{E}^{0,0}\{W_t W_s\} = t \wedge s \qquad (A.1.4)$$

for all $t, s \geqslant 0$. Conversely, any mean-zero Gaussian process satisfying (A.1.4) is a standard Brownian motion.

Let $\mathcal{F}_t = \sigma\{W_s,\ 0 \leqslant s \leqslant t\} \subset \mathcal{F}$ be the σ-algebra generated by all events of the form $\{a \leqslant W_s < b\}$, $0 \leqslant s \leqslant t$, $a < b$, that is, events depending only on the behaviour of the Brownian motion up to time t. Then $\mathcal{F}_s \subset \mathcal{F}_t$, whenever $s \leqslant t$. The family $\{\mathcal{F}_t\}_{t\geqslant 0}$ is called the *canonical filtration* generated by the Brownian motion.

A random variable $\tau : \Omega \to [0, \infty]$ is called a *stopping time* with respect to the filtration $\{\mathcal{F}_t\}_{t\geqslant 0}$ if $\tau^{-1}([0,t]) := \{\omega \in \Omega \colon \tau(\omega) \leqslant t\} \in \mathcal{F}_t$ for all $t > 0$. A typical example of a stopping time is the *first-exit time*

$$\tau_A = \inf\{t > 0 \colon W_t \notin A\} \qquad (A.1.5)$$

from an open[1] set $A \ni 0$: Indeed, the knowledge of the Brownian motion up to time t suffices to decide whether or not $\tau \leqslant t$. By contrast, the last time $\sup\{t \leqslant 1 \colon W_t \in A\}$ the Brownian motion is in a (measurable) set A, is an example of a random time which is *no* stopping time, as in order to decide whether $\tau(\omega) \leqslant t$ holds we would have "to look into the future" beyond time t. With a stopping time τ, we associate the *pre-τ σ-algebra* $\mathcal{F}_\tau$, defined by

$$A \in \mathcal{F}_\tau \quad \text{if and only if} \quad A \cap \{\tau \leqslant t\} \in \mathcal{F}_t \text{ for all } t \geqslant 0. \qquad (A.1.6)$$

A stochastic process having the so-called strong Markov property can be "restarted" at any stopping time τ:

---

[1] For a general Borel-measurable set A, the first-exit time τ_A is still a stopping time, provided the canonical filtration is completed by the null sets.

Theorem A.1.2 (Strong Markov property). *For any bounded measurable function $f : \mathbb{R} \to \mathbb{R}$, and any stopping time τ satisfying $\mathbb{P}\{\tau < \infty\} = 1$,*

$$\mathbb{E}^{0,0}\{f(W_{\tau+t}) \mid \mathcal{F}_\tau\} = \mathbb{E}^{0,W_\tau}\{f(W_t)\} . \qquad (A.1.7)$$

A k-dimensional standard Brownian motion is a vector $(W_t^{(1)}, \ldots, W_t^{(k)})$ of k independent one-dimensional standard Brownian motions, and the underlying probability space can thus be chosen as the k-fold product space of the space $(\Omega, \mathcal{F}, \mathbb{P})$ of $\{W_t\}_{t \geqslant 0}$.

A.2 Stochastic Integrals

In order to give a meaning to a stochastic differential equation which is generally understood as an integral equation, one starts by defining integrals of the form

$$\int_0^t h(s,\omega)\, \mathrm{d}W_s , \qquad (A.2.1)$$

where the integrand may depend on the realisation of the Brownian motion only up to time s, that is, $\omega \mapsto h(s,\omega)$ is assumed to be measurable with respect to $\mathcal{F}_s$ for each s. Since the Brownian motion is almost surely nowhere differentiable, and thus not of bounded variation, the integral (A.2.1) cannot be defined as a Riemann–Stieltjes integral. The following construction is due to Itô. As usual in integration theory, one starts by defining the integral for elementary integrands, and then extends the definition to a larger class of integrands by a limiting procedure.

Definition A.2.1 (Elementary function). *Fix a constant $T > 0$. A function $e : [0,T] \times \Omega \to \mathbb{R}$ is called* elementary *if there is a (deterministic) partition $0 = t_0 < t_1 < \cdots < t_N = T$ of the time interval $[0,T]$ such that, for all $s \in [0,T]$,*

$$e(s,\omega) = \sum_{k=0}^{N-1} e_k(\omega) \mathbf{1}_{[t_k, t_{k+1})}(s) , \qquad (A.2.2)$$

where $e_k : \Omega \mapsto \mathbb{R}$ is $\mathcal{F}_{t_k}$-measurable for all k.

Definition A.2.2 (Itô integral for elementary functions). *Let $e : [0,T] \times \Omega \to \mathbb{R}$ be elementary. Then the Itô integral of e on $[0,T]$ is defined by*

$$\int_0^T e(s)\, \mathrm{d}W_s = \int_0^T e(s,\omega)\, \mathrm{d}W_s(\omega) := \sum_{k=0}^{N-1} e_k(\omega) \bigl[W_{t_{k+1}}(\omega) - W_{t_k}(\omega) \bigr] . \qquad (A.2.3)$$

Proposition A.2.3 (Itô isometry). *Let e be an elementary function of the form (A.2.2), such that $e_k \in L^2(\mathbb{P})$ for all k. Then*

$$\mathbb{E}\left\{ \left(\int_0^T e(s)\, \mathrm{d}W_s \right)^2 \right\} = \int_0^T \mathbb{E}\{e(s)^2\}\, \mathrm{d}s . \qquad (A.2.4)$$

The Itô isometry, which is an isometry between the two Hilbert spaces $L^2(\lambda_{[0,T]} \otimes \mathbb{P})$ and $L^2(\mathbb{P})$,[2] allows to extend the definition of the Itô integral to the class of all functions $h : [0,T] \times \Omega \to \mathbb{R}$ such that

- $(t,\omega) \mapsto h(t,\omega)$ is jointly measurable, i.e., measurable with respect to $\mathcal{B}([0,T]) \otimes \mathcal{F}$, where $\mathcal{B}([0,T])$ denotes the Borel σ-algebra on the time interval $[0,T]$;
- h is $\mathcal{F}_t$-adapted, that is, $\omega \mapsto h(t,\omega)$ is $\mathcal{F}_t$-measurable for each $t \in [0,T]$;
- h satisfies the integrability assumption

$$\int_0^T \mathbb{E}\{h(s)^2\}\,\mathrm{d}s < \infty . \tag{A.2.5}$$

Any h of this form can be approximated by a sequence of elementary functions $\{e^{(n)}\}_{n \geqslant 1}$, in the sense that

$$\lim_{n \to \infty} \int_0^T \mathbb{E}\{(h(s) - e^{(n)}(s))^2\}\,\mathrm{d}s = 0 . \tag{A.2.6}$$

Proposition A.2.3 shows that the limit

$$\lim_{n \to \infty} \int_0^T e^{(n)}(s)\,\mathrm{d}W_s =: \int_0^T h(s)\,\mathrm{d}W_s \tag{A.2.7}$$

exists in $L^2(\mathbb{P})$, and does not depend on the particular choice of the approximating sequence $\{e^{(n)}\}_{n \geqslant 1}$.

Definition A.2.4 (Itô integral). *The limit (A.2.7) is called the* Itô integral *of h on $[0,T]$. For any interval $[a,b] \subset [0,T]$, one sets*

$$\int_a^b h(s)\,\mathrm{d}W_s := \int_0^T \mathbf{1}_{[a,b]}(s) h(s)\,\mathrm{d}W_s . \tag{A.2.8}$$

Obviously, the Itô integral defines a linear, additive functional on the space of admissible integrands. Moreover, the random variable $X_t = \int_0^t h(s)\,\mathrm{d}W_s$ is $\mathcal{F}_t$-measurable, has zero expectation, and variance

$$\mathrm{Var}\{X_t\} = \int_0^t h(s)^2\,\mathrm{d}s , \tag{A.2.9}$$

and there exists a continuous version of the stochastic process $\{X_t\}_{t \geqslant 0}$, so that we may assume that $X_t(\omega)$ depends continuously on t. Furthermore, the stochastic process $\{X_t\}_{t \geqslant 0}$ is a martingale, meaning that X_t is $\mathcal{F}_t$-measurable for all t and $\mathbb{E}\{X_t | \mathcal{F}_s\} = X_s$ holds, whenever $s \leqslant t$.

---

[2] Here $\lambda_{[0,T]}$ denotes Lebesgue measure on the interval $[0,t]$.

Remark A.2.5. The definition of the Itô integral can be extended to integrands satisfying only

$$\mathbb{P}\left\{\int_0^T h(s)^2 \, ds < \infty\right\} = 1 \qquad (A.2.10)$$

instead of (A.2.7) by defining the stochastic integral as the limit in probability of stochastic integrals of elementary functions, but then the properties involving expectations, in particular (A.2.9), may no longer hold true.

Remark A.2.6. Multidimensional Itô integrals can be defined component-wise in the obvious way.

Consider now the process

$$X_t = X_0 + \int_0^t f(s) \, ds + \int_0^t g(s) \, dW_s, \qquad (A.2.11)$$

where $f, g : [0, T] \times \Omega \to \mathbb{R}$ satisfy the same measurability assumptions as h above, and

$$\mathbb{P}\left\{\int_0^T |f(s)| \, ds < \infty\right\} = \mathbb{P}\left\{\int_0^T g(s)^2 \, ds < \infty\right\} = 1. \qquad (A.2.12)$$

For simplicity, we assume here that the initial value X_0 is not random. One may include the case of random initial values, independent of the Brownian motion, by adequately enlarging the filtration $\{\mathcal{F}_t\}_{t \geq 0}$.

Theorem A.2.7 (Itô formula). *Let $u : \mathbb{R} \times [0, T] \to \mathbb{R}$ be twice continuously differentiable in its first argument, and continuously differentiable in its second argument. Then the process $Y_t = u(X_t, t)$ satisfies*

$$Y_t = u(X_0, 0) + \int_0^t \left[\frac{\partial u}{\partial t}(X_s, s) + \frac{\partial u}{\partial x}(X_s, s)f(s) + \frac{1}{2}\frac{\partial^2 u}{\partial x^2}(X_s, s)g(s)^2\right] ds$$

$$+ \int_0^t \frac{\partial u}{\partial x}(X_s, s)g(s) \, dW_s. \qquad (A.2.13)$$

It is customary to write (A.2.11) and (A.2.13) in differential form as

$$dX_t = f(t) \, dt + g(t) \, dW_t \qquad (A.2.14)$$

$$dY_t = \frac{\partial u}{\partial t} dt + \frac{\partial u}{\partial x} dX_t + \frac{1}{2}\frac{\partial^2 u}{\partial x^2}(dX_t)^2$$

$$= \left[\frac{\partial u}{\partial t} + \frac{\partial u}{\partial x}f(t) + \frac{1}{2}\frac{\partial^2 u}{\partial x^2}g(t)^2\right] dt + \frac{\partial u}{\partial x}g(t) \, dW_t. \qquad (A.2.15)$$

The first two terms on the right-hand side of (A.2.15) are natural, since they also exist in the deterministic case. The additional term containing the second derivative of u with respect to x is interpreted as reflecting the fact that "$(dX_t)^2 = g(t)^2(dW_t)^2 = g(t)^2 dt$".

In the case of an n-dimensional stochastic processes $(X_t^{(1)}, \ldots, X_t^{(n)})$, defined component-wise, with the help of the same Brownian motion,

$$dX_t^{(i)} = f^{(i)}(t)\,dt + g^{(i)}(t)\,dW_t , \qquad (A.2.16)$$

and a function $u : \mathbb{R}^n \times [0,T] \to \mathbb{R}$, satisfying the analogous assumptions, the real-valued stochastic process $Y_t = u(X_t^{(1)}, \ldots, X_t^{(n)}, t)$ satisfies

$$dY_t = \frac{\partial u}{\partial t}\,dt + \sum_{i=1}^n \frac{\partial u}{\partial x_i}\,dX_t^{(i)} + \frac{1}{2}\sum_{i,j=1}^n \frac{\partial^2 u}{\partial x_i \partial x_j} g^{(i)}(t)g^{(j)}(t)\,dt . \qquad (A.2.17)$$

Finally, should the $X_t^{(i)}$, $i = 1, \ldots, n$, involve different, independent Brownian motions $W_t^{(j)}$, $j = 1, \ldots, n$, then $Y_t t = u(X_t^{(1)}, \ldots, X_t^{(n)}, t)$ is calculated according to the *general scheme*

$$dY_t = \frac{\partial u}{\partial t}\,dt + \sum_{i=1}^n \frac{\partial u}{\partial x_i}\,dX_t^{(i)} + \frac{1}{2}\sum_{i,j=1}^n \frac{\partial^2 u}{\partial x_i \partial x_j}(dX_t^{(i)})(dX_t^{(j)}) , \qquad (A.2.18)$$

using the substitution rules $(dt)(dt) = (dt)(dW_t^{(i)}) = (dW_t^{(i)})(dt) = 0$ and $(dW_t^{(i)})(dW_t^{(j)}) = \delta_{ij} dt$.

Example A.2.8. A few classical examples where the Itô formula allows to compute specific stochastic integrals are the following.

1. The relation $d(W_t^2) = dt + 2W_t\,dW_t$ implies

$$\int_0^t W_s\,dW_s = \frac{1}{2}W_t^2 - \frac{1}{2}t . \qquad (A.2.19)$$

2. The relation $d(tW_t) = W_t\,dt + t\,dW_t$ yields the integration-by-parts formula

$$\int_0^t s\,dW_s = tW_t - \int_0^t W_s\,ds . \qquad (A.2.20)$$

3. Let $dX_t = h(t)\,dW_t - \frac{1}{2}h(t)^2\,dt$. Then $Y_t = e^{X_t}$ satisfies the equation

$$dY_t = h(t)Y_t\,dW_t . \qquad (A.2.21)$$

In particular, $Y_t = e^{W_t - t/2}$ obeys the equation $dY_t = Y_t\,dW_t$. The *exponential martingale* Y_t is thus to be considered as the exponential of the Brownian motion W_t. It is called the *Doléans exponential* of W_t.

A.3 Strong Solutions

Having defined stochastic integrals, we can finally give a meaning to stochastic differential equations of the form

$$\mathrm{d}x_t = f(x_t, t)\,\mathrm{d}t + F(x_t, t)\,\mathrm{d}W_t, \tag{A.3.1}$$

where $f, F : \mathbb{R} \times [0, T] \to \mathbb{R}$ are jointly measurable deterministic functions of x and t.

Definition A.3.1 (Strong solution). *A stochastic process $\{x_t\}_{t \in [0,T]}$ is called a* strong solution *of the SDE (A.3.1) with initial condition x_0 if*

- $\{x_t\}_{t \in [0,T]}$ *is $\{\mathcal{F}_t\}_{t \in [0,T]}$-adapted;*
- *The integrability condition*

$$\mathbb{P}\left\{\int_0^T |f(x_s, s)|\,\mathrm{d}s < \infty\right\} = \mathbb{P}\left\{\int_0^T F(x_s, s)^2\,\mathrm{d}s < \infty\right\} = 1 \tag{A.3.2}$$

is satisfied;
- *For any $t \in [0, T]$,*

$$x_t = x_0 + \int_0^t f(x_s, s)\,\mathrm{d}s + \int_0^t F(x_s, s)\,\mathrm{d}W_s \tag{A.3.3}$$

holds with probability 1.

The following result is obtained by Picard iterations, as in the deterministic case.

Theorem A.3.2 (Existence and uniqueness of strong solutions). *Assume that the following conditions hold for all $t \in [0, T]$:*

- *Local Lipschitz condition: For any compact $A \subset \mathbb{R}$, there is a constant K_A such that*

$$|f(x, t) - f(y, t)| + |F(x, t) - F(y, t)| \leqslant K_A |x - y| \tag{A.3.4}$$

for all $x, y \in A$ and all $t \in [0, T]$.
- *Bounded-growth condition: There is a constant L such that*

$$xf(x, t) + F(x, t)^2 \leqslant L^2(1 + x^2) \tag{A.3.5}$$

for all $x, y \in \mathbb{R}$ and all $t \in [0, T]$.

Then the SDE (A.3.1) admits an almost surely continuous strong solution $\{x_t\}_{t \in [0,T]}$. This solution is pathwise unique, in the sense that if $\{x_t\}_{t \in [0,T]}$ and $\{y_t\}_{t \in [0,T]}$ are two almost surely continuous solutions, then

$$\mathbb{P}\left\{\sup_{0 \leqslant t \leqslant T} |x_t - y_t| > 0\right\} = 0. \tag{A.3.6}$$

We denote by $\mathbb{P}^{t_0,x_0}$ the law of the solution of (A.3.1), starting in x_0 at some time $t_0 \in [0,T]$, and by $\mathbb{E}^{t_0,x_0}$ expectations with respect to $\mathbb{P}^{t_0,x_0}$. The solution x_t satisfies the strong Markov property, that is, for any almost surely finite stopping time τ, and any bounded measurable function $f : \mathbb{R} \to \mathbb{R}$,

$$\mathbb{E}^{0,0}\{f(x_t) \mid \mathcal{F}_\tau\} = \mathbb{E}^{0,x_\tau}\{f(x_t)\} \quad \text{on the set } \{\tau \leq t\}. \tag{A.3.7}$$

Example A.3.3 (Linear SDE). Let $a, b, F : [0,T] \to \mathbb{R}$ be bounded, measurable deterministic functions. Then the linear SDE

$$dx_t = [a(t)x_t + b(t)]\,dt + F(t)\,dW_t \tag{A.3.8}$$

admits a strong solution. Let $\alpha(t,s) = \int_s^t a(u)\,du$. Itô's formula shows that the solution with initial condition $x_{t_0} = x_0$ can be written as

$$x_t = x_0 e^{\alpha(t,t_0)} + \int_{t_0}^t e^{\alpha(t,s)} b(s)\,ds + \int_{t_0}^t e^{\alpha(t,s)} F(s)\,dW_s\,. \tag{A.3.9}$$

In the particular case $a(t) \equiv -\gamma$, $b(t) \equiv 0$, $F(t) \equiv 1$, its solution

$$x_t = x_0 e^{-\gamma(t-t_0)} + \int_{t_0}^t e^{-\gamma(t-s)}\,dW_s \tag{A.3.10}$$

is called an *Ornstein–Uhlenbeck process*.

A.4 Semigroups and Generators

Consider, for simplicity, an autonomous, n-dimensional SDE

$$dx_t = f(x_t)\,dt + F(x_t)\,dW_t\,, \tag{A.4.1}$$

admitting a pathwise unique strong solution $\{x_t\}_{t \geq 0}$ for any initial condition $(0,x)$. One can associate with it a family $\{T_t\}_{t \geq 0}$ of linear operators, acting on bounded, measurable functions $\varphi : \mathbb{R}^n \to \mathbb{R}$, as

$$T_t\varphi(x) = \mathbb{E}^{0,x}\{\varphi(x_t)\}\,. \tag{A.4.2}$$

The Markov property, together with time homogeneity, implies that the $\{T_t\}_{t \geq 0}$ form a semigroup:

$$T_t T_s = T_{t+s} \quad \forall t, s \geq 0\,. \tag{A.4.3}$$

Each linear operator T_t has the following properties:

1. It preserves the constant functions.
2. It maps non-negative functions to non-negative functions.
3. It is a contraction with respect to the supremum norm.

4. It maps the set of bounded, continuous functions into itself (*weak Feller property*).
5. Under additional assumptions, for instance if the matrix $F(x)F(x)^{\mathrm{T}}$ is positive definite for all x, it maps bounded, measurable functions to continuous ones for $t > 0$ (*strong Feller property*).

The *(infinitesimal) generator* of the semigroup is defined as the operator L such that
$$L\varphi(x) = \lim_{t \to 0} \frac{T_t\varphi(x) - \varphi(x)}{t}, \tag{A.4.4}$$
for all φ for which this limit exists for all x. Itô's formula, and the fact that stochastic integrals have zero expectation, imply that the generator is the differential operator
$$L = \sum_{i=1}^{n} f_i(x)\frac{\partial}{\partial x_i} + \frac{1}{2}\sum_{i,j=1}^{n} d_{ij}(x)\frac{\partial^2}{\partial x_i \partial x_j}, \tag{A.4.5}$$
where $d_{ij}(x)$ are the matrix elements of $D(x) := F(x)F(x)^{\mathrm{T}}$. Indeed,
$$T_t\varphi(x) - \varphi(x) = \mathbb{E}^{0,x}\{\varphi(x_t)\} - \varphi(x) = \mathbb{E}^{0,x}\left\{\int_0^t L\varphi(x_s)\,\mathrm{d}s\right\}. \tag{A.4.6}$$

Note that $u(x,t) = T_t\varphi(x)$ satisfies the partial differential equation
$$\frac{\partial}{\partial t}u(x,t) = Lu(x,t), \tag{A.4.7}$$
called *Kolmogorov's backward equation*. Under suitable regularity assumptions, (A.4.7) admits, for each y, a particular solution with initial condition $u(x,0) = \delta_y(x)$. Denote this solution by $p(y,t|x,0)$. In that case,
$$\mathbb{P}^{0,x}\{x_t \in A\} = T_t 1_A(x) = \int_A p(y,t|x,0)\,\mathrm{d}y, \tag{A.4.8}$$
which shows that $y \mapsto p(y,t|x,0)$ is the density of the transition probability of the Markov process defined by $x \mapsto x_t$.

The semigroup $\{T_t\}_{t \geqslant 0}$ admits a dual semigroup $\{S_t\}_{t \geqslant 0}$, acting on σ-finite measures on $\mathbb{R}^n$, as
$$S_t\mu(A) = \int_{\mathbb{R}^n} \mathbb{P}^{0,x}\{x_t \in A\}\mu(\mathrm{d}x) =: \mathbb{P}^{0,\mu}\{x_t \in A\}. \tag{A.4.9}$$

Its generator is the formal adjoint of L, given by
$$L^*\varphi(y) = -\sum_{i=1}^{n} \frac{\partial}{\partial y_i}\bigl(f_i(y)\varphi(y)\bigr) + \frac{1}{2}\sum_{i,j=1}^{n} \frac{\partial^2}{\partial y_i \partial y_j}\bigl(d_{ij}(y)\varphi(y)\bigr). \tag{A.4.10}$$

For probability measures admitting a sufficiently smooth density ρ, the time-evolved density $\rho(y,t) := S_t \rho(y)$ satisfies *Kolmogorov's forward equation*, or *Fokker–Planck equation*,

$$\frac{\partial}{\partial t}\rho(y,t) = L^*\rho(y,t) \,, \qquad (A.4.11)$$

and so does the transition-probability density $p(y,t|x,0)$, considered as a function of (y,t). If the stochastic process x_t admits an invariant density ρ_0, then it obeys the equation $L^*\rho_0 = 0$.

Example A.4.1 (Gradient system). Consider an SDE of the form

$$\mathrm{d}x_t = -\nabla U(x)\,\mathrm{d}x + \sigma\,\mathrm{d}W_t \,, \qquad (A.4.12)$$

where U is a twice differentiable potential, and σ is a constant. The generator and its adjoint are given by

$$L = -\nabla U \cdot \nabla + \frac{\sigma^2}{2}\Delta \quad \text{and} \quad L^* = \Delta U + \nabla U \cdot \nabla + \frac{\sigma^2}{2}\Delta \,. \qquad (A.4.13)$$

If the potential grows sufficiently quickly at infinity, the stochastic process admits an invariant density given by

$$\rho_0(x) = \frac{1}{\mathcal{N}}\mathrm{e}^{-2U(x)/\sigma^2} \,, \qquad (A.4.14)$$

where $\mathcal{N}$ is the normalisation. In particular, the Ornstein–Uhlenbeck process (A.3.10) admits a Gaussian invariant density as the potential $U(x)$ is quadratic.

A.5 Large Deviations

Let $W_t^\sigma = \sigma W_t$ denote the scaled Brownian motion. By the Laplace method, $\mathbb{P}\{W_T^\sigma \in A\}$ behaves like $\exp\{-\inf_{x \in A} x^2/2\sigma^2 T\}$ for small σ, in the sense that

$$\lim_{\sigma \to 0} \sigma^2 \log \mathbb{P}\{W_T^\sigma \in A\} = -\frac{1}{2T}\inf_{x \in A} x^2 \qquad (A.5.1)$$

holds for all sufficiently regular sets A with non-empty interior. Similarly, the reflection principle shows that

$$\lim_{\sigma \to 0} \sigma^2 \log \mathbb{P}\Big\{\sup_{0 \leqslant t \leqslant T} W_t^\sigma > L\Big\} = -\frac{L^2}{2T} \,. \qquad (A.5.2)$$

The relations (A.5.1) and (A.5.2) are two examples of large-deviation estimates.

In the sequel, $\mathcal{C} = \mathcal{C}([0,T],\mathbb{R}^n)$ denotes the set of continuous functions from $[0,T]$ to $\mathbb{R}^n$, and $H_1 = H_1([0,T],\mathbb{R}^n)$ is the Sobolev space of absolutely continuous functions $\varphi : [0,T] \to \mathbb{R}^n$, possessing a square-integrable derivative and satisfying $\varphi(0) = 0$.

Definition A.5.1 (Large-deviation principle). *A functional $I : \mathcal{C} \to [0, \infty]$ is said to be a good rate function if it is lower semi-continuous and has compact level sets. A stochastic process $\{x_t^\sigma\}_{t \in [0,T]}$ is said to satisfy a large-deviation principle with rate function I if*

$$-\inf_{\Gamma^\circ} I \leqslant \liminf_{\sigma \to 0} \sigma^2 \log \mathbb{P}\{x^\sigma \in \Gamma\} \leqslant \limsup_{\sigma \to 0} \sigma^2 \log \mathbb{P}\{x^\sigma \in \Gamma\} \leqslant -\inf_{\overline{\Gamma}} I \tag{A.5.3}$$

holds for any measurable set of paths $\Gamma \subset \mathcal{C}$.

In many applications, the infima over the interior Γ° and the closure $\overline{\Gamma}$ coincide, and thus the limit of $\sigma^2 \log \mathbb{P}\{x^\sigma \in \Gamma\}$ exists and is equal to their common value.

Theorem A.5.2 (Schilder's Theorem). *The scaled Brownian motion $\{W_t^\sigma\}_{t \in [0,T]}$ satisfies a large-deviation principle with the good rate function*

$$I(\varphi) = \begin{cases} \dfrac{1}{2} \displaystyle\int_0^T \|\dot{\varphi}_t\|^2 \, dt \,, & \text{if } \varphi \in H_1, \\ +\infty \,, & \text{otherwise.} \end{cases} \tag{A.5.4}$$

The Euler–Lagrange equations show that the minimisers of I are linear functions $\varphi_t = ct$. This can be interpreted as meaning that the "most probable paths" in Γ are straight lines. For instance, in (A.5.1), $\Gamma = \{\varphi \colon \varphi_T \in A\}$, and the minimiser is $\varphi_t = (t/T) \inf_{x \in A} x$. In (A.5.2), one has $\Gamma = \{\varphi \colon \sup_{0 \leqslant t \leqslant T} \varphi_t > L\}$, and the minimiser is $\varphi_t = Lt/T$.

Consider now a general diffusion of the form

$$dx_t^\sigma = f(x_t^\sigma) \, dt + \sigma F(x_t^\sigma) \, dW_t \tag{A.5.5}$$

with an initial condition x_0. We assume that f and F satisfy the Lipschitz and bounded-growth conditions of Theorem A.3.2, and that the diffusion matrix $D(x) = F(x)F(x)^T$ is positive definite.[3] Then the analogue of Schilder's Theorem A.5.2 is the following.

Theorem A.5.3 (Large-deviation principle for solutions of SDEs). *The stochastic process $\{x_t^\sigma\}_{t \in [0,T]}$ satisfies a large-deviation principle with the good rate function*

$$J(\varphi) = \begin{cases} \dfrac{1}{2} \displaystyle\int_0^T (\dot{\varphi}_t - f(\varphi_t))^T D(\varphi_t)^{-1} (\dot{\varphi}_t - f(\varphi_t)) \, dt \,, & \text{if } \varphi \in x_0 + H_1, \\ +\infty \,, & \text{otherwise.} \end{cases} \tag{A.5.6}$$

---

[3]If the diffusion matrix happens to be positive semi-definite only, $\{x_t^\sigma\}_{t \in [0,T]}$ still satisfies a large-deviation principle with a good rate function J, but J is given by a variational principle and can generally not be represented in an explicit form like (A.5.6).

The minimisers of J satisfy Euler–Lagrange equations, which are equivalent to the Hamilton equations deriving from the Hamiltonian

$$H(\varphi, \psi) = \frac{1}{2}\psi^T D(\varphi)\psi + f(\varphi)^T \psi . \tag{A.5.7}$$

Here $\psi = D(\varphi)^{-1}(\dot\varphi - f(\varphi))$ is the conjugate momentum of φ, so that the rate function takes the form

$$J(\varphi) = \frac{1}{2}\int_0^T \psi_t^T D(\varphi_t)\psi_t \, dt . \tag{A.5.8}$$

A.6 The Exit Problem

Consider again the family of diffusions

$$dx_t^\sigma = f(x_t^\sigma)\, dt + \sigma F(x_t^\sigma)\, dW_t , \tag{A.6.1}$$

with f and F satisfying the same conditions as in the previous sections. In addition, we assume that the diffusion matrix $D(x) = F(x)F(x)^T$ is uniformly positive definite. Let $\mathcal{D} \subset \mathbb{R}^n$ be a bounded, open set with smooth boundary, and fix an initial condition $x_0^\sigma = x_0$ in $\mathcal{D}$. The aim of the exit problem is to determine the distributions of the *first-exit time*

$$\tau^\sigma = \tau_\mathcal{D}^\sigma = \inf\{t > 0 \colon x_t^\sigma \notin \mathcal{D}\} \tag{A.6.2}$$

of x^σ from $\mathcal{D}$, and of the *first-exit location* $x_{\tau^\sigma}^\sigma \in \partial\mathcal{D}$. In principle, the laws of τ^σ and $x_{\tau^\sigma}^\sigma$ can be determined by solving PDEs, involving the generator $L = L^\sigma$ of the diffusion.

Proposition A.6.1. *Let $c : \mathcal{D} \to (-\infty, 0]$, $g : \mathcal{D} \to \mathbb{R}$ and $\psi : \partial\mathcal{D} \to \mathbb{R}$ be bounded, continuous functions, and let $y_t^\sigma = \int_0^t c(x_s^\sigma)\, ds$. Then*

$$u(x) = \mathbb{E}^{0,x}\left\{\psi(x_{\tau^\sigma}^\sigma)e^{y_{\tau^\sigma}^\sigma} - \int_0^{\tau^\sigma} g(x_t^\sigma)e^{y_t^\sigma}\, dt\right\} \tag{A.6.3}$$

is the unique solution of the Dirichlet problem

$$\begin{cases} L^\sigma u(x) + c(x)u(x) = g(x) & \text{for } x \in \mathcal{D}, \\ u(x) = \psi(x) & \text{for } x \in \partial\mathcal{D}. \end{cases} \tag{A.6.4}$$

Two special cases are of particular interest for the exit problem.

1. $u(x) = \mathbb{E}^{0,x}\{\tau^\sigma\}$ is the unique solution of the PDE

$$\begin{cases} L^\sigma u(x) = -1 & \text{for } x \in \mathcal{D}, \\ u(x) = 0 & \text{for } x \in \partial\mathcal{D}. \end{cases} \tag{A.6.5}$$

2. For a continuous function $\psi: \partial \mathcal{D} \to \mathbb{R}$, $v(x) = \mathbb{E}^{0,x}\{\psi(x^\sigma_{\tau^\sigma})\}$ is the unique solution of the PDE

$$\begin{cases} L^\sigma v(x) = 0 & \text{for } x \in \mathcal{D}, \\ v(x) = \psi(x) & \text{for } x \in \partial \mathcal{D}. \end{cases} \quad (A.6.6)$$

Solving equation (A.6.6) for a sufficiently large set of functions ψ allows to determine the law of first-exit locations. This, however, is possible only in very specific cases.

An alternative approach is the theory of large deviations, which yields partial information on the exit problem, without the need to solve a PDE. Indeed, let

$$V(x, y; s) = \inf\{J(\varphi) \colon \varphi \in \mathcal{C}([0, s], \mathbb{R}^n), \varphi_0 = x, \varphi_s = y\} \quad (A.6.7)$$

be the "cost" required to bring the system from x to y in time s. Then Theorem A.5.3 implies

$$\lim_{\sigma \to 0} \sigma^2 \log \mathbb{P}^{0,x}\{\tau^\sigma \leqslant t\} = \inf\{V(x, y; s) \colon s \in [0, t], y \notin \mathcal{D}\}. \quad (A.6.8)$$

In order to obtain more precise information, one needs to make additional assumptions. Consider the case where $\overline{\mathcal{D}}$ is contained in the basin of attraction of an isolated, asymptotically stable equilibrium point $x^\star$ of the deterministic system $\dot{x} = f(x)$. The function

$$V(x^\star, y) = \inf_{t>0} V(x^\star, y; t) \quad (A.6.9)$$

is called the *quasipotential* of the system.[4] An important rôle is played by its infimum

$$\overline{V} = \inf_{y \in \partial \mathcal{D}} V(x^\star, y) \quad (A.6.10)$$

on the boundary of the domain $\mathcal{D}$.

Theorem A.6.2. *For all initial conditions $x \in \mathcal{D}$ and all $\delta > 0$,*

$$\lim_{\sigma \to 0} \mathbb{P}^{0,x}\left\{e^{(\overline{V}-\delta)/\sigma^2} < \tau^\sigma < e^{(\overline{V}+\delta)/\sigma^2}\right\} = 1 \quad (A.6.11)$$

and

$$\lim_{\sigma \to 0} \sigma^2 \log \mathbb{E}^{0,x}\{\tau^\sigma\} = \overline{V}. \quad (A.6.12)$$

Moreover, for any closed subset $N \subset \partial \mathcal{D}$ satisfying $\inf_{y \in N} V(x^\star, y) > \overline{V}$,

$$\lim_{\sigma \to 0} \mathbb{P}^{0,x}\{x^\sigma_{\tau^\sigma} \in N\} = 0. \quad (A.6.13)$$

If $y \mapsto V(x^\star, y)$ has a unique minimum $y^\star$ on $\partial \mathcal{D}$, then, for all $\delta > 0$,

$$\lim_{\sigma \to 0} \mathbb{P}^{0,x}\{\|x^\sigma_{\tau^\sigma} - y^\star\| < \delta\} = 1. \quad (A.6.14)$$

---

[4] Note that, so far, we dealt with a finite time horizon T only. The rate functions $J = J_{[0,T]}$ depend on T, and the infimum in (A.6.9) actually involves different rate functions for different t.

This theorem shows that sample paths are likely to leave $\mathcal{D}$ near the point on the boundary with the lowest value of quasipotential, in a mean time of order $e^{\overline{V}/\sigma^2}$. This does not completely characterise the distribution of τ^σ. However, Day has shown that this distribution tends to an exponential one in the limit of vanishing noise, that is

$$\lim_{\sigma \to 0} \mathbb{P}^{0,x}\{\tau^\sigma > s\mathbb{E}^{0,x}\{\tau^\sigma\}\} = e^{-s} \,. \tag{A.6.15}$$

Example A.6.3 (Gradient system – continued). Consider again the case $f(x) = -\nabla U(x)$ and $F(x) = \mathbb{1}$. If $x^\star$ is a non-degenerate minimum of the potential U, and y is such that $U(y) \leqslant \min_{z \in \partial \mathcal{D}} U(z)$, then the function φ with $\lim_{t \to -\infty} \varphi_t = x^\star$, $\varphi_T = y$, and going against the deterministic flow, i.e., $\dot\varphi_t = \nabla U(\varphi_t)$ is the unique minimiser of the rate function, contributing to the quasipotential $V(x^\star, y)$.[5] The resulting value for the quasipotential is simply given by twice the potential difference,

$$V(x^\star, y) = 2\bigl(U(y) - U(x^\star)\bigr) \,. \tag{A.6.16}$$

The expected first-exit time is thus of the order e^{2H/σ^2}, where H is the minimal potential difference to be overcome to leave $\mathcal{D}$. This result is known as *Arrhenius' law*.

Many generalisations of the above results exist. In particular, for gradient systems, the subexponential behaviour of the expected time needed to leave a potential well, by crossing a saddle, was first determined by Eyring [Eyr35] and Kramers [Kra40]. The geometry of well and saddle enter into the prefactor in the following way:

$$\mathbb{E}^{0,x_0}\{\tau^\sigma\} \simeq \frac{2\pi}{\lambda_1(z)} \sqrt{\frac{|\det \partial_{xx} U(z)|}{\det \partial_{xx} U(x^\star)}} \, e^{2[U(z) - U(x^\star)]/\sigma^2} \tag{A.6.17}$$

holds in the weak-noise limit $\sigma \to 0$, where $\partial_{xx} U(z)$ and $\partial_{xx} U(x^\star)$ denote the Hessian matrices of the potential function U at the bottom $x^\star$ of the well and at the saddle z, respectively, and $\lambda_1(z)$ is the unique negative eigenvalue of $\partial_{xx} U(z)$.

Bibliographic Comments

Brownian motion, discovered by Brown in the 1830s, was first described and investigated mathematically by Bachelier [Bac00], Einstein [Ein05], Smoluchowski [vS06], Langevin [Lan08], Wiener [Wie23], Ornstein and Uhlenbeck [UO30]. The stochastic integral was introduced by Paley, Wiener and

---

[5] Note that the paths φ_t with $\dot\varphi_t = -\nabla U(\varphi_t)$, i.e., moving *with* the deterministic flow, yield a vanishing rate function.

Zygmund [PWZ33] for non-random integrands, and by Itô [Itô44] in the general case presented here. Nowadays, there is a broad choice of literature on stochastic integration, also with respect to more general processes than Brownian motion, and on stochastic differential equations. We have mainly drawn on the monographs by McKean [McK69], Karatzas and Shreve [KS91] and Øksendal [Øks85], which provide detailed information on stochastic differential equations from a mathematical point of view. A more physical approach can be found in [HL84], [Ris89] or [vK81], for example.

Classical references on the general theory of large deviations are the monographs by Deuschel and Stroock [DS89] and by Dembo and Zeitouni [DZ98]. Large-deviation results for solutions of stochastic differential equations have first been established by Wentzell and Freidlin in a series of papers including [VF69, VF70]. The theory is explained in detail in their monograph [FW98]. As a first introduction, we also recommend the corresponding part of the excellent lecture notes by Varadhan [Var80].

The earliest insights into the exit problem are due to Arrhenius [Arr89], Eyring [Eyr35] and Kramers [Kra40]. The mathematical results by Wentzell and Freidlin have been extended in several directions. Asymptotic expansions of exit probabilities have been considered by Azencott [Aze85], and by Fleming and James [FJ92]. In the case of gradient systems, very precise results relating expected first-exit times, capacities, and small eigenvalues of the diffusion have been obtained by Bovier, Eckhoff, Gayrard and Klein [BEGK04, BGK05], see also the recent developments by Helffer, Klein and Nier [HKN04, HN05].

The dynamics near a hyperbolic fixed point has been considered by Kifer [Kif81] and Day [Day95]. Day has in particular extensively studied the exit from a domain with characteristic boundary (i.e., a boundary that is invariant under the deterministic dynamics, instead of being attracted by the equilibrium point) [Day90a, Day92, Day94, Day96]. See also related results by Maier and Stein [MS93, MS97].

B
Some Useful Inequalities

B.1 Doob's Submartingale Inequality and a Bernstein Inequality

Definition B.1.1 (Submartingale). *Let $(\Omega, \mathcal{F}, \mathbb{P})$ be probability space with a filtration $\{\mathcal{F}_t\}_{t\geqslant 0}$ of $\mathcal{F}$, and let $\{M_t\}_{t\geqslant 0}$ be an $\{\mathcal{F}_t\}_{t\geqslant 0}$-adapted, integrable stochastic process on that probability space, i.e., M_t is $\mathcal{F}_t$-measurable and $\mathbb{E}|M_t| < \infty$ for all t. The stochastic process $\{M_t\}_{t\geqslant 0}$ is called a submartingale, if $\mathbb{E}\{M_t|\mathcal{F}_s\} \geqslant M_s$ for all $s \leqslant t$.*

The following classical estimate allows to estimate the probability that a positive submartingale exceeds a level during the time interval $[0, t]$ with the help of its expected value at the endpoint.

Lemma B.1.2 (Doob's submartingale inequality). *Let $\{M_t\}_{t\geqslant 0}$ be a positive submartingale with (right-)continuous paths. Then, for any $L > 0$ and $t > 0$,*

$$\mathbb{P}\left\{\sup_{0\leqslant s\leqslant t} M_s \geqslant L\right\} \leqslant \frac{1}{L}\mathbb{E}\{M_t\}. \tag{B.1.1}$$

As a consequence of the preceding inequality, we obtain the following estimate for stochastic integrals with deterministic integrands.

Lemma B.1.3 (Bernstein-type inequality). *Let $\varphi(u)$ be a Borel-measurable deterministic function such that*

$$\Phi(t) = \int_0^t \varphi(u)^2 \, du \tag{B.1.2}$$

exists. Then

$$\mathbb{P}\left\{\sup_{0\leqslant s\leqslant t}\int_0^s \varphi(u)\,dW_u \geqslant \delta\right\} \leqslant \exp\left\{-\frac{\delta^2}{2\Phi(t)}\right\}. \tag{B.1.3}$$

Proof. Let P denote the left-hand side of (B.1.3). For any $\gamma > 0$, we have

$$P = \mathbb{P}\left\{\sup_{0 \leq s \leq t} \exp\left\{\gamma \int_0^s \varphi(u)\,dW_u\right\} \geq e^{\gamma\delta}\right\} \leq \mathbb{P}\left\{\sup_{0 \leq s \leq t} M_s \geq e^{\gamma\delta - \frac{\gamma^2}{2}\Phi(t)}\right\}, \tag{B.1.4}$$

where

$$M_s = \exp\left\{\int_0^s \gamma\varphi(u)\,dW_u - \frac{1}{2}\int_0^s \gamma^2 \varphi(u)^2\,du\right\} \tag{B.1.5}$$

is an (exponential) martingale, satisfying $\mathbb{E}\{M_t\} = \mathbb{E}\{M_0\} = 1$, which implies by Doob's submartingale inequality, that

$$\mathbb{P}\left\{\sup_{0 \leq s \leq t} M_s \geq \lambda\right\} \leq \frac{1}{\lambda}\mathbb{E}\{M_t\} = \frac{1}{\lambda}. \tag{B.1.6}$$

This gives us

$$P \leq e^{-\gamma\delta + \frac{\gamma^2}{2}\Phi(t)}, \tag{B.1.7}$$

and we obtain the result by optimising (B.1.7) over γ. □

B.2 Using Tail Estimates

Consider a random variable τ which has a density with respect to Lebesgue measure. Given an upper bound on the *tails* of the distribution of τ, we want to estimate $\mathbb{E}\{1_{[0,t)}(\tau)g(\tau)\}$ for sufficiently well-behaved functions g. Integration by parts allows to express the expectation with the help of the distribution function, and enables us to use the bound on the tails of the distribution. The following lemma provides such an estimate, without assuming the existence of a density.

Lemma B.2.1 (Using tail estimates). *Let $\tau \geq 0$ be a random variable satisfying $F_\tau(s) = \mathbb{P}\{\tau < s\} \geq G(s)$ for some continuously differentiable function G. Then,*

$$\mathbb{E}\{1_{[0,t)}(\tau)g(\tau)\} \leq g(s)\big[F_\tau(s) - G(s)\big]\Big|_{s=0}^{t} + \int_0^t g(s)G'(s)\,ds \tag{B.2.1}$$

holds for all $t > 0$, and all piecewise differentiable, non-decreasing functions g.

Proof. First note that, whenever g is differentiable on an interval (s,t), Fubini's theorem allows to write

$$\int_s^t g'(u)\mathbb{P}\{\tau \geq u\}\,du = \mathbb{E}\left\{\int_{s \wedge \tau}^{t \wedge \tau} g'(u)\,du\right\}$$

$$= \mathbb{E}\{g(t \wedge \tau)\} - \mathbb{E}\{g(s \wedge \tau)\} \tag{B.2.2}$$

$$= \mathbb{E}\{1_{[s,t)}(\tau)g(\tau)\} + g(t)\mathbb{P}\{\tau \geq t\} - g(s)\mathbb{P}\{\tau \geq s\}.$$

Then, integration by parts then shows

$$\mathbb{E}\{1_{[s_0,t)}(\tau)g(\tau)\} = \int_s^t g'(u)\big[1 - F_\tau(u)\big]\,\mathrm{d}u - g(u)\big[1 - F_\tau(u)\big]\Big|_{u=s}^t$$

$$\leqslant \int_s^t g(u)G'(u)\,\mathrm{d}u + g(u)\big[F_\tau(u) - G(u)\big]\Big|_{u=s}^t, \quad (\text{B.2.3})$$

where we have used $F_\tau(u) \geqslant G(u)$. By additivity, the same estimate holds on any interval $(0, t)$, which proves (B.2.1). □

B.3 Comparison Lemma

In the context of stochastic differential equations, the following variant of Gronwall's lemma is particularly useful, as it does not require differentiability.

Lemma B.3.1 (Gronwall's inequality). *Let $\phi, \psi : [0, \infty) \to [0, \infty)$ be a continuous functions, satisfying*

$$\phi(t) \leqslant L + K \int_0^t \psi(s)\phi(s)\,\mathrm{d}s \quad \text{for all } t \geqslant 0, \quad (\text{B.3.1})$$

with constants $K, L \geqslant 0$. Then,

$$\phi(t) \leqslant K \exp\left\{L \int_0^t \psi(s)\,\mathrm{d}s\right\} \quad \text{for all } t \geqslant 0. \quad (\text{B.3.2})$$

The following lemma allows to compare the relative position of solutions to two *different* stochastic differential equations, where initial conditions and drift coefficients can be different, while the diffusion coefficients are assumed to be the same.

Lemma B.3.2 (Comparison lemma). *Let*

$$x_t^{(i)} = x_0^{(i)} + \int_0^t f_i(x_s^{(i)}, s)\,\mathrm{d}s + \int_0^t F(x_s^{(i)}, s)\,\mathrm{d}W_s, \quad i = 1, 2, \quad (\text{B.3.3})$$

denote the solutions of two one-dimensional stochastic differential equations, where we assume that the drift and diffusion coefficients are Lipschitz continuous,[1] uniformly in t. If both,

- *the initial conditions are almost surely ordered: $x_0^{(1)} \leqslant x_0^{(2)}$ $\mathbb{P}$-almost surely,*
- *the drift coefficients are ordered: $f_1(x,t) \leqslant f_2(x,t)$ for all (x,t),*

then

$$\mathbb{P}\{x_t^{(1)} \leqslant x_t^{(2)} \forall t\} = 1. \quad (\text{B.3.4})$$

---
[1] The assumption of Lipschitz continuity of drift and diffusion coefficient in the preceding lemma can actually be relaxed, see [KS91, Proposition 2.18 of Section 5.2].

B.4 Reflection Principle

Consider a Brownian motion W_t, starting in some $-b < 0$. We are interested in its first-passage time $\tau_0 = \inf\{t > 0\colon W_t \geqslant 0\}$ at level $x = 0$. Using the strong Markov property and starting Brownian motion "afresh" at time τ_0 shows

$$\begin{aligned}
&\mathbb{P}^{0,-b}\{\tau_0 < t\}\\
&= \mathbb{P}^{0,-b}\{\tau_0 < t, W_t \geqslant 0\} + \mathbb{P}^{0,-b}\{\tau_0 < t, W_t < 0\} \quad\quad\text{(B.4.1)}\\
&= \mathbb{E}^{0,-b}\{1_{\{\tau_0<t\}}\mathbb{P}^{\tau_0,0}\{W_t \geqslant 0\}\} + \mathbb{E}^{0,-b}\{1_{\{\tau_0<t\}}\mathbb{P}^{\tau_0,0}\{W_t < 0\}\} .
\end{aligned}$$

The distribution of the Brownian sample path, starting in 0, being invariant under reflection at level $x = 0$, immediately implies *André's reflection principle* for Brownian motion,

$$\mathbb{P}^{0,-b}\{\tau_0 < t\} = 2\mathbb{E}^{0,-b}\{1_{\{\tau_0<t\}}\mathbb{P}^{\tau_0,0}\{W_t \geqslant 0\}\} = 2\mathbb{P}^{0,-b}\{W_t \geqslant 0\} , \quad\text{(B.4.2)}$$

which allows to express the distribution of the first-passage time to a constant level with the help of a Gaussian distribution. Also note that differentiating the right-hand side with respect to t yields an exact expression for the first-passage density; cf. Appendix C on first-passage densities to non-constant boundaries.

The preceding argument only uses the strong Markov property and the symmetry of the distribution of the sample paths with respect to the level to be crossed. Thus the following generalisation is immediate.

Lemma B.4.1 (Reflection principle). *Let x_t be a (real-valued) strong Markov process with continuous sample paths. Assume that the distribution of x_t, if started in 0, is symmetric, i.e., that the processes x_t and $-x_t$ (both starting in 0) have the same distribution in path space. Then, when starting in $-b < 0$, the first-passage time τ_0 at the level $x = 0$, satisfies*

$$\mathbb{P}^{0,-b}\{\tau_0 < t\} = 2\mathbb{P}^{0,-b}\{x_t \geqslant 0\} . \quad\quad\text{(B.4.3)}$$

Remark B.4.2. We will typically apply the reflection principle for processes x_t, which are given as the solution of a SDE

$$\mathrm{d}x_t = f(x_t, t)\,\mathrm{d}t + F(x_t, t)\,\mathrm{d}W_t , \quad\quad x_0 = -b < 0 . \quad\quad\text{(B.4.4)}$$

Then the symmetry assumption of the preceding lemma amounts to assuming $f(-x, t) = -f(x, t)$ and $|F(x, t)| = |F(-x, t)|$ for all x and t. Note that the requirement of the drift coefficient being odd is in particular satisfied for linear drift coefficients $f(x, t) = a(t)x$.

C

First-Passage Times for Gaussian Processes

C.1 First Passage through a Curved Boundary

Let $v(t)$ be continuously differentiable on $[0, \infty)$ and satisfy $v(0) = 0$ and $v'(t) \geq v_0 > 0$ for all $t \geq 0$. Consider the Gaussian process

$$z_t = \sigma \int_0^t \sqrt{v'(s)} \, \mathrm{d}W_s , \qquad (\text{C.1.1})$$

whose variance is $\sigma^2 v(t)$. We will consider z_t as a Markov process and introduce the notation $\mathbb{P}^{s,x}\{z_t \in \cdot\} = \mathbb{P}\{z_t \in \cdot \,|\, z_s = x\}$, $t > s$, for its transition probabilities. Their densities are given by

$$\begin{aligned} y \mapsto \varphi(t, y | s, x) &:= \frac{\partial}{\partial y} \mathbb{P}^{s,x}\{z_t < y\} \\ &= \frac{1}{\sigma} \frac{1}{\sqrt{2\pi v(t,s)}} e^{-(y-x)^2 / 2\sigma^2 v(t,s)} , \end{aligned} \qquad (\text{C.1.2})$$

where $v(t, s) = v(t) - v(s)$.

Let $d(t)$ be continuously differentiable on $[0, \infty)$ and satisfy $d(0) > 0$. The object of the "first-passage problem" is to determine the density

$$\psi(t) = \frac{\mathrm{d}}{\mathrm{d}t} \mathbb{P}^{0,0}\{\tau < t\} \qquad (\text{C.1.3})$$

of the first-passage time $\tau = \inf\{s > 0 : z_s > d(s)\}$. For the sake of brevity, we will call $\psi(t)$ the *first-passage density of z_t to $d(t)$*. Note that in the case of a constant boundary $d(t) \equiv d_0$, the first-passage density $\psi(t)$ can be obtained exactly from the reflection principle (Lemma B.4.1). For Brownian motion, the problem of determining the first-passage density to a straight line, i.e., a linear boundary, is equivalent to finding the density of an inverse Gaussian random variable, and in case the boundary is declining, the exact expression is known.[1] The case of general boundaries is often referred to as *curved boundary*.

For curved boundaries, the first-passage density can be estimated with the help of certain integral equations satisfied by $\psi(t)$. Let $\phi(t) = \varphi(t, d(t)|0, 0)$ denote the value of the density of z_t at $d(t)$ and let $\phi(t|s) = \varphi(t, d(t)|s, d(s))$ denote the transition density at $y = d(t)$ for paths starting at time s in $x = d(s)$. For any $y > d(t)$, the Markov property enables us to write

$$\mathbb{P}^{0,0}\{z_t \geqslant y\} = \int_0^t \mathbb{P}^{s,d(s)}\{z_t \geqslant y\} \psi(s) \, ds \, . \tag{C.1.4}$$

Differentiating with respect to y and taking the limit $y \searrow d(t)$ yields

$$\phi(t) = \int_0^t \phi(t|s) \psi(s) \, ds \, . \tag{C.1.5}$$

A second integral equation satisfied by $\psi(t)$ can be obtained by differentiating (C.1.4) twice with respect to y, before again taking the limit $y \searrow d(t)$. Combing the resulting equation with (C.1.5) yields

$$\psi(t) = b_0(t)\phi(t) - \int_0^t \tilde{b}(t,s)\phi(t|s)\psi(s) \, ds \, , \tag{C.1.6}$$

where

$$b_0(t) = v'(t) \left[\frac{d(t)}{v(t)} - \frac{d'(t)}{v'(t)} \right] = -v(t) \frac{\partial}{\partial t}\left(\frac{d(t)}{v(t)}\right), \tag{C.1.7}$$

$$\tilde{b}(t,s) = v'(t) \left[\frac{d(t,s)}{v(t,s)} - \frac{d'(t)}{v'(t)} \right] = -v(t,s) \frac{\partial}{\partial t}\left(\frac{d(t,s)}{v(t,s)}\right), \tag{C.1.8}$$

with $d(t,s) = d(t) - d(s)$. In the particular case of a standard Brownian motion, that is for $\sigma = 1$ and $v(t) \equiv t$, Equation (C.1.6) has been established in [Dur92, Appendix by D. Williams]. The general case is easily obtained from the fact that $z_t = \sigma W_{v(t)}$ in distribution.

Equation (C.1.6) suggests that the first-passage density can be written in the form

$$\psi(t) = \frac{1}{\sigma} c(t) e^{-d(t)^2/2\sigma^2 v(t)} \, , \tag{C.1.9}$$

where $c(t)$ is a subexponential prefactor. In fact, the following bound on $c(t)$ follows immediately from (C.1.5) and (C.1.6).

Lemma C.1.1. *Let $c_0(t) = b_0(t)/\sqrt{2\pi v(t)}$. Then (C.1.9) holds with*

$$|c(t) - c_0(t)| \leqslant \frac{1}{\sqrt{2\pi v(t)}} \sup_{0 \leqslant s \leqslant t} |\tilde{b}(t,s)| \, . \tag{C.1.10}$$

---

[1] If the boundary is linear but ascending instead of declining, the first-passage distribution is improper in the sense, that there is positive probability of *not* hitting the boundary at all.

Remark C.1.2. Note that Lemma C.1.1 does not require $v'(t)$ to be bounded away from zero as t varies.

If, for instance, $v(t)$ and $d(t)$ are twice continuously differentiable, then $s \mapsto \tilde{b}(t,s)$ is easily seen to be bounded, and thus $c(t)$ behaves like $v(t)^{-3/2}$ near $t=0$.

Taking advantage of the fact that σ is a small parameter, we can compute an asymptotic expansion for $c(t)$, valid on a finite, but exponentially long time interval. Writing $\tilde{c}(t,s) = \tilde{b}(t,s)/\sqrt{2\pi v(t,s)}$, we see from (C.1.6) that $c(t)$ must be a fixed point of the operator

$$(\mathcal{T}c)(t) = c_0(t) - \frac{1}{\sigma}\int_0^t \tilde{c}(t,s)c(s)e^{-r(t,s)/2\sigma^2}\,\mathrm{d}s\;, \qquad \text{(C.1.11)}$$

where

$$r(t,s) = \frac{d(s)^2}{v(s)} - \frac{d(t)^2}{v(t)} + \frac{d(t,s)^2}{v(t,s)} = \frac{v(t)v(s)}{v(t,s)}\left[\frac{d(s)}{v(s)} - \frac{d(t)}{v(t)}\right]^2. \qquad \text{(C.1.12)}$$

Remark C.1.3. The exponent $r(t,s)$ is non-negative and vanishes for $s=t$. If $r(t,s)$ does not vanish anywhere else, then the main contribution to the integral in (C.1.11) comes from s close to t. In the generic case $\partial_s r(t,t) \neq 0$, the integral is at most of order σ^2. If the functions involved are sufficiently smooth, one easily sees that the integral is of order σ^3. If $r(t,s)$ vanishes in a quadratic minimum in $s=t$ or elsewhere, then the integral is at most of order σ.

The most probable path reaching z at time t is represented by a straight line in the (v,z)-plane, cf. Section A.5. Thus $r(t,s)$ vanishes for some $s \neq t$ if and only if the most probable path reaching $d(t)$ as already reached $d(s)$. In that case, there exists a time $u \in (0,t)$ such that the tangent to the curve $(v(s),d(s))_{s\geqslant 0}$ at $(v(u),d(u))$ goes through the origin, i.e., $(d(u)/v(u))' = 0$. This situation can be excluded under a convexity assumption on $d(t)$, which is automatically satisfied in the adiabatic case.

The following lemma establishes the existence and some properties of a fixed point of (C.1.11).

Lemma C.1.4. *Assume that there are constants $\Delta, M_1, M_2 > 0$ such that the conditions*

$$d(s)v'(s) - v(s)d'(s) \geqslant \Delta v'(s)\bigl(1+\sqrt{v(s)}\bigr)\;, \qquad \text{(C.1.13)}$$
$$|\tilde{c}(t,s)| \leqslant M_1\;, \qquad \text{(C.1.14)}$$
$$M_2 v'(s) \geqslant 1 + v(s)\;, \qquad \text{(C.1.15)}$$

hold for all $0 \leqslant s \leqslant t$. Then (C.1.9) holds with a prefactor $c(t)$ satisfying

$$|c(t)-c_0(t)| \leqslant \frac{\epsilon}{1-\epsilon}\frac{1+v(t)^{3/2}}{v(t)^{3/2}}\sup_{0\leqslant s\leqslant t}\left|\frac{v(s)^{3/2}}{1+v(s)^{3/2}}c_0(s)\right|, \qquad \text{(C.1.16)}$$

whenever

$$\epsilon := 2M_1 \left(\frac{e^{-\Delta^2/4\sigma^2}}{\sigma} t + \frac{4M_2\sigma}{\Delta^2} \right) < 1 . \qquad (C.1.17)$$

Proof. We shall prove that $\mathcal{T}$ is a contraction on the Banach space $\mathcal{X}$ of continuous functions $c : [0, t] \to [0, \infty)$, equipped with the norm

$$\|c\| = \sup_{0 \leqslant s \leqslant t} \left| \frac{v(s)^{3/2}}{1 + v(s)^{3/2}} c(s) \right| . \qquad (C.1.18)$$

For any two functions $c_1, c_2 \in \mathcal{X}$, we have by (C.1.14)

$$|\mathcal{T}c_2(t) - \mathcal{T}c_1(t)| \leqslant \|c_2 - c_1\| \frac{M_1}{\sigma} \int_0^t \frac{1 + v(s)^{3/2}}{v(s)^{3/2}} e^{-r(t,s)/2\sigma^2} ds . \qquad (C.1.19)$$

Using Assumption (C.1.13), we obtain

$$\frac{d(s)}{v(s)} - \frac{d(t)}{v(t)} = \int_s^t \frac{d(u)v'(u) - v(u)d'(u)}{v(u)^2} du$$

$$\geqslant \Delta \left[\frac{v(t,s)}{v(t)v(s)} + 2 \frac{\sqrt{v(t)} - \sqrt{v(s)}}{\sqrt{v(t)v(s)}} \right] , \qquad (C.1.20)$$

and thus

$$r(t,s) \geqslant \Delta^2 \left[\frac{v(t,s)}{v(t)v(s)} + 4 \frac{(\sqrt{v(t)} - \sqrt{v(s)})^2}{v(t,s)} \right] \geqslant \Delta^2 \frac{v(t,s)}{v(t)} \left[1 + \frac{1}{v(s)} \right] . \qquad (C.1.21)$$

For the sake of brevity, we restrict our attention to the case $v(t) > 2$. We split the integral in (C.1.19) at times s_1 and s_2 defined by $v(s_1) = 1$ and $v(s_2) = v(t)/2$. By (C.1.15), the integral on the first interval is bounded by

$$2 \int_0^{s_1} \frac{1}{v(s)^{3/2}} e^{-\Delta^2/4\sigma^2 v(s)} ds \leqslant \frac{4M_2\sigma}{\Delta} \int_{\Delta^2/4\sigma^2}^{\infty} \frac{e^{-y}}{\sqrt{y}} dy \leqslant \frac{8M_2\sigma^2}{\Delta^2} e^{-\Delta^2/4\sigma^2} . \qquad (C.1.22)$$

The second part of the integral is smaller than $2te^{-\Delta^2/4\sigma^2}$ because $r(t,s) \geqslant \Delta^2/2$ for $s_1 < s < s_2$, while the last part is bounded by

$$\int_{s_2}^t e^{-\Delta^2 v(t,s)/2\sigma^2 v(t)} ds \leqslant \frac{2^{3/2} M_2}{v(t)} \int_{s_2}^t v'(s) e^{-\Delta^2 v(t,s)/2\sigma^2 v(t)} ds \leqslant \frac{2^{5/2} M_2 \sigma^2}{\Delta^2} . \qquad (C.1.23)$$

This shows that $\mathcal{T}$ is a contraction with contraction constant ϵ, and the result follows by bounding $\|c - c_0\| = \|\mathcal{T}^n c - \mathcal{T}0\|$ by a geometric series. Here 0 denotes the function which is zero everywhere. $\square$

Remark C.1.5. The proof shows that if, in addition to (C.1.13)–(C.1.15), the bound is satisfied for all s such that $v(t)/2 \leq v(s) \leq v(t)$, then (C.1.16) holds with

$$\epsilon = 2M_1 \left(\frac{4M_2\sigma}{\Delta^2} + \frac{t}{\sigma} \right) e^{-\Delta^2/4\sigma^2} + \frac{8M_1 M_2'\sigma}{\Delta^2}. \tag{C.1.25}$$

This result is useful in certain situations involving bifurcations, in which one may obtain a substantially smaller value of M_2' than of M_2.

Corollary C.1.6. *Let the assumptions of Lemma C.1.4 be satisfied and assume in addition that there exists a constant $M_3 > 0$ such that*

$$d(t)v'(t) - v(t)d'(t) \leq M_3(1 + v(t)^{3/2}) \qquad \text{for all } t \geq 0. \tag{C.1.26}$$

Then

$$c_0(t)\left[1 - \frac{\epsilon}{1-\epsilon}\frac{M_2 M_3}{\Delta}\right] \leq c(t) \leq c_0(t)\left[1 + \frac{\epsilon}{1-\epsilon}\frac{M_2 M_3}{\Delta}\right] \tag{C.1.27}$$

holds for all $t > 0$ such that $\epsilon = \epsilon(t) < 1$, where ϵ is defined by (C.1.17).

Proof. The proof follows directly from the bounds (C.1.26) and

$$\frac{1 + v(t)^{3/2}}{v(t)^{3/2}} \frac{1}{c_0(t)} \leq \frac{\sqrt{2\pi} M_2}{\Delta}, \tag{C.1.28}$$

the latter being a consequence of (C.1.13) and (C.1.15). □

C.2 Small-Ball Probabilities for Brownian Motion

For d-dimensional standard Brownian motion W_t, the distribution function of the first-exit time $\tau_r = \tau_{B(0,r)}$ from a centred sphere $B(0,r)$ of radius r can be expressed with the help of an infinite series, see [CT62, Theorem 2].

Theorem C.2.1. *Fix $r > 0$. Then, for any $t > 0$,*

$$\mathbb{P}\{\tau_r > t\} = \mathbb{P}\{\sup_{0 \leq s \leq t} \|W_s\| < r\} = \sum_{l=1}^{\infty} \xi_{d,l} e^{-q_{d,l}^2 t/2r^2}, \tag{C.2.1}$$

where $q_{d,l}$, $l \geq 1$, are the positive roots of the Bessel function J_ν, for $\nu = d/2 - 1$, and the prefactors are given by

$$\xi_{d,l} = \frac{1}{2^{\nu-1}\Gamma(\nu+1)} \frac{q_{d,l}^{\nu-1}}{J_{\nu+1}(q_{d,l})}. \tag{C.2.2}$$

We apply the preceding theorem to one-dimensional Brownian motion, i.e., for $d = 1$. In that case, the positive roots of J_ν are $q_{1,l} = (2l-1)\pi/2$, and we find $J_{\nu+1}(q_{1,l}) = \frac{2}{\pi}(2l-1)^{-1/2}(-1)^{l-1}$. Thus the series (C.2.1) is alternating, and the summands decrease in absolute value. This immediately implies the following estimate, which is useful for "small balls", namely for small r.

Corollary C.2.2. *For $d = 1$ and any $r > 0$,*

$$\mathbb{P}\{\sup_{0 \leqslant s \leqslant 1} |W_s| < r\} \leqslant \frac{4}{\pi} e^{-\pi^2/8r^2} . \qquad (\text{C.2.3})$$

Bibliographic Comments

Section C.1: First-passage problems for diffusions are important in many applications, and, consequently, the study of first-passage densities has received a lot of attention. As explicit analytic solutions are available only in a few special situations, most work for nonlinear boundaries aims at the derivation of integral equations for first-passage densities [Dur85, GNRS89] and their approximate solutions. Asymptotic expansions have, for instance, been derived for the passage of Brownian motion at a sufficiently smooth, one-sided boundary in [Fer83, Dur92], and for more general diffusions in [ST96]. Tangent approximation and other image methods have been used in [Str67, Dan69, Fer82, Ler86, Dan96, LRD02], e.g., yielding approximations either in the limit $t \to \infty$ or for increasingly remote boundaries. For numerical techniques in determining first-passage times see, for instance, [GS99, GSZ01].

The set-up considered in Section C.1 is tailored to applications for which ϵ, defined in (C.1.17), is small, so that we obtain approximations which hold uniformly for t from a large region. The application in Section 3.1.2 corresponds to passage of Brownian motion to a boundary which approaches a multiple of the square-root boundary as $t \to \infty$. For $h/\sigma \to \infty$, this boundary becomes increasingly remote.

Section C.2: Among other applications, estimates for small-ball probabilities are of importance in establishing rate-of-convergence results for functional laws of the iterated logarithm, and many, much more general, results than presented here have been obtained. These include studies for different norms, such as *weighted* sup-norms [Mog74, Shi96], L^p-norms [DV77], Hölder norms [BR92], [KLS95], Sobolev norms [LS99], as well as treatment of larger classes of stochastic processes, in particular for Gaussian processes with stationary increments [MR95], and Gaussian fields [SW95]. In addition, balls with moving centre have been studied. For a general overview, see [LS01].

References

[AC95]   Muktish Acharyya and Bikas K. Chakrabarti. Response of Ising systems to oscillating and pulsed fields: Hysteresis, ac, and pulse susceptibility. *Phys. Rev. Letters*, 52:6550–6568, 1995.

[Arn92] Vladimir I. Arnol'd. *Ordinary differential equations*. Springer Textbook. Springer-Verlag, Berlin, 1992. Translated from the third Russian edition by Roger Cooke.

[Arn94] V. I. Arnol'd, editor. *Dynamical systems. V*, volume 5 of *Encyclopaedia of Mathematical Sciences*. Springer-Verlag, Berlin, 1994. Bifurcation theory and catastrophe theory, A translation of *Current problems in mathematics. Fundamental directions. Vol. 5 (Russian)*, Akad. Nauk SSSR, Vsesoyuz. Inst. Nauchn. i Tekhn. Inform., Moscow, 1986, Translation by N. D. Kazarinoff.

[Arn98] Ludwig Arnold. *Random Dynamical Systems*. Springer-Verlag, Berlin, 1998.

[Arn01] Ludwig Arnold. Hasselmann's program revisited: The analysis of stochasticity in deterministic climate models. In Peter Imkeller and Jin-Song von Storch, editors, *Stochastic Climate Models*, volume 49 of *Progress in Probability*, pages 141–158, Boston, 2001. Birkhäuser.

[Arr89] Svante Arrhenius. *J. Phys. Chem.*, 4:226, 1889.

[Att98] S. Attal. Classical and quantum stochastic calculus: survey article. In *Quantum probability communications*, QP-PQ, X, pages 1–52. World Sci. Publishing, River Edge, NJ, 1998.

[Aze85] Robert Azencott. Petites perturbations aléatoires des systèmes dynamiques: développements asymptotiques. *Bull. Sci. Math. (2)*, 109:253–308, 1985.

[Bac00] L. Bachelier. Théorie de la spéculation. *Annales Scientifiques de l'Ecole Normale Supérieure*, 17:21, 1900.

[Bae95] Claude Baesens. Gevrey series and dynamic bifurcations for analytic slow-fast mappings. *Nonlinearity*, 8(2):179–201, 1995.

[BEGK04] Anton Bovier, Michael Eckhoff, Véronique Gayrard, and Markus Klein. Metastability in reversible diffusion processes. I. Sharp asymptotics for capacities and exit times. *J. Eur. Math. Soc. (JEMS)*, 6(4):399–424, 2004.

[Bel60]   Richard Bellman. *Introduction to Matrix Analysis*. McGraw–Hill, New York, 1960.
[Ben91]   E. Benoît, editor. *Dynamic Bifurcations*, Berlin, 1991. Springer-Verlag.
[Ber85]   M. V. Berry. Classical adiabatic angles and quantal adiabatic phase. *J. Phys. A*, 18(1):15–27, 1985.
[Ber90]   M. V. Berry. Histories of adiabatic quantum transitions. *Proc. Roy. Soc. London Ser. A*, 429(1876):61–72, 1990.
[Ber98]   Nils Berglund. *Adiabatic Dynamical Systems and Hysteresis*. PhD thesis, EPFL, 1998.
[BG02a]   Nils Berglund and Barbara Gentz. Beyond the Fokker–Planck equation: Pathwise control of noisy bistable systems. *J. Phys. A*, 35(9):2057–2091, 2002.
[BG02b]   Nils Berglund and Barbara Gentz. The effect of additive noise on dynamical hysteresis. *Nonlinearity*, 15(3):605–632, 2002.
[BG02c]   Nils Berglund and Barbara Gentz. Metastability in simple climate models: Pathwise analysis of slowly driven Langevin equations. *Stoch. Dyn.*, 2:327–356, 2002.
[BG02d]   Nils Berglund and Barbara Gentz. Pathwise description of dynamic pitchfork bifurcations with additive noise. *Probab. Theory Related Fields*, 122(3):341–388, 2002.
[BG02e]   Nils Berglund and Barbara Gentz. A sample-paths approach to noise-induced synchronization: Stochastic resonance in a double-well potential. *Ann. Appl. Probab.*, 12:1419–1470, 2002.
[BG03]   Nils Berglund and Barbara Gentz. Geometric singular perturbation theory for stochastic differential equations. *J. Differential Equations*, 191:1–54, 2003.
[BG04]   Nils Berglund and Barbara Gentz. On the noise-induced passage through an unstable periodic orbit I: Two-level model. *J. Statist. Phys.*, 114:1577–1618, 2004.
[BG05a]   Nils Berglund and Barbara Gentz. On the noise-induced passage through an unstable periodic orbit II: The general case. In preparation, 2005.
[BG05b]   Nils Berglund and Barbara Gentz. Universality of first-passage and residence-time distributions in non-adiabatic stochastic resonance. *Europhys. Letters*, 70:1–7, 2005.
[BGK05]   Anton Bovier, Véronique Gayrard, and Markus Klein. Metastability in reversible diffusion processes. II. Precise asymptotics for small eigenvalues. *J. Eur. Math. Soc. (JEMS)*, 7(1):69–99, 2005.
[BH04]   Dirk Blömker and Martin Hairer. Multiscale expansion of invariant measures for SPDEs. *Comm. Math. Phys.*, 251(3):515–555, 2004.
[BK99]   Nils Berglund and Hervé Kunz. Memory effects and scaling laws in slowly driven systems. *J. Phys. A*, 32(1):15–39, 1999.
[BK04]   Victor Bakhtin and Yuri Kifer. Diffusion approximation for slow motion in fully coupled averaging. *Probab. Theory Related Fields*, 129(2):157–181, 2004.
[Blö03]   Dirk Blömker. Amplitude equations for locally cubic nonautonomous nonlinearities. *SIAM J. Appl. Dyn. Syst.*, 2(3):464–486 (electronic), 2003.

References

[Bog56] N. N. Bogoliubov. *On a new form of adiabatic perturbation theory in the problem of particle interaction with a quantum field*. Translated by Morris D. Friedman, 572 California St., Newtonville 60, Mass., 1956.

[BPSV83] Roberto Benzi, Giorgio Parisi, Alfonso Sutera, and Angelo Vulpiani. A theory of stochastic resonance in climatic change. *SIAM J. Appl. Math.*, 43(3):565–578, 1983.

[BR92] P. Baldi and B. Roynette. Some exact equivalents for the Brownian motion in Hölder norm. *Probab. Theory Related Fields*, 93(4):457–484, 1992.

[BSV81] Roberto Benzi, Alfonso Sutera, and Angelo Vulpiani. The mechanism of stochastic resonance. *J. Phys. A*, 14(11):L453–L457, 1981.

[Car81] Jack Carr. *Applications of centre manifold theory*, volume 35 of *Applied Mathematical Sciences*. Springer-Verlag, New York, 1981.

[CCRSJ81] B. Caroli, C. Caroli, B. Roulet, and D. Saint-James. On fluctuations and relaxation in systems described by a one-dimensional Fokker-Planck equation with a time-dependent potential. *Phys. A*, 108(1):233–256, 1981.

[Ces94] Paola Cessi. A simple box model of stochastically forced thermohaline flow. *J. Phys. Oceanogr.*, 24:1911–1920, 1994.

[CF94] Hans Crauel and Franco Flandoli. Attractors for random dynamical systems. *Probab. Theory Related Fields*, 100(3):365–393, 1994.

[CF98] Hans Crauel and Franco Flandoli. Additive noise destroys a pitchfork bifurcation. *J. Dynam. Differential Equations*, 10(2):259–274, 1998.

[CFJ98] Mee H. Choi, R. F. Fox, and P. Jung. Quantifying stochastic resonance in bistable systems: Response vs residence-time distribution functions. *Phys. Rev. E*, 57(6):6335–6344, 1998.

[Chu48] Kai Lai Chung. On the maximum partial sums of sequences of independent random variables. *Trans. Amer. Math. Soc.*, 64:205–233, 1948.

[Cro75] J. Croll. *Climate and time in their geological relations: A theory of secular changes in the Earth's climate*. D. Appleton and Co., New York, 1875.

[CT62] Z. Ciesielski and S. J. Taylor. First passage times and sojourn times for Brownian motion in space and the exact Hausdorff measure of the sample path. *Trans. Amer. Math. Soc.*, 103:434–450, 1962.

[Dan69] H. E. Daniels. The minimum of a stationary Markov process superimposed on a U-shaped trend. *J. Appl. Probability*, 6:399–408, 1969.

[Dan96] Henry E. Daniels. Approximating the first crossing-time density for a curved boundary. *Bernoulli*, 2(2):133–143, 1996.

[Day83] Martin V. Day. On the exponential exit law in the small parameter exit problem. *Stochastics*, 8:297–323, 1983.

[Day90a] Martin Day. Large deviations results for the exit problem with characteristic boundary. *J. Math. Anal. Appl.*, 147(1):134–153, 1990.

[Day90b] Martin V. Day. Some phenomena of the characteristic boundary exit problem. In *Diffusion processes and related problems in analysis, Vol. I (Evanston, IL, 1989)*, volume 22 of *Progr. Probab.*, pages 55–71. Birkhäuser Boston, Boston, MA, 1990.

[Day92] Martin V. Day. Conditional exits for small noise diffusions with characteristic boundary. *Ann. Probab.*, 20(3):1385–1419, 1992.

[Day94]   Martin V. Day. Cycling and skewing of exit measures for planar systems. *Stoch. Stoch. Rep.*, 48:227–247, 1994.

[Day95]   Martin V. Day. On the exit law from saddle points. *Stochastic Process. Appl.*, 60:287–311, 1995.

[Day96]   Martin V. Day. Exit cycling for the van der Pol oscillator and quasipotential calculations. *J. Dynam. Differential Equations*, 8(4):573–601, 1996.

[DD91]   Francine Diener and Marc Diener. Maximal delay. In *Dynamic bifurcations (Luminy, 1990)*, volume 1493 of *Lecture Notes in Math.*, pages 71–86. Springer, Berlin, 1991.

[DGM$^+$01]   M. I. Dykman, B. Golding, L. I. McCann, V. N. Smelyanskiy, D. G. Luchinsky, R. Mannella, and P. V. E. McClintock. Activated escape of periodically driven systems. *Chaos*, 11:587–594, 2001.

[dH04]   F. den Hollander. Metastability under stochastic dynamics. *Stochastic Process. Appl.*, 114(1):1–26, 2004.

[DJC93]   W. Dansgaard, S. J. Johnsen, and H. B. Clausen. Evidence for general instability of past climate from a 250-kyr ice-core record. *Nature*, 364:218–220, 1993.

[DR96]   Freddy Dumortier and Robert Roussarie. Canard cycles and center manifolds. *Mem. Amer. Math. Soc.*, 121(577), 1996. With an appendix by Cheng Zhi Li.

[DS89]   Jean-Dominique Deuschel and Daniel W. Stroock. *Large deviations.* Academic Press, Boston, 1989. Reprinted by the American Mathematical Society, 2001.

[DT92]   Deepak Dhar and Peter B. Thomas. Hysteresis and self-organized criticality in the $O(N)$ model in the limit $N \to \infty$. *J. Phys. A*, 25:4967–4984, 1992.

[Dur85]   J. Durbin. The first-passage density of a continuous Gaussian process to a general boundary. *J. Appl. Probab.*, 22:99–122, 1985.

[Dur92]   J. Durbin. The first-passage density of the Brownian motion process to a curved boundary. *J. Appl. Prob.*, 29:291–304, 1992. with an appendix by D. Williams.

[DV77]   M. D. Donsker and S. R. S. Varadhan. On laws of the iterated logarithm for local times. *Comm. Pure Appl. Math.*, 30(6):707–753, 1977.

[DZ98]   Amir Dembo and Ofer Zeitouni. *Large deviations techniques and applications*, volume 38 of *Applications of Mathematics*. Springer-Verlag, New York, second edition, 1998.

[Ein05]   Albert Einstein. On the movement of small particles suspended in stationary liquids required by the molecular-kinetic theory of heat. *Ann. Phys.*, 17:549–560, 1905.

[EPRB99]   J.-P. Eckmann, C.-A. Pillet, and L. Rey-Bellet. Non-equilibrium statistical mechanics of anharmonic chains coupled to two heat baths at different temperatures. *Comm. Math. Phys.*, 201(3):657–697, 1999.

[ET82]   J.-P. Eckmann and L. E. Thomas. Remarks on stochastic resonance. *J. Phys. A*, 15:L261–L266, 1982.

[Eyr35]   H. Eyring. The activated complex in chemical reactions. *Journal of Chemical Physics*, 3:107–115, 1935.

[Fen79]   Neil Fenichel. Geometric singular perturbation theory for ordinary differential equations. *J. Differential Equations*, 31(1):53–98, 1979.

References

[Fer82]   Brooks Ferebee. The tangent approximation to one-sided Brownian exit densities. *Z. Wahrsch. Verw. Gebiete*, 61(3):309–326, 1982.

[Fer83]   Brooks Ferebee. An asymptotic expansion for one-sided Brownian exit densities. *Z. Wahrsch. Verw. Gebiete*, 63(1):1–15, 1983.

[Fit61]   R. FitzHugh. Impulses and physiological states in models of nerve membrane. *Biophys. J.*, 1:445–466, 1961.

[FJ92]   W. H. Fleming and M. R. James. Asymptotic series and exit time probabilities. *Ann. Probab.*, 20(3):1369–1384, 1992.

[FKM65]   G. W. Ford, M. Kac, and P. Mazur. Statistical mechanics of assemblies of coupled oscillators. *J. Mathemical Phys.*, 6:504–515, 1965.

[Fox89]   Ronald F. Fox. Stochastic resonance in a double well. *Phys. Rev. A*, 39:4148–4153, 1989.

[Fre00]   Mark I. Freidlin. Quasi-deterministic approximation, metastability and stochastic resonance. *Physica D*, 137:333–352, 2000.

[Fre01]   Mark I. Freidlin. On stable oscillations and equilibriums induced by small noise. *J. Statist. Phys.*, 103:283–300, 2001.

[FS03]   Augustin Fruchard and Reinhard Schäfke. Bifurcation delay and difference equations. *Nonlinearity*, 16(6):2199–2220, 2003.

[FW98]   M. I. Freidlin and A. D. Wentzell. *Random Perturbations of Dynamical Systems*. Springer-Verlag, New York, second edition, 1998.

[GBS97]   Guillermo H. Goldsztein, Fernando Broner, and Steven H. Strogatz. Dynamical hysteresis without static hysteresis: scaling laws and asymptotic expansions. *SIAM J. Appl. Math.*, 57(4):1163–1187, 1997.

[GH90]   John Guckenheimer and Philip Holmes. *Nonlinear Oscillations, Dynamical Systems, and Bifurcations of Vector Fields*. Springer-Verlag, New York, 1990. Reprint.

[GHJM98] Luca Gammaitoni, Peter Hänggi, Peter Jung, and Fabio Marchesoni. Stochastic resonance. *Rev. Mod. Phys.*, 70:223–287, 1998.

[GMSS+89] L. Gammaitoni, E. Menichella-Saetta, S. Santucci, F. Marchesoni, and C. Presilla. Periodically time-modulated bistable systems: Stochastic resonance. *Phys. Rev. A*, 40:2114–2119, 1989.

[GNRS89] V. Giorno, A. G. Nobile, L. M. Ricciardi, and S. Sato. On the evaluation of first-passage-time probability densities via nonsingular integral equations. *Adv. in Appl. Probab.*, 21(1):20–36, 1989.

[GR02]   A. Ganopolski and S. Rahmstorf. Abrupt glacial climate changes due to stochastic resonance. *Phys. Rev. Let.*, 88(3), 2002.

[Gra53]   I. S. Gradšteĭn. Application of A. M. Lyapunov's theory of stability to the theory of differential equations with small coefficients in the derivatives. *Mat. Sbornik N. S.*, 32(74):263–286, 1953.

[GS97]   P.M. Grootes and M. Stuiver. Oxygen 18/16 variability in greenland snow and ice with 10^3 to 10^5-year time resolution. *Journal of Geophysical Research*, 102:26455–26470, 1997.

[GS99]   Maria Teresa Giraudo and Laura Sacerdote. An improved technique for the simulation of first passage times for diffusion processes. *Comm. Statist. Simulation Comput.*, 28(4):1135–1163, 1999.

[GSZ01]   Maria Teresa Giraudo, Laura Sacerdote, and Cristina Zucca. A Monte Carlo method for the simulation of first passage times of diffusion processes. *Methodol. Comput. Appl. Probab.*, 3(2):215–231, 2001.

[GT84]   R. Graham and T. Tél. Existence of a potential for dissipative dynamical systems. *Phys. Rev. Letters*, 52:9–12, 1984.

[GT85] R. Graham and T. Tél. Weak-noise limit of Fokker–Planck models and nondifferentiable potentials for dissipative dynamical systems. *Phys. Rev. A*, 31:1109–1122, 1985.

[Hab79] Richard Haberman. Slowly varying jump and transition phenomena associated with algebraic bifurcation problems. *SIAM J. Appl. Math.*, 37(1):69–106, 1979.

[Hän02] Peter Hänggi. Stochastic resonance in biology: How noise can enhance detection of weak signals and help improve biological information processing. *Chemphyschem*, 3(3):285–290, 2002.

[Has76] K. Hasselmann. Stochastic climate models. Part I. Theory. *Tellus*, 28:473–485, 1976.

[HH52] A. L. Hodgkin and A. F. Huxley. A quantitative description of ion currents and its applications to conduction and excitation in nerve membranes. *J. Physiol. (Lond.)*, 117:500–544, 1952.

[HI02] Samuel Herrmann and Peter Imkeller. Barrier crossings characterize stochastic resonance. *Stoch. Dyn.*, 2(3):413–436, 2002. Special issue on stochastic climate models.

[HI05] Samuel Herrmann and Peter Imkeller. The exit problem for diffusions with time-periodic drift and stochastic resonance. *Ann. Appl. Probab.*, 15(1A):36–68, 2005.

[HKN04] Bernard Helffer, Markus Klein, and Francis Nier. Quantitative analysis of metastability in reversible diffusion processes via a Witten complex approach. *Mat. Contemp.*, 26:41–85, 2004.

[HL84] Werner Horsthemke and René Lefever. *Noise-induced transitions*. Springer-Verlag, Berlin, 1984.

[HN79] H. Hasegawa and T. Nakagomi. Semiclassical laser theory in the stochastic and thermodynamic frameworks. *J. Stat. Phys.*, 21:191–214, 1979.

[HN05] Bernard Helffer and Francis Nier. *Hypoelliptic estimates and spectral theory for Fokker-Planck operators and Witten Laplacians*, volume 1862 of *Lecture Notes in Mathematics*. Springer-Verlag, Berlin, 2005.

[IP02] P. Imkeller and I. Pavlyukevich. Model reduction and stochastic resonance. *Stoch. Dyn.*, 2(4):463–506, 2002.

[Itô44] Kiyosi Itô. Stochastic integral. *Proc. Imp. Acad. Tokyo*, 20:519–524, 1944.

[Izh00] Eugene M. Izhikevich. Neural excitability, spiking and bursting. *Internat. J. Bifur. Chaos Appl. Sci. Engrg.*, 10(6):1171–1266, 2000.

[JGB+03] Wolfram Just, Katrin Gelfert, Nilüfer Baba, Anja Riegert, and Holger Kantz. Elimination of fast chaotic degrees of freedom: on the accuracy of the Born approximation. *J. Statist. Phys.*, 112(1-2):277–292, 2003.

[JGRM90] Peter Jung, George Gray, Rajarshi Roy, and Paul Mandel. Scaling law for dynamical hysteresis. *Phys. Rev. Letters*, 65:1873–1876, 1990.

[JH89] Peter Jung and Peter Hänggi. Stochastic nonlinear dynamics modulated by external periodic forces. *Europhys. Letters*, 8:505–510, 1989.

[JH91] Peter Jung and Peter Hänggi. Amplification of small signals via stochastic resonance. *Phys. Rev. A*, 44:8032–8042, 1991.

[JKK96] C. K. R. T. Jones, Tasso J. Kaper, and Nancy Kopell. Tracking invariant manifolds up to exponentially small errors. *SIAM J. Math. Anal.*, 27(2):558–577, 1996.

[JKP91]	A. Joye, H. Kunz, and Ch.-Ed. Pfister. Exponential decay and geometric aspect of transition probabilities in the adiabatic limit. *Ann. Physics*, 208(2):299–332, 1991.
[JL98]	Kalvis M. Jansons and G. D. Lythe. Stochastic calculus: application to dynamic bifurcations and threshold crossings. *J. Statist. Phys.*, 90(1–2):227–251, 1998.
[Jon95]	Christopher K. R. T. Jones. Geometric singular perturbation theory. In *Dynamical systems (Montecatini Terme, 1994)*, pages 44–118. Springer, Berlin, 1995.
[JP91]	A. Joye and Ch.-Ed. Pfister. Full asymptotic expansion of transition probabilities in the adiabatic limit. *J. Phys. A*, 24(4):753–766, 1991.
[Kha66]	R. Z. Khasminskii. A limit theorem for solutions of differential equations with random right-hand side. *Teor. Veroyatnost. i Primenen.*, 11:390–406, 1966.
[Kif81]	Yuri Kifer. The exit problem for small random perturbations of dynamical systems with a hyperbolic fixed point. *Israel J. Math.*, 40(1):74–96, 1981.
[Kif01a]	Yuri Kifer. Averaging and climate models. In *Stochastic climate models (Chorin, 1999)*, volume 49 of *Progr. Probab.*, pages 171–188. Birkhäuser, Basel, 2001.
[Kif01b]	Yuri Kifer. Stochastic versions of Anosov's and Neistadt's theorems on averaging. *Stoch. Dyn.*, 1(1):1–21, 2001.
[Kif03]	Yuri Kifer. L^2 diffusion approximation for slow motion in averaging. *Stoch. Dyn.*, 3(2):213–246, 2003.
[KKE91]	Thomas B. Kepler, Michael L. Kagan, and Irving R. Epstein. Geometric phases in dissipative systems. *Chaos*, 1(4):455–461, 1991.
[KLS95]	J. Kuelbs, W. V. Li, and Qi Man Shao. Small ball probabilities for Gaussian processes with stationary increments under Hölder norms. *J. Theoret. Probab.*, 8(2):361–386, 1995.
[Kol00]	Vassili N. Kolokoltsov. *Semiclassical analysis for diffusions and stochastic processes*, volume 1724 of *Lecture Notes in Mathematics*. Springer-Verlag, Berlin, 2000.
[Kra40]	H. A. Kramers. Brownian motion in a field of force and the diffusion model of chemical reactions. *Physica*, 7:284–304, 1940.
[KS91]	Ioannis Karatzas and Steven E. Shreve. *Brownian motion and stochastic calculus*, volume 113 of *Graduate Texts in Mathematics*. Springer-Verlag, New York, second edition, 1991.
[KS01]	M. Krupa and P. Szmolyan. Extending geometric singular perturbation theory to nonhyperbolic points—fold and canard points in two dimensions. *SIAM J. Math. Anal.*, 33(2):286–314 (electronic), 2001.
[KST04]	Yohai Kaspi, Roiy Sayag, and Eli Tziperman. A "triple sea-ice state" mechanism for the abrupt warming and synchronous ice sheet collapses during Heinrich events. *Paleoceanography*, 19:PA3004, 1–12, 2004.
[Kus99]	R. Kuske. Probability densities for noisy delay bifurcations. *J. Statist. Phys.*, 96(3–4):797–816, 1999.
[Lan08]	Paul Langevin. Sur la théorie du mouvement brownien. *C. R. Acad. Sci. (Paris)*, 146:530–533, 1908.
[Ler86]	Hans Rudolf Lerche. *Boundary crossing of Brownian motion*, volume 40 of *Lecture Notes in Statistics*. Springer-Verlag, Berlin, 1986. Its relation to the law of the iterated logarithm and to sequential analysis.

[LG97]   Don S. Lemons and Anthony Gythiel. Paul Langevin's 1908 paper "On the Theory of Brownian Motion". *Am. J. Phys.*, 65(11):1079–1081, 1997.

[Lon93]   André Longtin. Stochastic resonance in neuron models. *J. Stat. Phys.*, 70:309–327, 1993.

[LP90]   W. S. Lo and Robert A. Pelcovits. Ising model in a time-dependent magnetic field. *Phys. Rev. A*, 42:7471–7474, 1990.

[LRD02]   Violet S. F. Lo, Gareth O. Roberts, and Henry E. Daniels. Inverse method of images. *Bernoulli*, 8(1):53–80, 2002.

[LRH00a]   J. Lehmann, P. Reimann, and P. Hänggi. Surmounting oscillating barriers. *Phys. Rev. Lett.*, 84(8):1639–1642, 2000.

[LRH00b]   Jörg Lehmann, Peter Reimann, and Peter Hänggi. Surmounting oscillating barriers: Path-integral approach for weak noise. *Phys. Rev. E*, 62(5):6282–6303, 2000.

[LS75]   N. R. Lebovitz and R. J. Schaar. Exchange of stabilities in autonomous systems. *Studies in Appl. Math.*, 54(3):229–260, 1975.

[LS77]   N. R. Lebovitz and R. J. Schaar. Exchange of stabilities in autonomous systems. II. Vertical bifurcation. *Studies in Appl. Math.*, 56(1):1–50, 1976/77.

[LS99]   Wenbo V. Li and Qi-Man Shao. Small ball estimates for Gaussian processes under Sobolev type norms. *J. Theoret. Probab.*, 12(3):699–720, 1999.

[LS01]   W. V. Li and Q.-M. Shao. Gaussian processes: inequalities, small ball probabilities and applications. In *Stochastic processes: theory and methods*, volume 19 of *Handbook of Statist.*, pages 533–597. North-Holland, Amsterdam, 2001.

[LSCDM05] S. Lorito, D. Schmitt, G. Consolini, and P. De Michelis. Stochastic resonance in a bistable geodynamo model. *Astron. Nachr.*, 326(3/4):227–230, 2005.

[Mar77]   Philippe A. Martin. On the stochastic dynamics of Ising models. *J. Statist. Phys.*, 16(2):149–168, 1977.

[Mar79]   Philippe A. Martin. *Modèles en mécanique statistique des processus irréversibles*. Springer-Verlag, Berlin, 1979.

[Mar00]   Jochem Marotzke. Abrupt climate change and thermohaline curculation: Mechanisms and predictability. *PNAS*, 97(4):1347–1350, 2000.

[May91]   I. D. Mayergoyz. *Mathematical Models of Hysteresis*. Springer-Verlag, New York, 1991.

[McK69]   H. P. McKean, Jr. *Stochastic integrals*. Probability and Mathematical Statistics, No. 5. Academic Press, New York, 1969.

[ME84]   Paul Mandel and Thomas Erneux. Laser Lorenz equations with a time-dependent parameter. *Phys. Rev. Letters*, 53:1818–1820, 1984.

[Mil41]   M. Milankovitch. *Kanon der Erdbestrahlung und seine Anwendung auf das Eiszeitenproblem*. Académie Royale Serbe. Éditions Speciales, Tome CXXXIII. Section des Sciences Mathématiques et Naturelles, Tome 33. Publisher unknown, Belgrade, 1941.

[MKKR94]   E. F. Mishchenko, Yu. S. Kolesov, A. Yu. Kolesov, and N. Kh. Rozov. *Asymptotic methods in singularly perturbed systems*. Monographs in Contemporary Mathematics. Consultants Bureau, New York, 1994. Translated from the Russian by Irene Aleksanova.

References

[ML81]   C. Morris and H. Lecar. Voltage oscillations in the barnacle giant muscle fiber. *Biophys. J.*, pages 193–213, 1981.

[MNZ93]   Jack W. Macki, Paolo Nistri, and Pietro Zecca. Mathematical models for hysteresis. *SIAM Rev.*, 35(1):94–123, 1993.

[Mog74]   A. A. Mogul′skiĭ. Small deviations in the space of trajectories. *Teor. Verojatnost. i Primenen.*, 19:755–765, 1974.

[Mon02]   Adam Hugh Monahan. Stabilisation of climate regimes by noise in a simple model of the thermohaline circulation. *Journal of Physical Oceanography*, 32:2072–2085, 2002.

[MPO94]   Frank Moss, David Pierson, and David O'Gorman. Stochastic resonance: tutorial and update. *Internat. J. Bifur. Chaos Appl. Sci. Engrg.*, 4(6):1383–1397, 1994.

[MR80]   E. F. Mishchenko and N. Kh. Rozov. *Differential equations with small parameters and relaxation oscillations*. Plenum Press, New York, 1980.

[MR95]   Ditlev Monrad and Holger Rootzén. Small values of Gaussian processes and functional laws of the iterated logarithm. *Probab. Theory Related Fields*, 101(2):173–192, 1995.

[MS93]   Robert S. Maier and D. L. Stein. Escape problem for irreversible systems. *Physical Review E*, 48(2):931–938, 1993.

[MS96]   Robert S. Maier and D. L. Stein. Oscillatory behavior of the rate of escape through an unstable limit cycle. *Phys. Rev. Lett.*, 77(24):4860–4863, 1996.

[MS97]   Robert S. Maier and Daniel L. Stein. Limiting exit location distributions in the stochastic exit problem. *SIAM J. Appl. Math.*, 57:752–790, 1997.

[MS01]   Robert S. Maier and D. L. Stein. Noise-activated escape from a sloshing potential well. *Phys. Rev. Letters*, 86(18):3942–3945, 2001.

[MW89]   Bruce McNamara and Kurt Wiesenfeld. Theory of stochastic resonance. *Phys. Rev. A*, 39:4854–4869, 1989.

[MW95]   Frank Moss and Kurt Wiesenfeld. The benefits of background noise. *Scientific American*, 273:50–53, 1995.

[NAY62]   J. S. Nagumo, S. Arimoto, and S. Yoshizawa. An active pulse transmission line simulating nerve axon. *Proc. IRE*, 50:2061–2070, 1962.

[Nay73]   Ali Hasan Nayfeh. *Perturbation methods*. John Wiley & Sons, New York-London-Sydney, 1973. Pure and Applied Mathematics.

[Neĭ87]   A. I. Neĭshtadt. Persistence of stability loss for dynamical bifurcations I. *Differential Equations*, 23:1385–1391, 1987.

[Neĭ88]   A. I. Neĭshtadt. Persistence of stability loss for dynamical bifurcations II. *Differential Equations*, 24:171–176, 1988.

[Nei95]   A. I. Neishtadt. On calculation of stability loss delay time for dynamical bifurcations. In *XIth International Congress of Mathematical Physics (Paris, 1994)*, pages 280–287. Internat. Press, Cambridge, MA, 1995.

[Nen80]   G. Nenciu. On the adiabatic theorem of quantum mechanics. *J. Phys. A*, 13:L15–L18, 1980.

[NN81]   C. Nicolis and G. Nicolis. Stochastic aspects of climatic transitions—additive fluctuations. *Tellus*, 33(3):225–234, 1981.

[NP03]   E. Neumann and A. Pikovsky. Slow–fast dynamics in Josephson junctions. *Eur. Phys. J. B*, 34:293–303, 2003.

References

[NSASG98] Alexander Neiman, Alexander Silchenko, Vadim Anishchenko, and Lutz Schimansky-Geier. Stochastic resonance: Noise-enhanced phase coherence. *Phys. Rev. E*, 58:7118–7125, 1998.

[NST96] A. I. Neishtadt, C. Simó, and D. V. Treschev. On stability loss delay for a periodic trajectory. In *Nonlinear dynamical systems and chaos (Groningen, 1995)*, volume 19 of *Progr. Nonlinear Differential Equations Appl.*, pages 253–278. Birkhäuser, Basel, 1996.

[Øks85] Bernt Øksendal. *Stochastic differential equations*. Springer-Verlag, Berlin, 1985.

[Olb01] Dirk Olbers. A gallery of simple models from climate physics. In *Stochastic climate models (Chorin, 1999)*, volume 49 of *Progr. Probab.*, pages 3–63. Birkhäuser, Basel, 2001.

[O'M74] Robert E. O'Malley, Jr. *Introduction to singular perturbations*. Academic Press [A subsidiary of Harcourt Brace Jovanovich, Publishers], New York-London, 1974. Applied Mathematics and Mechanics, Vol. 14.

[O'M91] Robert E. O'Malley, Jr. *Singular perturbation methods for ordinary differential equations*, volume 89 of *Applied Mathematical Sciences*. Springer-Verlag, New York, 1991.

[PM04] Khashayar Pakdaman and Denis Mestivier. Noise induced synchronization in a neuronal oscillator. *Phys. D*, 192(1-2):123–137, 2004.

[Pon57] L. S. Pontryagin. Asymptotic behavior of solutions of systems of differential equations with a small parameter in the derivatives of highest order. *Izv. Akad. Nauk SSSR. Ser. Mat.*, 21:605–626, 1957.

[PR60] L. S. Pontryagin and L. V. Rodygin. Approximate solution of a system of ordinary differential equations involving a small parameter in the derivatives. *Soviet Math. Dokl.*, 1:237–240, 1960.

[PRK01] Arkady Pikovsky, Michael Rosenblum, and Jürgen Kurths. *Synchronization, a universal concept in nonlinear sciences*, volume 12 of *Cambridge Nonlinear Science Series*. Cambridge University Press, Cambridge, 2001.

[PS02] André Paul and Michael Schulz. Holocene climate variability on centennial-to-millenial time scales: 2. Internal and forced oscillations as possible causes. In G. Wefer, W. Berger, Behre K.-E., and E. Jansen, editors, *Climate development and history of the North Atlantic realm*, pages 55–73, Berlin Heidelberg, 2002. Springer-Verlag.

[PWZ33] R. E. A. C. Paley, N. Wiener, and A. Zygmund. Note on random functions. *Math. Z.*, 37:647–668, 1933.

[Rah95] Stefan Rahmstorf. Bifurcations of the Atlantic thermohaline circulation in response to changes in the hydrological cycle. *Nature*, 378:145–149, 1995.

[Rah03] Stefan Rahmstorf. Timing of abrupt climate change: A precise clock. *Geophysical Research Letters*, 30(10):17–1–17–4, 2003.

[Rao92] Madan Rao. Comment on "Scaling law for dynamical hysteresis". *Phys. Rev. Letters*, 68:1436–1437, 1992.

[RBT00] Luc Rey-Bellet and Lawrence E. Thomas. Asymptotic behavior of thermal nonequilibrium steady states for a driven chain of anharmonic oscillators. *Comm. Math. Phys.*, 215(1):1–24, 2000.

References

[RBT02] Luc Rey-Bellet and Lawrence E. Thomas. Exponential convergence to non-equilibrium stationary states in classical statistical mechanics. *Comm. Math. Phys.*, 225(2):305–329, 2002.

[Ris89] H. Risken. *The Fokker-Planck equation*, volume 18 of *Springer Series in Synergetics*. Springer-Verlag, Berlin, second edition, 1989. Methods of solution and applications.

[RKP90] Madan Rao, H. K. Krishnamurthy, and Rahul Pandit. Magnetic hysteresis in two model spin systems. *Phys. Rev. B*, 42:856–884, 1990.

[Sch89] Björn Schmalfuß. Invariant attracting sets of nonlinear stochastic differential equations. In Heinz Langer and Volker Nollau, editors, *Markov processes and control theory (Gaußig, 1988)*, volume 54 of *Math. Res.*, pages 217–228, Berlin, 1989. Akademie-Verlag.

[Sch02a] Michael Schulz. On the 1470-year pacing of Dansgaard–Oeschger warm events. *Paleoceanography*, 17(2):4–1–4–9, 2002.

[Sch02b] Michael Schulz. The tempo of climate change during Dansgaard–Oeschger interstadials and its potential to affect the manifestation of the 1470-year climate cycle. *Geophysical Research Letters*, 29(1):2–1–2–4, 2002.

[SD93] A. M. Somoza and R. C. Desai. Kinetics of systems with continuous symmetry under the effect of an external field. *Phys. Rev. Letters*, 70:3279–3282, 1993.

[SHA91] J. B. Swift, P. C. Hohenberg, and Guenter Ahlers. Stochastic Landau equation with time-dependent drift. *Phys. Rev. A*, 43:6572–6580, 1991.

[Shi73] M. A. Shishkova. Examination of one system of differential equations with a small parameter in highest derivatives. *Soviet Math. Dokl.*, 14:384–387, 1973.

[Shi96] Z. Shi. Small ball probabilities for a Wiener process under weighted sup-norms, with an application to the supremum of Bessel local times. *J. Theoret. Probab.*, 9(4):915–929, 1996.

[SL77] Herbert Spohn and Joel L. Lebowitz. Stationary non-equilibrium states of infinite harmonic systems. *Comm. Math. Phys.*, 54(2):97–120, 1977.

[SMM89] N. G. Stocks, R. Manella, and P. V. E. McClintock. Influence of random fluctuations on delayed bifurcations: The case of additive white noise. *Phys. Rev. A*, 40:5361–5369, 1989.

[SNA95] B. Shulgin, A. Neiman, and V. Anishchenko. Mean switching frequency locking in stochastic bistable systems driven by a periodic force. *Phys. Rev. Letters*, 75:4157–4160, 1995.

[SP99] Kotaro Sakai and W. Richard Peltier. A dynamical systems model of the Dansgaard-Oeschger oscillation and the origin of the bond cycle. *Journal of Climate*, 12:2238–2255, 1999.

[SPT02] Michael Schulz, André Paul, and Axel Timmermann. Relaxation oscillators in concert: A framework for climate change at millennial timescales during the late Pleistocene. *Geophysical Research Letters*, 29(24):46–1–46–4, 2002.

[SRN98] S. W. Sides, P. A. Rikvold, and M. A. Novotny. Stochastic hysteresis and resonance in a kinetic Ising system. *Phys. Rev. E*, 57:6512–6533, 1998.

[SRT04] Jianzhong Su, Jonathan Rubin, and David Terman. Effects of noise on elliptic bursters. *Nonlinearity*, 17(1):133–157, 2004.

References

[SS98] Roberto H. Schonmann and Senya B. Shlosman. Wulff droplets and the metastable relaxation of kinetic Ising models. *Comm. Math. Phys.*, 194(2):389–462, 1998.

[ST96] L. Sacerdote and F. Tomassetti. On evaluations and asymptotic approximations of first-passage-time probabilities. *Adv. in Appl. Probab.*, 28:270–284, 1996.

[Sto61] Henry Stommel. Thermohaline convection with two stable regimes of flow. *Tellus*, 13:224–230, 1961.

[Str67] Volker Strassen. Almost sure behavior of sums of independent random variables and martingales. In *Proc. Fifth Berkeley Sympos. Math. Statist. and Probability (Berkeley, Calif., 1965/66)*, pages Vol. II: Contributions to Probability Theory, Part 1, pp. 315–343. Univ. California Press, Berkeley, Calif., 1967.

[SW95] Q.-M. Shao and D. Wang. Small ball probabilities of Gaussian fields. *Probab. Theory Related Fields*, 102(4):511–517, 1995.

[Tal99] Peter Talkner. Stochastic resonance in the semiadiabatic limit. *New Journal of Physics*, 1:4.1–4.25, 1999.

[TdO90] Tânia Tomé and Mário J. de Oliveira. Dynamic phase transition in the kinetic Ising model under a time-dependent oscillating field. *Phys. Rev. A*, 41:4251–4254, 1990.

[TGST03] Axel Timmermann, Hezi Gildor, Michael Schulz, and Eli Tziperman. Coherent resonant millennial-scale climate oscillations triggered by massive meltwater pulses. *Journal of Climate*, 16:2569–2585, 2003.

[Tih52] A. N. Tihonov. Systems of differential equations containing small parameters in the derivatives. *Mat. Sbornik N. S.*, 31:575–586, 1952.

[TL04] Peter Talkner and Jerzy Łuczka. Rate description of Fokker-Planck processes with time-dependent parameters. *Phys. Rev. E (3)*, 69(4):046109, 10, 2004.

[TSM88] M. C. Torrent and M. San Miguel. Stochastic-dynamics characterization of delayed laser threshold instability with swept control parameter. *Phys. Rev. A*, 38:245–251, 1988.

[Tuc89] Henry C. Tuckwell. *Stochastic Processes in the Neurosciences*. SIAM, Philadelphia, PA, 1989.

[UO30] G. E. Uhlenbeck and L. S. Ornstein. On the theory of the Brownian motion. *Phys. Rev.*, 36:823, 1930.

[Var80] S. R. S. Varadhan. *Lectures on diffusion problems and partial differential equations*, volume 64 of *Tata Institute of Fundamental Research Lectures on Mathematics and Physics*. Tata Institute of Fundamental Research, Bombay, 1980. With notes by Pl. Muthuramalingam and Tara R. Nanda.

[VBAC+01] P. Vélez-Belchí, A. Alvarez, P. Colet, J. Tintoré, and R. L. Haney. Stochastic resonance in the thermohaline circulation. *Geophysical Research Letters*, 28(10):2053–2056, 2001.

[VBK95] Adelaida B. Vasil'eva, Valentin F. Butuzov, and Leonid V. Kalachev. *The boundary function method for singular perturbation problems*, volume 14 of *SIAM Studies in Applied Mathematics*. Society for Industrial and Applied Mathematics (SIAM), Philadelphia, PA, 1995. With a foreword by Robert E. O'Malley, Jr.

[vdP20] B. van der Pol. A theory of the amplitude of free and forced triode vibration. *Radio. Rev.*, 1:701, 1920.

[vdP26]	B. van der Pol. On relaxation oscillation. *Phil. Mag.*, 2:978–992, 1926.
[vdP27]	B. van der Pol. Forced oscillations in a circuit with non-linear resistance. (Reception with reactive triode). *Phil. Mag.*, 3:64–80, 1927.
[VF69]	A. D. Ventcel' and M. I. Freĭdlin. Small random perturbations of a dynamical system with stable equilibrium position. *Dokl. Akad. Nauk SSSR*, 187:506–509, 1969.
[VF70]	A. D. Ventcel' and M. I. Freĭdlin. Small random perturbations of dynamical systems. *Uspehi Mat. Nauk*, 25(1 (151)):3–55, 1970.
[vK81]	N. G. van Kampen. *Stochastic processes in physics and chemistry*. North-Holland Publishing Co., Amsterdam, 1981. Lecture Notes in Mathematics, 888.
[vK04]	N. G. van Kampen. A new approach to noise in quantum mechanics. *J. Statist. Phys.*, 115(3–4):1057–1072, 2004.
[vS06]	M. von Smoluchowski. *Ann. Phys. (Leipzig)*, 21:756, 1906.
[Was87]	Wolfgang Wasow. *Asymptotic Expansions for Ordinary Differential Equations*. Dover Publications Inc., New York, 1987. Reprint of the 1976 edition.
[Wie23]	Norbert Wiener. Differential space. *Journal of Math. and Physics*, 2:131–174, 1923.
[WJ98]	Kurt Wiesenfeld and Fernan Jaramillo. Minireview of stochastic resonance. *Chaos*, 8:539–548, 1998.
[WM95]	Kurt Wiesenfeld and Frank Moss. Stochastic resonance and the benefits of noise: from ice ages to crayfish and SQUIDs. *Nature*, 373:33–36, 1995.
[WSB04]	T. Wellens, V. Shatokhin, and A. Buchleitner. Stochastic resonance. *Reports on Progress in Physics*, 67:45–105, 2004.
[ZMJ90]	Ting Zhou, Frank Moss, and Peter Jung. Escape-time distributions of a periodically modulated bistable system with noise. *Phys. Rev. A*, 42:3161–3169, 1990.
[ZZ95]	Fan Zhong and Jinxiu Zhang. Renormalization group theory of hysteresis. *Phys. Rev. Letters*, 75:2027–2030, 1995.
[ZZL95]	Fan Zhong, Jinxiu Zhang, and Xiao Liu. Scaling of hysteresis in the Ising model and cell-dynamical systems in a linearly varying external field. *Phys. Rev. E*, 52:1399–1402, 1995.

List of Symbols and Acronyms

$a^\star(y)$, $a^\star(t)$	linearisation at equilibrium	
$a(t)$	linearisation at $\bar{x}(t,\varepsilon)$	
a_0	lower bound for $|a(t)|$	
$b(x,y)$	nonlinear term	
$b(t), b_0(t)$	prefactors of $\psi(t)$	Appendix C
$c, c_0, c_1, \ldots$	constants (usually small)	
c	heat capacity	Section 4.1.1
$c_k(t)$	coefficients of Floquet mode	Section 4.1.2
$c(t), c_0(t)$	prefactors of $\psi(t)$	Appendix C
dx	Lebesgue measure	
d	minimal distance to $\partial\mathcal{D}$	Section 3.1
d	size of x-domain	Section 3.3
$d(t)$	level crossed by z_t	Section 3.1.1
$d(t,s)$	$d(t) - d(s)$	Appendix C
$d_1(t), d_2(t)$	boundaries of $\partial\mathcal{D}$	Section 3.1
e	Napier's (or Neper's) number	
$e(t,\omega)$	elementary function	Appendix A.2
$f(x,y)$	fast vector field	
$g(x,y)$	slow vector field	
$g(t,s)$	integrand	Theorem 3.2.8
h	width of $\mathcal{B}(h)$	Section 3.1.1
$h(y)$	auxiliary function	Examples 2.1.2, 2.1.10
$h_1(y,t), h_2(t)$	change of variables	Lemma 3.1.2
$h_\pm(t)$	potential-well depths	Section 4.1.2
i	imaginary unit $\sqrt{-1}$	
i, j, k, ℓ	integers (summation indices)	
m, n, q	integers (dimensions)	
$n(\varphi,y)$	normal vector to periodic orbit	Section 2.3.1
$p(y,t|x,s)$	transition probability	
p, q	scaling exponents	Definition 2.2.6

$p_k(x), p_k(x,t)$	eigenfunctions of $L^*, L^*(t)$	Section 4.1.2
$p(t\|s)$	first-passage-time distribution	Section 4.1.4
$q_t(s)$	exit probability	Theorem 3.2.8
$q(t)$	residence-time distribution	Section 4.1.4
$q_k(x,t)$	Floquet modes	Section 4.1.2
r	degree of differentiability	Section 2.2.4
r	distance to periodic orbit	Section 2.3.1
$r(h/\sigma, t, \varepsilon)$	remainder	Section 3.1.1
$r_\pm(t)$	transition rates	Section 4.1.4
s	fast time	
t	slow time	
u, v	auxiliary time variables	
$v(t), v^\star(t)$	variance	Section 3.1.1
$v(t,s)$	$v(t) - v(s)$	Appendix C
x	fast variable	
$x^\star(y)$	slow manifold	Section 2.1.1
$x^\star(t)$	equilibrium branch	Section 2.1.1
$\bar{x}(y,\varepsilon), \bar{x}(t,\varepsilon)$	adiabatic manifold or solution	Theorem 2.1.8
(x_0, y_0)	initial condition	
$(\hat{x}, \hat{y})$	bifurcation point	Section 2.2.1
(x_c, t_c)	avoided bifurcation point	Section 4.2
y	slow variable	
y, z	auxiliary dynamic variables	
$A(y), A^\star(y)$	Jacobian matrix (fast variables)	
A, A_c	driving amplitude, critical value	Chapter 4
$B(y), C(y)$	Jacobian matrix (slow variables)	Chapter 5
C	constant	Example 2.2.10
$C_{h/\sigma}(t,\varepsilon)$	prefactor	Section 3.1
$C_A(t,s)$	autocorrelation function	Section 4.1.2
$D(x)$	diffusion matrix	Appendix A
$D_k, \widehat{D}_k$	prefactors of $P_k(h), Q_k(h)$	Section 5.1.2
$E(T), E_0$	emissivity	Section 4.1.1
F	freshwater flux	Section 6.2.1
$F(x,t), F(x,y)$	diffusion coefficient (of dx_t)	Chapter 3
$F_0(t), F_1(y,t)$	expansion of F around $\bar{x}$	Section 3.1
F_-	lower bound on F	Chapter 3
$G(x,y)$	diffusion coefficient (of dy_t)	Chapter 5
$G_0(y), G_1(x,y)$	expansion of G	Section 5
H	potential barrier height	Section 4.1
I	interval	
$I(\varphi), J(\varphi)$	rate functions	Appendix A
K, L, M	(large) constants	
K	solar radiation modulation	Section 4.1.1

List of Symbols and Acronyms   265

L, L^*	generator, adjoint	Appendix A.4
$L(X)$	$AX + XB$	Lemma 5.1.2
N	cardinality of a partition	
N	normalisation	Section 4.1
$P_{\lambda T}(\theta)$	periodic part of $p(t\|s)$	Section 4.1.4
$P_k(h), Q_k(h)$	exit probabilities on $[0, kT]$	Section 5.1.2
$Q(\Delta \rho)$	mass exchange	Example 2.1.5
Q	solar constant	Section 4.1.1
$Q_t(s)$	exit probability	Theorem 3.2.8
$Q_\lambda(t)$	periodic part of $q(t)$	Section 4.1.4
$Q_s(u)$	matrix related to covariance	Section 5.1.1
R	remainder	
$R(\varphi, r, y, \varepsilon)$	radial fast vector field	Section 2.3.1
$R_{\text{in}}, R_{\text{out}}$	solar radiation	Section 4.1.1
$R_{\text{SN}}(\sigma)$	signal-to-noise ratio	Section 4.1.2
S	transformation matrix	
S	salinity	Section 6.2.1
$S_A(\omega)$	power spectrum	Section 4.1.2
$S(t, s)$	entry of principal solution	Section 5.1
S_t, T_t	semigroups	Appendix A.4
$T, T(y)$	period	Section 2.3.1
T	size of time interval	Chapter 3
T	temperature	Section 4.1.1
$T_{\text{K}}^0, T_{\text{K}}(\sigma)$	Kramers time	Section 4.1.4
$U(x)$	potential	Examples 2.1.3, 2.1.9
$U(t, s)$	principal solution	Section 2.3.1
$V(x, y)$	Lyapunov function	Theorem 2.1.7
$V(t, s)$	principal solution	Section 5.1
$V(x, y)$	quasipotential	Appendix A
$\overline{V}$	minimum of quasipotential	Appendix A
$W(Y)$	a row vector	Section 2.2.5
W_t	Wiener process	
$W(x)$	weight of scalar product	Section 4.1.2
$X, X^\star(y, \varepsilon)$	solution of Lyapunov equation	Section 5.1
$\overline{X}(y, \varepsilon)$	asymptotic covariance matrix	Section 5.1
X_t	Itô integral	Appendix A
Y_t	particular slow solution	Section 2.2.5
$Y(t, Y_0)$	covariance (slow variables)	Section 5.1.4
$Z^*(y), \overline{Z}(y, \varepsilon)$	covariance of $\xi^0(\eta^0)^{\text{T}}$	Section 5.1.4
$\mathcal{A}^\tau(h)$	concentration domain of paths	Section 3.4.3
$\mathcal{A}(y)$	Jacobian matrix	Section 5.1.4
$\mathcal{B}(h)$	concentration domain of paths	Section 3.1.1

List of Symbols and Acronyms

$\mathcal{C}^r$	r-times continuously differentiable functions	
$\mathcal{D}, \mathcal{D}_0$	open sets (domains)	
$\mathcal{F}, \{\mathcal{F}_t\}_t$	σ-algebra, canonical filtration	
$\mathcal{F}_0(y)$	diffusion coefficient	Section 5.1.4
$\mathcal{J}(v)$		Section 5.1.4
$\mathcal{K}(\delta), \mathcal{K}(\kappa)$	drift-dominated region	Section 3.2.2
$\mathcal{M}, \mathcal{M}_\pm$	slow manifolds	Definition 2.1.1
$\mathcal{M}_\varepsilon, \widehat{\mathcal{M}}$	adiabatic manifolds	Theorem 2.1.8
$\mathcal{N}$	neighbourhood	
$\mathcal{O}(x)$	Landau symbol: $\|\mathcal{O}(x)\| \leqslant K\|x\|$ for $\|x\| \leqslant \delta$	
$o_x(1)$	Landau symbol: $\lim_{x \to 0} o_x(1) = 0$	
$\mathcal{P}$	Newton polygon	Section 2.2.4
$\mathcal{Q}(t)$	matrix related to covariance	Section 5.1.4
$\mathcal{S}(h)$	diffusion-dominated region	Section 3.2.1
$\mathcal{S}_\omega(t,s)$	entry of principal solution	Section 5.3.1
$\mathcal{T}$	contraction	Appendix C
$\mathcal{U}(t,s)$	principal solution	Section 5.1.4
$\mathcal{V}_\omega(t,s)$	principal solution	Section 5.3.1
$\mathcal{X}$	Banach space	Appendix C
$\bar{\mathcal{Z}}(t)$	asymptotic covariance matrix	Section 5.1.4
$\mathbb{C}$	complex numbers	
$\mathbb{E}\{\cdot\}, \mathbb{E}^{x_0,y_0}\{\cdot\}$	expectation; with initial value	
$\mathbb{E}^\mu\{\cdot\}$	expectation, initial distribution μ	
$\mathbb{N}$	positive integers	
$\mathbb{N}_0$	non-negative integers	
$\mathbb{P}\{\cdot\}, \mathbb{P}^{x_0,y_0}\{\cdot\}$	probability; with initial value	
$\mathbb{Q}$	rational numbers	
$\mathbb{R}$	real numbers	
$\mathbb{R}^n$	n-dimensional space	
$\mathbb{R}^{n \times m}$	n by m real matrices	
$\mathbb{Z}$	integers	
$\alpha(t,s)$	integral of $a(u)$ over $[s,t]$	
$\alpha(T)$	albedo	Section 4.1.1
α_k, β_k	integral of $x p_k(x)$	Section 4.1.2
α_T, α_S	thermal-expansion coefficients	Section 6.2.1
β	relaxation parameter	Section 4.1.1
γ	damping coefficient	Examples 2.1.3, 2.1.6
γ	spacing of partition	Section 3.1.1
γ	exponent, $0 < \gamma < 1$	Chapter 5

$\gamma^\star(y,s)$	periodic solution	Section 2.3.1
$\gamma(T)$	atmospheric response	Section 4.1.1
δ	small constant	
δ	size of neighbourhood	Section 3.2.2
δ	bifurcation imperfection	Section 4.2
ε	adiabatic parameter	
ε_0	maximal value of ε	
ϵ	contraction constant	Appendix C
$\zeta(y,\varepsilon)$	integral (scaling function)	Lemma 2.2.8
$\zeta(t)$	width of $\mathcal{B}(h)$	Section 3.1.1
$\hat\zeta(t)$	$\sup_{s:s\leqslant t}\zeta(s)$	Section 3.3.1
ζ_t	$\zeta_t^{\mathrm{T}}=(\xi_t^{\mathrm{T}},\eta_t^{\mathrm{T}})$	Section 5.1.4
η	parameter	
$\eta(\sigma)$	spectral power amplification	Section 4.1.2
η_t,η_t^0	$y_t-y_t^{\mathrm{det}}$, linear approximation	Chapter 5
θ	intermediate value	Theorem 2.2.9
θ	angle on periodic orbit	Section 2.3
$\theta(t)$	natural time	Section 4.1.4
$\vartheta(t,s)$	bound on $\|\mathcal{V}_\omega(t,s)\|$	Section 5.3.1
$\kappa,\kappa_\pm$	exponents	
λ	exponential time scale	Section 4.1.3
λ	Lyapunov exponent	Section 4.1.4
λ_k	eigenvalues of $-L$	Section 4.1.2
λ,μ	parameters	
ξ	deviation from periodic orbit	Section 2.3.1
ξ_t,ξ_t^0	$x_t-\bar x(y_t,\varepsilon)$, linear approx.	Chapter 5
ρ	σ'/σ	Chapter 5
$\rho(y)$	remainder	Example 2.2.10
$\rho(t)$	width of $\mathcal{S}(h)$	Section 3.2.1
ϱ	spacing of partition	Proposition 3.3.5
σ	noise intensity	
σ'	noise intensity, slow variables	Chapter 5
σ_c	critical noise intensity	Section 4.2
$\tau,\tau_{\mathcal{B}(h)},\ldots$	first-exit times	
τ_0^κ	first-passage time at 0	
φ	angle, auxiliary function	
ϕ_t	phase	Section 2.3.1
$\phi_k(x)$	eigenfunctions of L	Section 4.1.2
$\phi(t,x,\lambda)$	deterministic limiting function	Section 4.1.3
$\chi^{(1)}(t),\chi^{(2)}(t)$	growth measures of $V(t,0)$	Chapter 5
$\psi(t)$	arrival-phase density	Section 4.1.4
$\psi(t)$	first-passage density	Appendix C
$\psi^\kappa(t;s,y)$	first-passage density, initial cond. (s,y)	Section 3.2.2

268   List of Symbols and Acronyms

ω	realisation ($\omega \in \Omega$)	
$\omega(y)$	imaginary part of eigenvalue	Section 2.2.5
ω	frequency	Section 4.1.2
$\omega_i(t)^2$	curvature of potential	Section 4.1.2
$\Gamma^*(y,\varphi)$	periodic orbit, parametrized	Section 2.3.1
$\bar{\Gamma}(\theta,y,\varepsilon)$	invariant cylinder	Section 2.3.2
Δ	discriminant of a quadratic eq.	Example 2.2.10
Δ	spacing of partition	Theorem 3.2.2
$\Delta(t)$	difference of well depths	Section 4.1.2
ΔS	salinity difference	Section 6.2.1
ΔT	temperature difference	Section 6.2.1
$\Delta\rho$	density difference	Section 6.2.1
$\Theta(t)$	related to $\|U(t,u)\|$	Section 5.1.2
Λ	integral of $\lambda_1(t)$	Section 4.1.2
Ξ	auxiliary variable	Lemma 3.2.11
Ξ_t	submartingale	Lemma 5.1.10
$\Pi(t_0)$	bifurcation-delay time	Section 2.2.3, 2.2.5
Σ	covariance of $Q_s(t)\Upsilon_t^0$	Lemma 5.1.8
Υ_u	$U(s,u)\xi_u$	Section 5.1
$\Phi(x)$	normal law distribution function	
$\Phi_t(s,y)$	exit probability	Theorem 3.2.8
$\Phi(t), \widehat{\Phi}(t)$	related to $U(t,u), \mathcal{U}(t,u)$	Section 5.1
$\Psi(t)$	integral of $a(s)+i\omega(s)$ up to t	Section 2.2.5
$\Psi(t), \widehat{\Psi}(t)$	related to $U(t,u), \mathcal{U}(t,u)$	Section 5.1
Ω	probability space	
Ω	forcing frequency	Section 4.1.2
$\Omega_1(\varphi,r,y,\varepsilon)$	angular fast vector field	Section 2.3.1
$a \wedge b$	minimum of a and b	
$a \vee b$	maximum of a and b	
$\lceil a \rceil$	smallest integer $k \geqslant a$	
$\lfloor a \rfloor$	largest integer $k \leqslant a$	
$x := y$	x is defined by y	
$x =: y$	x is denoted y	
$x(t,\varepsilon) \asymp y(t,\varepsilon)$	x scales as y	Notation 2.2.1
$h \gg \sigma, \sigma \ll h$	$h = h(\sigma)$ is of larger order than σ: $h(\sigma)/\sigma \to \infty$ as $\sigma \to 0$	
$x \simeq y$	x is approximately y	
$x(\sigma) \cong y(\sigma)$	$x(\sigma)/y(\sigma) = 1 + \mathcal{O}(\sigma^2)$	Section 4.1.2
$x(\sigma) \sim y(\sigma)$	$\lim_{\sigma \to 0} \sigma^2 \log(x(\sigma)/y(\sigma)) = 1$	
$\|x\|$	Euclidean norm of x	
$\langle x,y \rangle$	scalar product of x and y	

List of Symbols and Acronyms

$\langle x, y \rangle_{W^{-1}}$	scalar prod. weighted by $W(x)^{-1}$ Section 4.1.2
$\|A\|$	operator norm: $\sup_{\|x\|=1} \|Ax\|$
$\|A\|_I$	$\sup_{t \in I} \|A(t)\|$
$\|A\|_\infty$	$\|A\|_I$, $I = \mathbb{R}$ or context-dep.
A^T	transposed matrix
$\partial_x f(x)$	derivative, Jacobian matrix
$\partial_{xx} f(x)$	2nd derivative, Hessian matrix
$\nabla f(x)$	gradient
$\mathbb{1}$	identity matrix
$1_A(x)$	indicator function on the set A

Acronyms

DO	Dansgaard–Oeschger
EMIC	Earth Model of Intermediate Complexity
GCM	General Circulation Model
GL	Ginzburg–Landau
ODE	Ordinary Differential Equation
PDE	Partial Differential Equation
RCLSJ	Resistive-Capacitive-Inductively Shunted Junction
SDE	Stochastic Differential Equation
SK	Suzuki–Kubo
SQUID	Superconducting Quantum Interference Device
SR	Stochastic Resonance
THC	Thermohaline Circulation

Index

σ-algebra 224
  pre-τ 224
action
  functional 4, 128
  potential 207
activation time 2
adiabatic
  limit 19, 120, 128
  manifold 24, 27, 145, 164, 205
  solution 51, 69, 90, 134, 148, 198
  theorem 48
Airy function 32
Aristotle's law 20
Arrhenius' law 236
associated system 17, 20, 28, 33, 45, 178, 213
assumption
  asymptotically stable periodic orbit 172
  avoided transcritical bifurcation 133
  Hopf bifurcation 185
  multidimensional bifurcation 178
  multidimensional stable case 144, 148
  saddle–node bifurcation 85
  slowly varying stable periodic orbit 175
  stable case 53
  symmetric pitchfork bifurcation 97
  unstable case 69
asymptotically stable
  equilibrium branch 53
  equilibrium point 5, 6, 209, 221, 235

  periodic orbit 45, 170, 172, 175
  slow manifold 19, 23, 54, 202
attractive
  locally 25, 28
autocorrelation function 116
averaged system 46, 190
averaging 45, 177
avoided bifurcation 28, 132, 139, 216

backward Kolmogorov equation 231
Balanus nubilus 208
barnacle 208
Bernoulli process 11, 141
Bernstein inequality 58, 239
Bessel
  function 247
  process 153
bifurcation
  avoided 28, 132, 139, 216
  concentration of paths 180
  delay 7, 11, 34, 43, 70, 98, 213
  diagram 19
  dynamic 27, 178
  Hopf 43, 185, 213
  pitchfork 8, 34, 42, 97, 184
  point 22, 27, 51, 178
  reduction 183
  saddle–node 8, 28, 84
  saddle–node of periodic orbits 213
  saddle–node-to-invariant-circle 209
  transcritical 41, 105
boundary
  characteristic 128
  curved 59, 60, 243, 244

272 Index

bounded-growth condition   66, 229
box model   21, 199, 200
branch
  equilibrium   11, 19, 22, 51, 68
Brownian motion   223
buffer time   44
bursters   212
bursting   209, 212

Cauchy's formula   26
Cayley–Hamilton theorem   147
centre manifold   27, 178
centre-manifold theorem   27
characteristic
  boundary   128
  exponent   45, 118, 128
  multiplier   45, 171
climate model   112, 199
comparison lemma   241
concentration
  of sample paths   53, 145, 166, 178
    near bifurcation point   180
condition
  bounded-growth   66, 229
  Lipschitz   229
controllability Grammian   147
Curie–Weiss model   215
curved boundary   59, 60, 243, 244
cycling   128

Dansgaard–Oeschger events   200, 204
delayed bifurcation   7, 11, 34, 43, 70, 98, 213
density
  first-passage   243
  invariant   3, 8, 232
distribution
  Gumbel   129
  residence-time   126, 206
Doléans exponential   228
Doob's submartingale inequality   58, 153, 154, 239
double-well potential   8, 97, 111, 113, 139
dynamic
  bifurcation   27, 178
  hysteresis   214
  phase transition   215

elementary function   225
energy-balance model   112
equation
  Euler–Lagrange   233, 234
  Fitzhugh–Nagumo   13, 208
  Fokker–Planck   3, 117, 232
  Ginzburg–Landau   215
  Hamilton   234
  Hodgkin–Huxley   13, 207
  Kolmogorov   3, 117, 231, 232
  Langevin   194, 215
  Lyapunov   146
  Morris–Lecar   208
  Riccati   32, 135
  Smoluchowski   194
  Suzuki–Kubo   215
equilibrium
  branch   11, 19, 22, 51
    asymptotically stable   53
    stable   29, 53
    tame   38
    unstable   29, 68
    with exponent q 37
  point
    asymptotically stable   209, 221
escape
  diffusion-dominated   71, 101, 189
  drift-dominated   78, 102
Euler Gamma function   129
Euler–Lagrange equation   233, 234
excitability   13, 209
exit problem   3, 127, 234
exponent
  characteristic   45, 118, 128
  Lyapunov   45, 128, 165, 170
  of exit probability   63, 78, 88, 100, 104, 106, 136
exponential martingale   228

fast
  system   17, 221
  time   6, 17, 24, 47, 112
  variable   17
Feller property   231
Fenichel's theorem   24, 69
first-exit
  location   4, 127, 234
  time   3, 53, 58, 62, 72, 224, 234
first-passage

density   243
  time   60, 126, 243
Fitzhugh–Nagumo equation   13, 208
Floquet theory   118, 170
Fokker–Planck equation   3, 117, 232
formula
  Cauchy   26
  Itô   55, 86, 145, 227
  Stirling   27
Fourier series   129
function
  Airy   32
  autocorrelation   116
  Bessel   247
  elementary   225
  Gamma   129
  Lyapunov   23
  rate   4

Gamma function   129
Gaussian
  martingale   58, 73, 152
  process   56, 71, 79, 166, 179, 187, 224, 243
generator   2, 117, 231
Ginzburg–Landau equation   215
good rate function   233
gradient system   5, 148, 232, 236
Gronwall's inequality   155, 241
Gumbel distribution   129

Hamilton equation   234
Hodgkin–Huxley equation   13, 207
Hopf bifurcation   43, 185, 213
hyperbolic slow manifold   19
hysteresis   7, 125
  cycles   10
  dynamic   214
  static area   215

ice age   112, 200, 204
implicit-function theorem   26, 29, 41, 43, 86
inequality
  Bernstein   58, 239
  Doob   58, 153, 154, 239
  Gronwall   155, 241
  Schwarz   67
integral

Itô   56, 225, 226
Riemann–Stieltjes   225
stochastic   225
integration by parts   66, 71, 228, 240
invariant
  circle   209
  density   3, 8, 232
  manifold   24, 47
  positively   3, 150
  set   3, 128, 150
  tube   175
Ising model   215
isometry
  Itô   225
Itô
  formula   55, 86, 145, 227
  integral   56, 225, 226
  isometry   225

Josephson junction   219

Kirchhoff's law   196, 219
Kolmogorov's
  backward equation   3, 231
  continuity theorem   223
  forward equation   3, 117, 232
Kramers' time   2, 54, 115, 206
Kronecker product   147

Langevin equation   194, 215
Laplace method   4, 39, 121, 232
large deviations   2, 59, 66, 124, 128, 232
large-deviation principle   233
law
  Arrhenius   236
  Kirchhoff   196, 219
  Stefan–Boltzmann   113
limit
  adiabatic   19, 120, 128
linear stochastic differential equation   56, 145, 230
Lipschitz condition   229
locally
  attractive   25, 28
  invariant   24
location
  first-exit   4, 127, 234
Lyapunov
  equation   146
  exponent   45, 128, 165, 170

function   23
time   6

macaroni   177
manifold
  adiabatic   24, 27, 145, 164, 205
  centre   27, 178
  invariant   47
  locally invariant   24
  slow   18, 144
    hyperbolic   19
    stable   19, 23, 144
    unstable   19
  stable   144
Markov
  process   1, 53, 224, 243
  property   61, 73, 93, 162, 224, 244
    strong   225, 230
martingale   226
  exponential   228
  Gaussian   58, 73, 152
metastability   1, 54, 59
model
  climate   112, 199
  Curie–Weiss   215
  energy-balance   112
  Ising   215
  Stommel   21, 200
moments of one-dimensional process   66
monodromy matrix   170
Morris–Lecar equation   208
multiplier
  characteristic   45, 171

neural dynamics   207
Newton polygon   37
noise-induced
  deviation   55
  transition   108, 111
non-autonomous stochastic differential equation   51, 56

orbit
  periodic   45, 170, 172, 173, 175, 176
Ornstein–Uhlenbeck process   9, 195, 203, 230, 232
oscillation
  relaxation   7, 22, 33

oscillator
  nonlinear   194
  van der Pol   12, 21, 33, 196
overdamped motion   20, 26, 51, 194

palm tree   201
periodic
  bursting   212
  orbit   45
    asymptotically stable   45, 170, 172, 175
    concentration of paths   173, 176
pitchfork bifurcation   8, 34, 97
  asymmetric   42, 108
  avoided   139
  multidimensional   184
  symmetric   34, 97
Poincaré map   134
Poisson point process   115
potential   20, 26, 51, 54, 59, 69, 85, 97, 194, 221
  $(\Phi^2)^2$ 215
  action   207
  double-well   8, 97, 111, 113, 139
  washboard   210
power spectrum   116
pre-τ σ-algebra   224
principal
  curvature   148
  solution   45, 146
principle
  large-deviation   233
  reflection   81, 94, 105, 232, 242
process
  Bernoulli   11, 141
  Bessel   153
  Gaussian   56, 71, 79, 166, 179, 187, 224, 243
  Markov   1, 53, 224, 243
  Ornstein–Uhlenbeck   9, 195, 203, 230, 232
  Poisson   115
  Wiener   52, 223
property
  Feller   231
  Markov   61, 73, 93, 162, 224, 244
  scaling   52, 224
  semigroup   230

quasipotential 128, 235
quasistatic regime 124

rate function 4
  good 233
reduced system 19, 24, 28, 43
  stochastic 164, 177, 182
reduction
  near bifurcation point 27, 182, 183
  near periodic orbit 174, 177
  to phase dynamics 174
  to slow variables 164, 178
reflection principle 81, 94, 105, 232, 242
regime
  large-amplitude 218
  quasistatic 124
  semiadiabatic 123
  small-amplitude 218
  stochastic resonance 114
  strong-noise 90, 101, 138, 218
  superadiabatic 123, 125
  synchronisation 10, 112, 131, 132
  weak-noise 96, 101, 135
relaxation oscillation 7, 22, 33
residence time 126, 206
Riccati equation 32, 135
Riemann–Stieltjes integral 225

saddle–node bifurcation 8, 28, 84
  of periodic orbits 213
  to invariant circle 209
sample paths 2
  approach 132
  concentration 53, 145, 166, 178
  of Brownian motion 223
scaling
  behaviour 30
  law 7, 30, 37
  property 52, 224
Schilder's theorem 233
Schwarz inequality 67
semiadiabatic regime 123
semigroup 230
signal-to-noise ratio 119
slow
  manifold 18, 144
    asymptotically stable 19, 23, 54, 144, 202

hyperbolic 19
  unstable 19
system 19
time 9, 19, 51
variable 17
  reduction 164, 178
slow–fast system 6
  deterministic 17
  stochastic 143
small-ball probability 71, 73, 247
Smoluchowski equation 194
Sobolev space 4, 232
solution
  principal 146
  strong 52, 229
space
  Sobolev 4, 232
spectral
  gap 2
  power amplification 119
  theory 2, 116
stability matrix 19
stable
  asymptotically 19, 53, 54, 202, 209, 221
  equilibrium branch 53
  manifold 23, 144
  structurally 28
static hysteresis area 215
Stefan–Boltzmann law 113
Stewart–McCumber parameter 220
Stirling's formula 27
stochastic
  integral 225
  resonance 10, 111, 206
  synchronisation 132
stochastic differential equation
  linear 56, 145, 230
  non-autonomous 51, 56
  reduced 164
  slow–fast 143
Stommel's box model 21, 200
stopping time 53, 93, 224
strong
  Feller property 231
  Markov property 225, 230
  solution 52, 229
structural stability 28
submartingale 153, 154, 239

superadiabatic regime   123, 125
Suzuki–Kubo equation   215
synchronisation
   noise-induced   132
   regime   10, 112, 131, 132
   stochastic   132
system
   associated   17, 20, 28, 33, 45, 178, 213
   fast   17, 221
   gradient   5, 232, 236
   reduced   19, 24, 28, 43
   slow   19
   slow–fast   6, 17

tame equilibrium branch   38
theorem
   adiabatic   48
   Cayley–Hamilton   147
   centre-manifold   27, 178
   concentration around a deterministic solution   168
   dynamic Hopf bifurcation   44
   dynamic pitchfork bifurcation   35
   dynamic saddle–node bifurcation   31
   Fenichel   24, 69
   implicit-function   26, 29, 41, 43, 86
   Itô   227
   Kolmogorov   223
   multidimensional bifurcation   180
   multidimensional stochastic stable case   149
   reduction near bifurcation point   183
   reduction to phase dynamics   174
   reduction to slow variables   164
   Schilder   233
   slowly varying periodic orbit   46
   stochastic avoided transcritical bifurcation   136, 138
   stochastic Hopf bifurcation   188, 189
   stochastic linear stable case   59
   stochastic motion far from equilibria   65
   stochastic motion near stable periodic orbit   173, 176
   stochastic nonlinear stable case   63
   stochastic pitchfork bifurcation   100–102, 104
   stochastic saddle–node bifurcation   88, 91, 96
   stochastic transcritical bifurcation   106, 107
   stochastic unstable case   72, 78
   strong Markov property   225
   strong solutions of SDE   229
   Tihonov   23, 56, 70
   Wiener–Khintchin   116
theory
   Floquet   118, 170
   large-deviation   2, 59, 66, 124, 128, 232
   spectral   2, 116
   Wentzell–Freidlin   4, 59, 124, 232
thermohaline circulation   200
Tihonov's theorem   23, 56, 70
time
   activation   2
   bifurcation delay   35, 44, 98
   buffer   44
   fast   6, 17, 24, 47, 112
   first-exit   3, 53, 58, 62, 72, 224, 234
   first-passage   60, 126, 243
   Kramers   2, 54, 115, 206
   Lyapunov   6
   residence   126, 206
   slow   9, 19, 51
   stopping   53, 93, 224
transcritical bifurcation   41, 105
   avoided   132
transformation of variables   55
transition
   noise-induced   108, 111
   probability   116, 231, 243

Union Jack   201
unstable
   equilibrium branch   68
   slow manifold   19

van der Pol oscillator   12, 21, 33, 196
variables
   fast   17
   slow   17

washboard potential   210
weak Feller property   231
Wentzell–Freidlin theory   4, 59, 124, 232, 235
Wiener process   52, 223
Wiener–Khintchin theorem   116